T0184767

Each World Resources Institute study represents a significant and timely treatment of a subject of public concern. WRI takes responsibility for choosing the study topics and guaranteeing its authors and researchers freedom of inquiry. It also solicits and responds to the guidance of advisory panels and expert reviewers. Unless otherwise stated, however, all the interpretations and findings set forth in WRI publications are those of the authors.

The World Resources Institute (WRI) is a policy research center created in late 1982 to help governments, international organizations, the private sector, and others address a fundamental question: How can societies meet basic human needs and nurture economic growth without undermining the natural resources and environmental integrity on which life, economic vitality, and international security depend?

The Institute's current program areas include tropical forests, biological diversity, sustainable agriculture, global energy futures, climate change, pollution and health, economic incentives for sustainable development, and resource and environmental information. Within these broad areas, two dominant concerns influence WRI's choice of projects and other activities:

The destructive effects of poor resource management on economic development and the alleviation of poverty in developing countries.

The new generation of globally important environmental and resource problems that threaten the economic and environmental interests of the United States and other industrial countries and that have not been addressed with authority in their laws.

Independent and nonpartisan, the World Resources Institute approaches the development and analysis of resource policy options objectively, with a strong grounding in the sciences. Its research is aimed at providing accurate information about global resources and population, identifying emerging issues, and developing politically and economically workable proposals. WRI's work is carried out by an interdisciplinary staff of scientists and policy experts augmented by a network of formal advisors, collaborators, and affiliated institutions in 30 countries.

WRI is funded by private foundations, United Nations and governmental agencies, corporations, and concerned individuals.

Public policies and the misuse of forest resources

Public policies and the misuse of forest resources

A World Resources Institute Book

Edited by

Robert Repetto
World Resources Institute

Malcolm Gillis
Duke University

The right of the
University of Cambridge
to print and sell
all manner of books
was granted by
Henry VIII in 1534.
The University has printed
and published continuously
since 1584.

CAMBRIDGE UNIVERSITY PRESS

Cambridge
New York   Port Chester   Melbourne   Sydney

CAMBRIDGE
UNIVERSITY PRESS

32 Avenue of the Americas, New York NY 10013-2473, USA

Cambridge University Press is part of the University of Cambridge.

It furthers the University's mission by disseminating knowledge in the pursuit of education, learning and research at the highest international levels of excellence.

www.cambridge.org
Information on this title: www.cambridge.org/9780521335744

Copyright © 1988 by World Resources Institute

This publication is in copyright. Subject to statutory exception and to the provisions of relevant collective licensing agreements, no reproduction of any part may take place without the written permission of Cambridge University Press.

First published 1988
Reprinted 1989, 1990

A catalogue record for this publication is available from the British Library

Library of Congress Cataloguing in Publication data

Public policies and the misuse of forest resources.
"A World Resources Institute book."
Includes index.
1. Forest policy – Developing countries – Case studies.
2. Deforestation – Environmental aspects – Developing countries – Case studies. I. Repetto, Robert C.
II. Gillis, Malcolm.   III. World Resources Institute.
HD9768.D44P82   1988
333.75'091724   87–33815

ISBN 978-0-521-34022-9 Hardback
ISBN 978-0-521-33574-4 Paperback

Cambridge University Press has no responsibility for the persistence or accuracy of URLs for external or third-party internet websites referred to in this publication, and does not guarantee that any content on such websites is, or will remain, accurate or appropriate.

For Rachel and Sarah
R. R.

For Bill, Elsie, Lillianette, and Ted
M. G.

Contents

vii

Contributors

Eufresina L. Boado
*Formerly of the Philippine
Forestry Bureau and ASEAN*

John O. Browder
Tulane University

Kong Fanwen and He Naihui
*Chinese Academy of Forestry Sciences
Beijing, China*

Malcolm Gillis
Duke University

Li Jinchang
*Research Center for Economic, Technological,
and Social Development under the State Council
of the People's Republic of China
Beijing, China*

Robert Repetto
World Resources Institute

Lester Ross
*J.D. Candidate at
the Harvard Law School*

Foreword

All beginning students learn that economics is about the efficient allocation of scarce resources. In that spirit, Robert Repetto, Malcolm Gillis, and the other contributors to this volume have shown that good economics can help stem the loss of the world's disappearing forest resources. Reversing the forces of tropical deforestation in the closing years of this century is of paramount importance. Rapid population growth, unemployment, and concentrated agricultural landholdings are fueling the drive to clear forested land for agriculture. Demands on the world's forests for timber, industrial raw materials, and fuel are growing. Even as these pressures mount, scientists are documenting with greater precision the roles that forests play in the control of erosion and floods, in the hydrological cycle, and in the survival of perhaps half of the planet's species.

This volume convincingly demonstrates that, on balance, government policies affecting the forest sector aggravate rather than counteract this pressure. Inappropriate tax and trade policies, distorted investment incentives, and shortsighted development priorities contribute to alarming waste of forest resources, and result in heavy economic and fiscal losses as well. In all ten countries covered by the studies underlying this book, such policies contribute to uneconomic and ecologically damaging exploitation.

Thus, *Public Policies and the Misuse of Forest Resources* shows how conflicts of interest between conservationists and developers are often more apparent than real. Indeed, an important message of this volume is that more reasonable policies can save both natural and financial resources. The policy changes that Gillis and Repetto outline in the closing chapter give specific content to the idea of sustainable development of forest resources. They deserve the broadest consideration and understanding because they are essential to the preservation of the world's remaining forests.

World Resources Institute sponsored the research leading up to this volume as part of a broader examination of policy changes by national and international agencies that can encourage sound resource use and sustainable economic growth. While such policy changes are urgent in agriculture, energy, and other sectors as well, forests are a special case. As the report *Tropical Forests: A Call for Action,* released in late 1985 by the

xi

WRI, the World Bank, and the United Nations Development Programme, made abundantly clear, more than a billion people in developing countries suffer from fuelwood scarcity. Moreover, the report shows, tropical forests are disappearing at a rate of more than 11 million hectares per year.

Managing the forests is the responsibility of local, national, and international authorities, but the actions of national governments in the next decade are critical. Money can grow on trees, but only if nations value, protect, and tend the forest's wealth.

WRI is grateful for the cooperation and collaboration of many international authorities in the preparation of this book. They are mentioned by the editors in the Acknowledgments. We would especially like to thank Li Jinchang, Kong Fanwen, and He Naihui of the People's Republic of China and Eufresina L. Boado of the Philippines for their chapters in this volume.

The World Resources Institute would like to express its special appreciation to the United States Agency for International Development, the World Bank, and the World Commission on Environment and Development for their support of this work.

James Gustave Speth
President
World Resources Institute

Acknowledgments

We wish to thank in particular the other authors of country case studies, whose scholarship and insights contributed so much to the study. In addition, many others assisted in the research, including Tinling Choong, Mark Dillenbeck, Ruth Griswold, Philip Huffman, Nancy Sheehan, William Spiller, and Craig Thomas. Hyacinth Billings, Esther Chambers, Elizabeth Gillis, Katherine Harrington, Ann Pence, and Karen Saxon contributed able and dedicated efforts in the lengthy production process.

We also wish to acknowledge helpful comments and contributions from Paul Aind, Michael Arnold, Peter Ashton, Peter Emerson, Julian Evans, Suzannah Hecht, Barin Ganguli, Carl Gallegos, Alan Grainger, Hans Gregersen, Roberto Lopez C., Norman Myers, Jeffrey Sayer, John Spears, William Beattie, Michael Roth, Roger Sedjo, and Marzuki Usman, and the support of WRI's advisory panel on the study of economic incentives for sustainable development: William Baumol of New York University, Gardner Brown, Jr., at the University of Washington, Anthony Fisher of the University of California at Berkeley, Shanta Devarajan at Harvard, Charles Howe at the University of Colorado, and Jeremy Warford at the World Bank. Colleagues at WRI, including William Burley, Peter Hazlewood, Jessica Mathews, Irving Mintzer, and Gus Speth, provided strong support. We are grateful for financial contributions for the underlying research from the U.S. Agency for International Development, the World Bank, and the World Commission on Environment and Development.

R. R.
M. G.

1 Overview

ROBERT REPETTO

Introduction

Threats to the world's forests are evoking responses at all levels, from villagers organizing to protect their woods to international summit meetings of world leaders. Many articles, books, and films have been produced documenting forest losses and the many threats – from peasant farmers, fuelgatherers, ranchers, herders, large-scale development projects, multinational companies, and atmospheric pollution. This book is different. It reports the results of an international research project that identified the impacts that governments, most of which are committed in principle to conservation and wise resource use, are themselves having on the forests under their stewardship through policies that inadvertently or intentionally aggravate the losses.

Such policies, by and large, were adopted for worthy objectives: industrial or agricultural growth, regional development, job creation, or poverty alleviation. But the study's important finding is that such objectives typically have not been realized, or have been attained only at excessive cost. The government policies identified in this book, both those usually identified as forestry policies and those impinging on the forestry sector from outside, have resulted in economic and fiscal losses while contributing to the depletion of forest resources.

Forestry policies, the terms on which potential users can exploit public forests, include harvesting fees, royalties, logging regulations, and administration of timber concessions with private loggers. Governments have typically sold off timber too cheaply, sacrificing public revenues and the undervalued non-timber benefits of the standing forest while encouraging rapid logging exploitation. The terms of many timber concession agreements and revenue systems have encouraged wasteful, resource-depleting logging.

Other government policies impinging significantly on the forest sector include tax and trade regimes, industrialization incentives, laws governing land tenure, and agricultural resettlement and development policies.

1

These frequently are biased against the preservation of standing forests and toward their exploitation or conversion to other land uses. Often, they tip the balance of incentives facing private parties to exploit or convert forest resources far faster and further than market forces would otherwise allow. This book shows that in several countries studied such biases have been sufficiently strong that forest resources have been destroyed for purposes that are intrinsically uneconomic.

Governments have become more aware in recent years of the threats and risks of forest depletion. In the Third World, most have reacted by adopting new measures to encourage reforestation, but few have modified existing policies that have aggravated forest depletion. As a result, in all the developing countries studied, on balance deforestation continues at significant rates.

While increased attention, investment, and research directed toward the solution of forest problems are now widely advocated, the special contribution of this book and the underlying research is in identifying the government policies that can be changed to reduce the wastage of forest without sacrificing other economic objectives and that must be changed to ensure that those other interventions will be effective.

The extent and rate of deforestation

Scientific evidence suggests that the world's forest area has declined by one-fifth, from about 5 to 4 billion hectares, from pre-agricultural times to the present. Temperate closed forests have evidently suffered the greatest cumulative losses, 32–35 percent, followed by subtropical woody savannahs and deciduous forests, 24–25 percent, and tropical climax forests, 15–20 percent. Over the entire period, tropical evergreen rain forests have suffered the smallest attrition, 4–6 percent, because until recently they were inaccessible and barely populated (Matthews 1983). Forests and woods still cover two-fifths of the earth's land surface (Table 1.1), an area three and a half times as large as that devoted to crops, and account for about 60 percent of the net biomass productivity of terrestrial ecosystems (Olson 1975). Just over half of the remaining closed and open forests are in the developing countries.

Since World War II, deforestation has shifted from the temperate zone to the tropics. In the industrialized North, rural out-migration and rising agricultural yields have allowed previously cultivated land to revert to woods, offsetting other sources of deforestation. Forest management poses acute policy issues, as industrialists, loggers, naturalists, hikers, and hunters urge their conflicting interests, but the forest area is stable.

Table 1.1. *Land use, 1850–1980*

	Area (million hectares)														Percentage change, 1850 to 1980
	1850	1860	1870	1880	1890	1900	1910	1920	1930	1940	1950	1960	1970	1980	
Ten regions															
Forests and woodlands	5,919	5,898	5,869	5,833	5,793	5,749	5,696	5,634	5,553	5,455	5,345	5,219	5,103	5,007	− 15
Grassland and pasture	6,350	6,340	6,329	6,315	6,301	6,284	6,269	6,260	6,255	6,266	6,293	6,310	6,308	6,299	− 1
Croplands	538	569	608	659	712	773	842	913	999	1,085	1,169	1,278	1,396	1,501	179
Tropical Africa															
Forests and woodlands	1,336	1,333	1,329	1,323	1,315	1,306	1,293	1,275	1,251	1,222	1,188	1,146	1,106	1,074	− 20
Grassland and pasture	1,061	1,062	1,064	1,067	1,070	1,075	1,081	1,091	1,101	1,114	1,130	1,147	1,157	1,158	9
Croplands	57	58	61	64	68	73	80	88	101	118	136	161	190	222	228
North Africa and Middle East															
Forests and woodlands	34	34	33	32	31	30	28	27	24	21	18	17	15	14	− 60
Grassland and pasture	1,119	1,119	1,118	1,117	1,116	1,115	1,113	1,112	1,108	1,103	1,097	1,085	1,073	1,060	− 5
Croplands	27	28	30	32	35	37	40	43	49	57	66	79	93	107	294
North America															
Forests and woodlands	971	968	965	962	959	954	949	944	941	940	939	939	941	942	− 3
Grassland and pasture	571	559	547	535	522	504	486	468	454	450	446	446	447	447	− 22
Croplands	50	65	80	95	110	113	156	179	196	201	206	205	204	203	309

(continued)

Table 1.1 (*continued*)

	Area (million hectares)														Percentage change, 1850 to 1980
	1850	1860	1870	1880	1890	1900	1910	1920	1930	1940	1950	1960	1970	1980	
Latin America															
Forests and woodlands	1,420	1,417	1,414	1,408	1,401	1,394	1,383	1,369	1,348	1,316	1,273	1,225	1,186	1,151	− 19
Grassland and pasture	621	623	625	627	630	634	638	646	655	673	700	730	751	767	23
Croplands	18	19	21	24	28	33	39	45	57	72	87	104	123	142	677
China															
Forests and woodlands	96	93	91	89	86	84	82	79	76	73	69	64	59	58	− 39
Grassland and pasture	799	799	798	798	797	797	797	796	796	794	793	789	784	778	− 3
Croplands	75	78	81	84	86	89	91	95	98	103	108	117	127	134	79
South Asia															
Forests and woodlands	317	315	311	307	303	299	294	289	279	265	251	235	210	180	− 43
Grassland and pasture	189	189	189	189	189	189	190	190	190	190	190	190	189	187	− 1
Croplands	71	73	77	81	85	89	93	98	108	122	136	153	178	210	196

															% change
Southeast Asia															
Forests and woodlands	252	252	251	251	250	249	248	247	246	244	242	240	238	235	− 7
Grassland and pasture	123	123	122	121	119	118	116	114	111	108	105	102	97	92	− 25
Croplands	7	7	8	10	12	15	18	21	25	30	35	40	47	55	670
Europe															
Forests and woodlands	160	158	157	157	156	156	155	155	155	154	154	156	161	167	4
Grassland and pasture	150	147	145	144	143	142	141	139	138	137	136	136	137	138	− 8
Croplands	132	136	140	142	143	145	146	147	149	150	152	151	145	137	4
USSR															
Forests and woodlands	1,067	1,060	1,052	1,040	1,027	1,014	1,001	987	973	961	952	945	940	941	− 12
Grassland and pasture	1,078	1,081	1,083	1,081	1,079	1,078	1,076	1,074	1,072	1,070	1,070	1,069	1,065	1,065	− 1
Croplands	94	98	103	118	132	147	162	178	194	208	216	225	233	233	147
Pacific															
Developed Countries															
Forests and woodlands	267	267	266	265	264	263	262	261	260	259	258	252	247	246	− 8
Grassland and pasture	638	638	638	637	635	634	632	630	629	627	625	617	609	608	− 5
Croplands	6	6	7	9	12	14	17	19	22	24	28	42	56	58	841

Source: International Institute for Environment and Development and World Resources Institute (1987: 272).

In the developing countries, the issues are more intense, because for hundreds of millions, the struggle is for survival. Growing rural populations invade the forests in search of land for their crops, fuel for cooking, and fodder for their animals. Governments impelled to raise foreign exchange earnings and employment, and to finance economic development programs, turn to the forests as a resource that can readily be exploited. Under this relentless assault, forests in the Third World are retreating. Every year more than 11 million hectares are cleared for other uses – 7.5 and 3.8 million hectares of closed and open forests, respectively – and in most developing countries, deforestation is accelerating. Between 1950 and 1983 the area of forest and woodland dropped 38 percent in Central America and 24 percent in Africa. Table 1.2 presents estimates of the annual losses of closed forests in developing countries. At these rates there are a number of countries – Nigeria, Ivory Coast, Paraguay, Costa Rica, Haiti, and El Salvador – where severely depleted forests would completely disappear within 30 years. Other countries, especially Indonesia, Brazil, and Colombia, have large reserves but are losing vast areas every year.

Large as they are, these figures show only a fraction of the losses, the area completely cleared to make way for other uses. Forests and woodlands are also deteriorating in quality. Each year over 4 million hectares of virgin tropical forests are harvested, becoming "secondary" forest (Melillo et al. 1985). Under prevailing practices, most of the mature stems of those few species with commercial value are removed (usually amounting to 10–20 percent of standing volume) but typically another 30–50 percent of the trees are destroyed or fatally damaged during logging, and the soil is sufficiently disturbed to impede regeneration, leaving a forest much diminished in quality (Guppy 1984).

In the open woodlands and savannahs of drier regions, where plant growth is slower, fuelwood and fodder demands are outstripping regeneration and accelerating devegetation as populations grow and tree stocks diminish. In countries of the Sahelian/Sudanian zone in Africa, for example, consumption now exceeds natural regeneration by 70 percent in Sudan, 75 percent in northern Nigeria, 150 percent in Ethiopia, and 200 percent in Niger (Anderson and Fishwick 1984). Worldwide, the FAO estimates that 1.5 of the 2 billion people who rely mostly on wood fuel are cutting wood faster than it is growing back (FAO 1983). Woodlands become progressively sparser and eventually disappear under such pressures.

Simple projections based on the growth of population and food demands, and inversely on the increase in agricultural yields, indicate future

Table 1.2. *Deforestation in tropical countries, 1981–1985*

Country	Closed forest area, 1980 (thousand hectares)	Annual rate of deforestation, 1981–85 (percent)	Area deforested annually (thousand hectares)
Group I			
Colombia	47,351	1.7	820
Mexico	47,840	1.2	595
Ecuador	14,679	2.3	340
Paraguay	4,100	4.6	190
Nicaragua	4,508	2.7	121
Guatemala	4,596	2.0	90
Honduras	3,797	2.4	90
Costa Rica	1,664	3.9	65
Panama	4,204	0.9	36
Malaysia	21,256	1.2	255
Thailand	10,375	2.4	252
Laos	8,520	1.2	100
Philippines	12,510	0.7	91
Nepal	2,128	3.9	84
Vietnam	10,810	0.6	65
Sri Lanka	2,782	2.1	58
Nigeria	7,583	4.0	300
Ivory Coast	4,907	5.9	290
Madagascar	12,960	1.2	150
Liberia	2,063	2.2	46
Angola	4,471	1.0	44
Zambia	3,390	1.2	40
Guinea	2,072	1.7	36
Ghana	2,471	0.9	22
Total	241,037	1.7	4,180
Group II			
Brazil	396,030	0.4	1,480
Peru	70,520	0.4	270
Venezuela	33,075	0.4	125
Bolivia	44,013	0.2	87
Indonesia	123,235	0.5	600
India	72,521	0.2	147
Burma	32,101	0.3	105
Kampuchea	7,616	0.3	25
Papua New Guinea	34,447	0.1	22
Zaire	105,975	0.2	182
Cameroon	18,105	0.4	80
Congo	21,508	0.1	22
Gabon	20,690	0.1	15
Total	979,836	0.3	3,160

(*continued*)

Table 1.2 (*continued*)

Country	Closed forest area, 1980 (thousand hectares)	Annual rate of deforestation, 1981–85 (percent)	Area deforested annually (thousand hectares)
Group III			
El Salvador	155	3.2	5
Jamaica	195	1.0	2
Haiti	58	3.4	2
Kenya	2,605	0.7	19
Guinea-Bissau	664	2.6	17
Mozambique	1,189	0.8	10
Uganda	879	1.1	10
Brunei	325	2.2	7
Rwanda	412	0.7	3
Benin	47	2.1	1
Total	6,529	1.2	76
Group IV			
Belize	1,385	0.6	9
Dominican Republic	685	0.6	4
Cuba	3,025	0.1	2
Trinidad and Tobago	368	0.3	1
Bangladesh	2,207	0.4	8
Pakistan	3,785	0.2	7
Bhutan	2,170	0.1	2
Tanzania	2,658	0.4	10
Ethiopia	5,332	0.2	8
Sierra Leone	798	0.8	6
Central African Republic	3,595	0.1	5
Somalia	1,650	0.2	4
Sudan	2,532	0.2	4
Equatorial Guinea	1,295	0.2	3
Togo	304	0.7	2
Total	31,789	0.2	75

Source: International Institute for Environment and Development and World Resources Institute (1986: 73).

deforestation at a declining rate that would reduce the tropical forest area by 10–20 percent by 2020 (Table 1.3). The projected deceleration is due to assumed reductions in population growth rates and inelastic food demands. However, feedbacks between logging and deforestation, migration of farmers into forest areas, and cash crop demands could easily produce much higher deforestation rates (Grainger 1987).

Table 1.3. *Projected deforestation scenarios, 1980–2020*

	High scenario					Low scenario				
	1980	1990	2000	2010	2020	1980	1990	2000	2010	2020
Deforestation rates[a] **(ha × 10⁶ per year)**										
Africa (1.1)	1.6	1.5	1.2	0.9	0.9	1.0	0.9	0.7	0.6	0.4
Asia-Pacific (1.3)	1.7	1.5	1.2	1.2	1.1	1.1	0.9	0.7	0.5	0.4
Latin America (3.1)	3.3	3.1	2.7	2.2	1.7	2.0	1.6	1.1	0.6	0.0
Humid tropics (5.6)	6.6	6.1	5.1	4.3	3.7	4.1	3.4	2.5	1.7	0.9
Forest area (ha × 10⁶)										
Africa	198.9	183.5	170.3	160.4	151.6	198.9	188.9	181.1	175.0	170.1
Asia-Pacific	239.4	222.8	209.5	197.5	185.8	239.4	228.9	220.8	214.8	210.2
Latin America	598.0	566.2	537.5	513.0	493.8	598.0	580.1	566.7	558.6	555.8
Humid tropics	1,036.3	972.6	917.2	870.8	831.1	1,036.3	997.9	968.7	948.5	936.1

[a]Lanly (1982) estimates for 1976–80 in parentheses.
Source: Grainger (1987).

The outlook for tropical timber is also unfavorable. Demand is not limited to subsistence needs in the developing countries but springs in large part from rapidly growing demands in richer countries of the North for exotic tropical hardwoods. Table 1.4 shows the rapid increase of tropical timber exports from developing countries between 1960 and 1984, mostly in the form of logs but increasingly as panels and sawn lumber. To supply European, Japanese, and American markets, countries exploited their mature forests for commercial species with little heed either for the pace or possibility of regeneration or for impairment of other forest functions and yields.

Forecasts of future timber harvest and supply predict a dramatic decline in harvests and exports due to depletion of commercial stands. According to the recent report of an international task force convened by the World Resources Institute, the World Bank, and the United Nations Development Programme, "By the end of the century, the 33 developing countries that are now net exporters of forest products will be reduced to fewer than 10, and total developing country exports of industrial forest products are predicted to drop from their current level of more than U.S. $7 billion to less than U.S. $2 billion" (World Resources Institute 1985: part 1, 10). The process is beginning in Asia, where once leading exporters like the Philippines have already virtually exhausted their lowland productive forests. But Africa and Latin America are seen as able to supply the market only for another few decades before their commercial stands are also depleted (Grainger 1987). Table 1.5 and Figure 1.1 depict future export trends predicted by one model of world production and trade in tropical timber. Although part of these losses in foreign exchange earnings will probably be offset by rising timber prices and increased domestic processing, the forecast nonetheless predicts a severe depletion of an extremely valuable natural resource within a single generation.

Tropical forests are slow to recover fully once disturbed, and although total primary productivity is high, the annual growth of the relatively few prized commercial species is relatively low in a heterogeneous stand. Although secondary tree species will quickly revegetate forest clearings unless soils have been severely depleted, canopy trees and large individuals emerging through the canopy may take a hundred years or more to mature, and the density of commercial stems will likely be lower than in the original stand (Richards 1973). Economists have therefore examined whether sustained yield management can be successfully justified on strict investment principles, especially since high transport costs and limited market demand for little-known varieties greatly reduces the stumpage value of residual species (Leslie 1987). The economic alternative is to

Table 1.4. *International trade in sawlogs, veneer logs, sawnwood, and wood-based panels*

	Trade (m³ × 1,000)				Annual increase (percent)		
	1961	1970	1979	1985	1961–70	1970–79	1979–85
Exports, world total	65,437	130,312	177,706	167,653	8	3	−1
Developing market economies	16,696	44,644	59,963	44,155	11	3	−5
Africa	5,551	8,128	7,427	5,238	4	−1	−6
Latin America	2,148	2,661	4,352	3,935	2	6	−2
Asia	8,971	33,381	47,393	33,444	16	4	−6
Imports, world total	64,750	126,508	178,150	167,524	8	4	−1
Developed market economies	56,343	107,505	142,133	127,387	7	3	−2
North America	12,791	19,880	34,707	43,332	5	6	4
West Europe	32,292	43,839	56,638	48,080	3	3	−3
Asia	10,021	42,658	49,685	34,433	17	2	−6
Developing economies	4,808	11,704	23,526	21,629	10	8	−1

Source: FAO, *Forest Production Yearbook*, various years.

Table 1.5. *Projected exports of logs and processed wood, from developing countries, 1980–2020* $(m^3 \times 10^6)$

	1980	1985	1990	1995	2000	2005	2010	2015	2020
Africa									
Base	7.9	7.7	11.4	15.8	32.8	33.7	21.6	25.2	8.1
High	7.9	7.0	12.1	20.4	57.2	22.0	15.3	5.7	6.0
Asia-Pacific									
Base	41.5	47.3	35.6	30.9	12.1	13.3	11.8	4.2	7.1
High	41.5	50.0	28.5	28.4	15.0	4.2	7.0	4.5	5.4
Latin America									
Base	1.2	8.4	32.2	51.7	77.0	104.0	55.4	33.2	20.7
High	1.2	9.5	46.3	63.9	73.2	64.5	33.9	10.8	4.0

Source: Grainger (1987).

harvest these mature forests, investing the proceeds elsewhere and converting the land to higher-yielding plantations and farms.

This alternative overlooks important arguments for sustained use management. First, tropical forests as a standing resource confer important ongoing benefits that improve conservation's economic return (Hartman 1976). For the 500 million forest dwellers worldwide, a wide variety of nuts, berries, game, fish, honey, and other foodstuffs are available. The values of these yields are badly underestimated, since they hardly ever register as market transactions and for the most part benefit weak cultural minorities. But resins, essential oils, medicinal substances, rattan, flowers, and a wide variety of other products flow from tropical forests into commercial channels. Although they are generally regarded as "minor" forest products and receive little promotion or development attention, their aggregate value is substantial. Exports of such products from Indonesia, for example, reached $125 million per year by the early 1980s, most of which represented employment income for those engaged in collection and trade (Chapter 2).

These products are a small fraction of the potential sustainable yield of the tropical forests, a potential arising from their biological richness (Myers 1984), only a tiny fraction of which has been investigated – let alone utilized. Ecuador, with a land area only 3 percent as large as Europe's, shelters 20–50 percent more plant species. A single hectare of tropical forest may contain 300 different trees, most of them represented by a single individual. The Amazon contains one-fifth of all bird species on earth, and at least eight times as many fish species as the Mississippi River system. A single tree in the Amazon was found to shelter 43 ant species

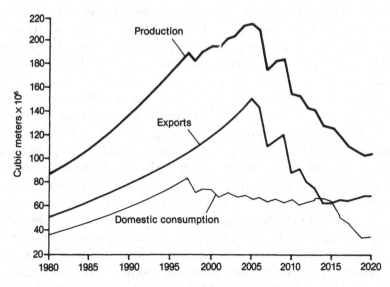

Figure 1.1. Projected production, trade, and domestic consumption of tropical timber, 1980–2020 (Grainger 1987)

from 26 genera, a greater diversity than exists in all the British Isles (Wilson 1987).

Tropical forests contribute genetic material for plant breeders that confers disease and pest resistance and other desirable properties to coffee, cocoa, bananas, pineapples, maize, rice, and many other crops. They contribute entirely new foods such as the mangosteen and the winged bean. Pyrethrins, rotenoids, and other insecticides and insect repellants have evolved in tropical plants in self-defense, while insect predators and parasites have been found to control at least 250 different agricultural pests (Myers 1984). Tropical plants underlie one-quarter of all prescription drugs sold in the United States (U.S. Congress, Office of Technology Assessment 1987). The pharmaceutical products from tropical plants include alkaloids such as quinine and others used in drugs to treat hypertension, childhood leukemia, and Hodgkin's disease; and plant steroids such as diosgenin, which comes from Mexican yams and West African calabar beans and is used in oral contraceptives.

Current uses are only a very minor fraction of the potential benefits. While a score or so of plants make up the large bulk of the human diet, 10,000 plant and animal foods are known to exist, mainly in the tropics. Forest-dwelling Indians in the Amazon know of 1,300 medicinal plants, including antibiotics, narcotics, abortifacients, contraceptives, anti-diarrheal agents, fungicides, anaesthetics, muscle relaxants, and many others, most of which have never been investigated by Western scientists. The sacrifice of these current and potential benefits as the tropics are deforested and endemic species (and indigenous knowledge of them) are lost is generally omitted from economic analyses of forest management options.

Second, the yields of alternative land uses have been greatly overestimated. The soils underlying 95 percent of the remaining tropical forests are infertile and easily degraded through erosion, laterization, or other processes if the vegetative cover is removed. Unlike temperate regions, where organic matter can build up in the soil, high temperatures and rainfalls quickly deplete nutrients in the soil, so that most of the entire nutrient stock is in the biomass itself, where it is quickly and efficiently recycled (Breunig 1985). The tropical forest ecosystem is highly adapted, to the extent that dense forests can even exist on what is essentially sand. Attempts to substitute monocultures, whether of annual or tree crops, typically encounter declining yields and invasion by pest and weed species (National Research Council 1982). Plantation crops that are most successful in the tropics are those such as tea, cocoa, and rubber that remove relatively little of the nutrient stock. Traditional long-fallow shifting cultivation systems also give the forest ecosystems ample time to recover, but most annual cropping efforts have proved to be economic as well as ecological failures. Agricultural settlement schemes and large-scale plantations have incurred economic losses and, if not covered by continuing government subsidies, have led to the eventual abandonment of large areas of degraded soils and impoverished biota (Buschbacher, Uhl, and Serrao 1987). These losses are rarely entered into the calculus of expected benefits and costs. Instead, as this book shows, governments often provide generous incentives to encourage conversion in the illusory expectation of higher returns and absorb the ensuing losses.

Finally, the economic benefits from timber harvests to the national economies of tropical countries are substantially overstated. The *gross* value of timber exports is the usual focus of attention. Overlooked are the large acknowledged and hidden outflows of profits gained by domestic and foreign timber concessionaires and the politicians and military officers who are often their silent partners. The net domestic benefits gained by the economies of tropical countries from depletion of their forest resources have been surprisingly small, as several of the following case studies document. Value-added in the forestry sector averaged 3.3 percent of GDP in a 1980 sample of African countries (Chapter 7), and this low figure substantially overstates its contribution to income: a substantial fraction of value-added should be regarded as depreciation of the forest sector's capital stock rather than as current income. In some countries, including the Philippines, annual revenues accruing to national treasuries from forest exploitation have not covered even the administrative and infrastructure costs incurred for timber harvesting (Chapter 4). Chapter 8 shows that the same is true in dozens of the national forests of the United States, where forestry policy and practice are usually presumed to be more highly developed. Therefore, the narrow economic case for "mining" tropical forests as an exhaustible resource is far from established, and there is no justification for exploiting the resource as wastefully and destructively as many countries have done. In fact, economic and ecological losses have gone hand in hand.

The reasons for deforestation in tropical countries

The reasons for the rapid deforestation taking place in the Third World are complex. Subsequent chapters show that some are rooted deeply in countries' development patterns: rapidly increasing populations, extreme concentration of landholdings that leave hundreds of millions in search of land, slow growth of job opportunities in both city and countryside. The Ivory Coast, where in-migration and natural increase produced population growth at 4.6 percent per year between 1965 and 1985 (by far the most rapid in the world), also experienced accelerating deforestation, from an annual rate of 2.4 percent per year between 1956 and 1965 to one of 7.3 percent annually between 1981 and 1985, the result of shifting cultivation, logging, and conversion to farms.

In countries where development opportunities for the majority of people have lagged, such as Ghana and the Philippines during the 1960s and 1970s, impoverished, often landless, rural households have moved into forested regions in search of available arable land. In Ghana, a nation of 11 million people, over nine million hectares had come under shifting cultivation by 1980, 40 percent of the total land area and eight times the area of remaining productive forest.

Further reasons include rules of land tenure in many countries that confer title to forest lands on parties who "improve" it by clearing the forest for some other use. Brazilian authorities have tended to settle disputed land claims by granting claimants title to areas that are multiples of the total forest area cleared for agricultural uses (Chapter 6). In the Malaysian state of Sabah, laws dating from the British colonial period make the state government the holder of all forestry property rights, vitiating the traditional rights of local communities, but permit any native person to obtain title to forest land by clearing and cultivating it (Chapter

3). In other countries, such as the Philippines, land claims predicated on forest clearance involve not only small-scale shifting cultivators, but also extensive livestock operations (Chapter 4).

In Ghana, rights to use the resources of the forests were governed by traditional communities, until taken over by the central government in the early 1970s. As a result, the forest has become even more vulnerable to open-access common property problems, because tribal heads no longer have any strong reason to try to limit shifting cultivation or timber operations (Chapter 7). In many other countries as well, the displacement of traditional communities exercising customary law over the forest actually weakened controls over the use of the resource.

In addition, some reasons stem directly from government policies in Third World countries toward forest exploitation, and toward industries that compete for the use of forest lands. These policies are the main focus of this report, which shows that better policies can conserve forest resources, raise economic welfare, and reduce fiscal burdens on governments.

Investment incentives, tax provisions, credit concessions, agricultural pricing policies, and the terms on which firms obtain access to the public forests are the means through which governments affect forest exploitation. In the developed countries, although the struggle is over priorities in the use of a stable forest resource, the policy mechanisms are much the same. In fact, throughout the world, governments largely determine how forests should be used.

In most Third World countries, and in all ten whose experiences are included in this volume, the large remaining natural forest areas are public lands. Governments have taken over authority and responsibility for managing them from indigenous communities, which traditionally used the forests in accordance with their own laws. According to a comprehensive FAO assessment, over 80 percent of the closed forest area in Third World countries is public land (Lanly 1982). Since central governments in many developing countries have had neither the means nor the local acceptability to enforce forestry regulations in remote regions, undermining local authority has sometimes opened up access to the resource for virtually uncontrolled exploitation.

In the industrialized countries as well, a substantial percentage of remaining forests are on public lands. In law and in fact, governments set the conditions for exploitation and use of the public forests. Even the use of private forests is greatly influenced by government policies. Because commercial forestry involves holding a growing asset for long periods, returns to private investors are sensitive to credit costs, inflation, taxes on land and capital assets, and other economic parameters greatly affected by government policy. Because some forest lands can be used for agricultural or other purposes, government policies that stimulate expansion of these competing land uses can do so at the expense of forest area.

The economic criterion for efficient forest management is getting the maximum total benefit from *all* the forests' various possible uses over the long run, discounting future benefits at an appropriate interest rate. This criterion implies that land areas should be devoted to the uses, forest or non-forest, that yield the greatest potential economic benefits – whether or not those benefits are reflected in market transactions. It implies that the various uses of each forest should be managed to produce the greatest total benefit. For example, applying the efficiency criterion means that land most valuable as a watershed protection forest would not be converted to crops; a forest most valuable as a recreational park would not be harvested for timber; or a forest containing immense mineral reserves would not be preserved as wilderness.

In countries endowed with large forest resources, government policies frequently ignore this criterion for efficient resource use. They overemphasize the timber harvest at the expense of other potential benefits. Even in managing timber on public lands and in creating incentives for use of private forests, they sacrifice much of the potential long-term benefit for a lesser transitory gain. Potential benefits from forest exploitation are dissipated in wasteful harvesting and processing or allowed to flow unnecessarily to stockholders of timber companies. Government policies also result in greater conversion of forest lands to agricultural and other uses than is economically warranted, with a loss in net benefits from the land. Despite official endorsements of conservation goals, government policies contribute significantly to the rapid deforestation now under way.

Because these policies are often implemented through manipulation of tax codes, public credits, and charges for the use of public lands, they typically create fiscal burdens for governments and taxpayers, along with the sacrifice of long-term economic welfare and the forest resource itself. Improving the policy framework offers many opportunities to promote more sustainable development in this sector: greater long-term economic benefits from forest use, more effective resource conservation, and reduced fiscal burdens on governments.

The case studies in this volume give examples of these policy opportunities. Country investigations undertaken in China, Indonesia, three separate regions of Malaysia, the Philippines, Brazil, Peru, Ghana, Liberia, the Ivory Coast, Gabon, and the United States were carried out either by local analysts and research institutions or by U.S. researchers with long-term local interests and experience. These countries represent different continents, different economic systems, different levels of development, and different ecological zones. The settings, the problems, and the opportunities differ among them. Yet, there are also surprising similarities.

Forest sector policies

Timber concessions

Many countries became independent still endowed with substantial public forests of mature commercial timber. These countries have generally promoted rapid depletion of those resources by conceding much of their economic value to logging interests that contract with the government for the timber on public lands. Typically, the first contractors were foreign firms, some with links to the countries from earlier colonial periods. British, French, American, and more recently, Japanese, Korean, and Taiwanese firms have actively sought and implemented concession agreements in tropical countries. Still more recently, logging firms from Malaysia and the Philippines have won sizeable concession areas in neighboring tropical countries (Chapter 2). Over time, domestic entrepreneurs have taken increasingly important positions in highly profitable timber extraction, either as local partners to foreign companies or as domestic concessionaires. In the Malaysian state of Sabah, for example, a domestic foundation will ultimately become the sole concession holder, with foreign and domestic firms limited to the role of logging contractors (Chapter 3).

The "stumpage value"[1] of an accessible virgin forest of commercial species is often considerable. This stumpage value is an economic rent, a value attributable not to any cost of production, but to the strength of market demand and favorable natural resource endowments and location. Table 1.6 presents estimates of the rents available in logging tropical forests, in U.S. dollars per cubic meter. For high-valued species, they were approximately $70–$100 per cubic meter of logs at the end of the 1970s; for low-valued and middle-valued species, about one-quarter to one-half as much.

Rent, by definition, is a value in excess of the total costs of bringing trees to market as logs or wood products, including the cost of attracting the necessary investment. That cost may include a risk premium that reflects uncertainties about future market and political conditions, so there are inevitably doubts about the exact magnitude of available rent. The-

Table 1.6. *Estimated rents in logging tropical forests, by country, species value, and time period (US$ per m³)*

Country and period	Value of species			
	Highest	Middle	Lowest	Overall
I. 1979				
Indonesia				85
Sabah, Malaysia				94
Philippines				69
Liberia	98	41		
II. 1973–1974				
Indonesia				45
Liberia	89	58	25	
Ivory Coast	47	31	17	
Gabon	89	54	22	
Cameroon				
Douala	61	32	14	
Pointe Noire	52	23	7	
Congo				
South	81	52	23	
North	69	42	13	
III. 1971–1972	79	28		
Ghana				

Sources: Chapters 2, 3, 4, and 7.

oretically, all rent can be captured by governments as a revenue source that stems from the country's advantageous natural resource assets. In practice, royalties, land rents, license fees, and various harvest taxes are all means of converting rent into government revenues. To the extent it is not captured, it remains as a source of greater than normal profits for the timber contractor, or as a cushion for defraying excess costs.

Most developing countries with large mature forests have failed to adopt forest revenue systems that come close to capturing these rents for the public treasury. While most countries have sought to raise the government's share over time, only some have been successful; in others, the real value of the government's revenues has been eroded by inflation, evasion, or poorly designed fiscal systems. Despite a variety of fees, royalties, taxes, and miscellaneous charges, total forest revenues have fallen far short of their potential in most exporting countries. Table 1.7 presents information from several country studies that document this shortfall. Column 5 shows that in the Philippines, the government captured only 16.5 percent of logging rents between 1979 and 1982, while Indonesia obtained

Table 1.7. *Government rent capture in tropical timber production (million U.S. dollars)*

(1) Country and period	(2) Potential rent from log harvest[a]	(3) Actual rent from log harvest[b]	(4) Official government rent capture[c]	(5) Col. 4 ÷ Col. 3 (%)	(6) Col. 4 ÷ Col. 2 (%)
Indonesia 1979–82	4,954	4,409	1,644	37.3	33.2
Sabah 1979–82	2,198	2,094	1,703	81.3	77.5
Ghana 1971–74	—	—	29	38.0	—
Philippines 1979–82	1,505	1,033	171	16.5	11.4

[a]Potential rent assumes that all harvested logs are allocated to uses (direct export, sawmills, plymills) that yield the largest net economic rent.
[b]Actual rent totals rents arising from the actual disposal of harvested logs.
[c]Rent capture totals timber royalties, export taxes, and other official fees and charges.
Sources: Chapters 2, 3, 4, and 7.

38 percent, as did Ghana in an earlier period. Of these countries, only Sabah's revenue system brought in a high percentage of potential revenues through aggressive taxation.

The result of this failure has almost invariably been a rush by private contractors for (aptly named) "timber concessions," because those contracts for timber harvests and leases of public forest lands offer high potential rates of return on investment. The greater the loss of potential public revenues, the greater the profit incentive to private investors. Entrepreneurs are induced to seek timber concessions before others sign agreements to exploit all the profitable areas.

This rent-seeking behavior has generated the "timber booms" experienced by many countries. Foreign and domestic entrepreneurs contract for harvesting rights on vast areas. Sometimes to forestall risks of contract renegotiation or revision, and sometimes because of contractual obligations imposed by governments, concessionaires quickly enter the forests to begin large-scale harvesting operations. In part because royalties and other forest revenue systems encourage it, concessionaires typically "high-grade" their tracts, taking the best specimens of the most highly valued species but disturbing extensive forest areas in the process (Chapter 2, Appendix A). Since access roads are vital to shifting cultivators and settlers, who rarely penetrate far into wilderness forests, loggers are quickly followed by migrants who complete the process of deforestation.

The Ivory Coast provides an extreme example. Despite *ad valorem* export taxes that ranged from 25 percent of f.o.b. value for low-valued species to 45 percent for highly prized varieties, the estimated rents left to concessionaires during the 1970s approximated $40 per cubic meter for the most valuable species, $30 for moderate valued species, and $20 for low-valued varieties (see Table 1.6). These were more than sufficient to stimulate rapid exploitation and depletion of the timber resource. Between 1965 and 1972 concession agreements assigned more than two-thirds of all productive forests to concessionaires within the space of seven years. Timber contractors have virtually exhausted the more valuable species, and shifting cultivators have moved in on their heels to clear the depleted forests. In 1985 Ivory Coast forests were only 22 percent of their extent 30 years earlier (Chapter 7). Ominously, the rents per cubic meter are much lower in the Ivory Coast than those available to concessionaires in countries such as Gabon, where considerable reserves still remain.

Detailed estimates made for Indonesia illustrate the size and disposition of rents (Chapter 2). Log exports from Sumatra and Kalimantan, two main concession areas, generated potential rents that averaged $62 per cubic meter exported between 1979 and 1982. This figure represents the difference between the logs' average export value and the total costs of harvesting and moving them to the ports (exclusive of taxes and fees). Total identifiable government revenues, including timber royalties, land taxes, reforestation fees, and other charges, averaged $28 per cubic meter. Thus, the government recaptured only 45 percent of the rents available from log exports.

Timber exported as sawn timber received even more favorable treatment. Due to lower tax rates intended to encourage local processing, the government captured only 21 percent of the rents generated by Indonesian logs exported in the form of sawn timber. Potential rents from logs used for plywood production were actually destroyed because production costs exceeded the price margin between log and plywood exports, so that, without tax concessions and investment incentives, plywood producers would have incurred losses. Between 1979 and 1982 the potential economic rents generated by log production, whether for further processing or direct export, exceeded $4.95 billion. Of this, the government's share, collected through official taxes and fees, was $1.64 billion. Five hundred million dollars of potential profits were lost because relatively high cost domestic processing generated negative economic returns. The remainder, $2.8 billion, was left to private parties. This represents an average of $700 million in rents annually. Largely because of such strong profit incentives as these, by 1985 the total area under more than 500 concession agreements or being awarded to applicants was 65.4 million hectares, 1.4 million hectares *more* than the total area of production forests in the country (Chapter 2).

In the Philippines, between 1979 and 1982, the forest sector generated rents in excess of a billion dollars. The potential rents were even larger, approximately $1.5 billion. The difference is the loss due to conversion of an increasing volume of exportable logs to plywood in inefficient mills. As discussed in greater detail below, the low conversion rates in Philippine mills implied that each log exported as plywood brought a lower net return over cost than the same log exported as sawn timber or without processing. By comparison, the government's total revenues over these years from export taxes and forest charges on log, timber, and plywood production was $170 million, less than 11 percent of potential rents and 16 percent of actual available rents (Chapter 4). The remainder, more than 800 million dollars, was retained by exploiters of the forest resource.

Moreover, although production cost estimates are unavailable with which to estimate aggregate rents for earlier periods, it appears that before 1979, when timber harvests in the Philippines were at a higher level, the government's share of rents generated by forest exploitation was even lower. From 1979 to 1982 forest charges and export taxes totaled 11 percent of the value of forest product exports. In the preceding five years, they came only to 8 percent.

The result in the Philippines was also a dramatic timber boom. Between 1960 and 1970 the area under concession agreements rose rapidly from about 4.5 to 10.5 million hectares. Many large U.S timber companies participated, along with increasing involvement by Philippine military, political, and traditional elites. Logging activity rose in step, peaking in the mid-1970s. By the mid-1980s virgin productive forests had virtually been logged out.

Many governments have reinforced these powerful incentives for timber exploitation with other provisions that tend to accelerate the process. Governments enter agreements with many concessionaires, not through a process of competitive bidding, which would increase the government's share of the rents, but on the basis of standard terms or individually negotiated agreements. Potential investors are thus led to rush into agreements before others take up all the favorable sites offered for exploitation.

Governments typically impose conditions on concessionaires requiring them to begin harvesting their sites within a stipulated time, and also limit agreements to periods much shorter than a single forest rotation: to 25, 20, 10, or 5 years, and even a single year. In the Malaysian state of Sabah, for example, half of all timber leases are for the regular 21-year term, but most of the remainder are for only 10 years, and 5 percent are for just one year (see Chapter 3). These conditions are imposed to prevent concessionaires from stockpiling leases, but often result in a much faster harvest schedule and less concern for future productivity than the private investor would choose if he owned the land and the timber outright and had to pay only an ordinary income tax. Licensees in Indonesia have often reentered the forests for a second cut long before the stand has had an adequate chance to recover, for example, in order to strip the concession of remaining merchantable timber before expiration of the concession.

Macroeconomic policies, such as exchange rate overvaluation and undervaluation, may affect timber rents. In Indonesia, for example, the rupiah was allowed to appreciate against the U.S. dollar by more than 50 percent between 1970 and 1977, before a currency devaluation in 1978 partially restored the earlier relationship. The cycle was repeated between 1979 and 1983. These episodes of currency overvaluation reduced private rents from timber exports. But, they may have more severely discouraged non-timber forest product exports, because their supply costs are determined mostly by domestic labor costs (Chapter 2).

Of course, countries sometimes maintain undervalued exchange rates, which have the opposite effect. Exports, including exports of forest products, are encouraged by payments to shippers in domestic currency in excess of the value of the foreign exchange they earn. Malaysia's currency has been consistently undervalued, a situation which has stimulated logging and large-scale conversion of logged-over areas to plantations of export tree crops, mainly rubber and palm oil (Chapter 3).

In addition, some governments increase the potential rents available to contractors on timber from public lands by assuming some of the costs of bringing timber to market. These include costs of constructing trunk roads, port facilities, and other infrastructure; administrative costs of surveying, marking, and grading timber to be sold; and costs due to the environmental side effects of timber operations.

In extreme cases, these budgetary subsidies can result in commercial harvest of timber that has negative rent, i.e., timber that is not worth the cost of marketing. In the United States, for example, the U.S. Forest Service supports logging on over 100 million acres of the national forests that are economically unfit for sustained timber production. It does so by selling timber at prices that do not cover its own growing, road building, harvesting, and selling expenses, with a cost to taxpayers of approximately $100 million a year (see Chapter 8). Although the Forest Service attempts to justify this policy for its benefits to wildlife and recreational users of the national forests, it is clear that commercial logging unduly takes precedence over other management objectives in most of the U.S. public forests because of these subsidies.

Governments affect the pace of deforestation both by the level of their charges for use of public forests and by the form those charges take. Differences in the structure of forest revenue systems can markedly alter the pattern and level of harvesting, as well as the distribution of revenues between government and concessionaire (Gray 1983). Most forest charges are based on the volume of timber removed, not the volume of merchantable timber in the tract. This practice, along with high transport costs and narrow market preferences for known species, encourages licensees to harvest highly selectively, taking only the stems of greatest value. The effects in tropical forests are serious: first, a larger area must be harvested to fulfill timber demand, opening up more virgin forest to squatters and shifting cultivators; second, the remaining stems and underlying soils are usually severely damaged by logging operations. In Sabah, between 45 and 74 percent of trees remaining after a harvesting operation are substantially damaged or destroyed (see Chapter 3); in Indonesia and the Philippines, estimates of damage fall in the same range (Chapters 2 and 4). If concessionaires re-enter the tract within 5–10 years to extract any valuable timber left before their licenses expire, as not infrequently occurs, they leave the forest virtually valueless and damaged badly enough that regeneration and recovery are uncertain. Figure 7.1 in Chapter 7 depicts the dramatic depletion of highly valued species, including ebony and mahogany, from the forests of the Ivory Coast, and the rising percentage of "other" low-valued trees in timber exports: between 1962 and 1978 these residual species rose from 15 to over 50 percent of the harvest as the prized woods became ever more scarce.

Inappropriately designed forest revenue systems encourage high-grading or "mining" of timber stands. Of the many forms timber charges take, flat charges per cubic meter harvested provide licensees the strongest incentive for high-grading, unless they are finely differentiated by species, grade, and site condition. *Ad valorem* royalties are better than flat charges but not by much. The reason is simple: trees with a stumpage value less than the forest charge are worthless to the licensee, and can be left or destroyed with impunity. The Philippine government charges licensees a relatively undifferentiated specific royalty that tends to encourage high-grading. Finely differentiated systems are beyond the administrative capabilities of most tropical countries, and are not widely used.[2] However, the Malaysian state of Sarawak imposes specific charges that vary considerably by species, with much lower rates on low-valued trees, and suffers only half as much residual tree damage from logging operations as either Sabah or Indonesia (see Chapter 3). If such differentiated systems are administratively infeasible, revenue systems based on income taxes or ground rents (area license fees) promote more complete utilization of the growing timber, because even inferior trees are more likely to have some positive stumpage value to the licensee.

At least as important as the stipulations of forestry concession agreements is their enforcement, which is inherently difficult. Forested regions are vast and remote. What happens there is far removed from public scrutiny. Even in countries that have used forest guards such as Liberia (Chapter 7), government agents are thinly spread, and forest exploitation provides ample funds with which to suborn them. In many countries, including the Philippines, Malaysia, and Indonesia, timber concessionaires have been closely linked to political or military leaders, making enforcement by mere forest agents difficult. When concession terms are not adequately enforced, licensees can cut costs and raise their returns, usually at the expense of the government and the forest resource base. In the Ivory Coast, for example, harvest methods were not even prescribed in concession agreements until 1972. Throughout the 1970s, as discussed in Chapter 7, forestry officials lacked the resources and the information required to determine and enforce annual allowable cuts, to administer obligatory removal of secondary species, and to verify the working programs of logging companies. As a result, concessionaires were not obliged to follow any particular technique of selection, or any particular cutting method.

In the Philippines, illegal cutting and timber smuggling have been widespread. During the 1970s restrictions on log exports were introduced to encourage local processing and to preserve timber resources. From 1976 on, for example, log exports were allowed only from certain regions, and in 1979 they were limited to 25 percent of the total annual allowable cut. The result was considerable log smuggling and underreporting of exports. Between 1977 and 1979, Japanese trade data recorded imports of 4.7 million cubic meters from the Philippines, but the Philippines' recorded exports to Japan, where three-fourths of forest products exports were sent, totaled only 4.1 million cubic meters, a discrepancy worth U.S. $70 million in export receipts. In 1980, after log export restrictions were tightened, the incentive to smuggle out logs became stronger. A glut of logs on the domestic market kept prices within the Philippines much below prices on the export market. Correspondingly, the discrepancy in reported exports widened: Japan imported 1.1 million cubic meters of logs from the Philippines, although only 0.5 million were recorded as Philippine exports. Indonesia is known to have experienced similar problems of log smuggling, underreporting of exports, and evasion of forestry stipulations and export bans (see Chapter 4).

Persistent, and sometimes extreme, currency overvaluation has been another reason for log smuggling from some Third World countries. By affording all exporters a return in domestic currency much less than the market value of the foreign exchange they earn, currency overvaluation discourages production for exports. In Ghana, between 1977 and 1980, the free market exchange rate varied between five and eight times the official exchange rate. Such extreme currency overvaluation effectively discouraged all export production, except that which could be smuggled out. For those years timber smuggled overland is estimated to have earned 80 percent as much foreign exchange as timber exported officially, but without contributing to tax receipts (see Chapter 7). Inadvertently, these policies have reduced government revenues and management control over the public forests along with export earnings.

Incentives for wood-processing industries

Log-exporting countries have had to struggle to establish local wood-processing industries, even though processing reduces the weight of the raw material and economizes on shipping costs. One important reason is that industrial countries typically have set tariffs much higher on imports of processed wood products than on logs, in order to protect wood manufacturing industries, as illustrated in Table 1.8. Such tariff escalation allows protected industries to compete successfully even if their labor and capital costs are much higher. Indeed, studies of wood-processing industries in Japan and Europe show that without trade protection their costs would be uncompetitively high (Contreras 1982). Removal of these discriminatory trade barriers would increase world trade and welfare.

In the past, to counteract these trade barriers and to stimulate investment in processing capacity that would create employment and additional value-added in wood industries, log-exporting countries have banned log exports, reduced or waived export taxes on processed wood, and offered substantial investment incentives to forest product industries. For example, Ghana used four different measures in the attempt to stimulate investment in domestic processing industries: log export bans were enacted but most were evaded; plywood and other wood products were exempted from export taxes (which were far less onerous than the overvaluation of the exchange rate); long-term loans for sawmills and plymills were granted at zero or negative real interest rates;[3] and, finally, a 50 percent rebate on income tax liabilities was given to firms that exported more than

Table 1.8. *Most favored nation (MFN) tariff levels for selected forest products: Australia, EEC, Japan, and United States (as of December 1985)*

CCCN tariff no.	General product description	Tariff rate[a] (percent)			
		Australia	EEC	Japan	United States
44.03	Wood in rough	0	0	0	0
44.05	Wood simply sawn	5	0, 4.1	0, 7, 10	0
44.09	Wood chips	0	0	0	0
44.13	Wood planed, grooved, etc.	0–15	0, 4.1	10	0, 4.4
44.14	Veneer	5	6.1	15	0
44.15	Plywood	30–40	10.4	15, 20	4.1–20
	Laminated lumber	15	11.1	20	1.9¢/lb + 3.4%
44.19 to 44.28	Manufacture of wood products	15	2.6–9.1	2.5–7.2	0–8
44.01 to 44.03	Furniture	30	5.6, 6.3	4.8	4.7–9.3
47.01	Wood pulp	0	0	0	0
48.01 to	Newsprint	7	5.4	3.9	0
48.15	Other paper and paperboard	6–14	4.1–12.8	5–12	0–3.3

[a]These are MFN rates. Special preferences may be available for specified supplying countries. Most products are eligible for GSP treatment. Non-tariff barriers may place limitations on some products.
Source: Bourke (1986).

25 percent of output (when the ordinary income tax rate stood at 55 percent of earnings). By 1982 these policies had created a considerable domestic industry consisting of 95 sawmills, 10 veneer and plywood plants, and 30 wood-products manufacturing plants (Chapter 7).

Similarly, in the Ivory Coast there have been generous incentives to companies for creation of wood-processing capacity. Companies that make approved investments can write off half the costs against existing income tax liabilities and are eligible for income tax holidays on subsequent production for seven to 11 years. These incentives, in addition to the large reductions in export taxes on exports of processed wood, explain the creation of a sizable but inefficient processing industry (see Chapter 7).

During the 1970s, Brazil offered very liberal subsidies to wood-processing industries in the Amazon, within the framework of an extraordinarily generous set of incentives to encourage investments in that region. First, firms were offered tax credits equal to their investments in approved Amazonian projects, up to 50 percent of their total income tax liability and 75 percent of the total project costs. In other words, firms could use money owed in taxes to invest in the Amazon. Between 1965 and 1983, about a half billion dollars of such funds were invested in wood-processing industries, 35 percent of all tax credit funds committed to Amazonian investments (see Chapter 6).

In addition, approved projects were eligible for partial or complete income tax holidays for periods up to 15 years. By 1983, the agency administering these incentive programs had granted tax holidays to 260 wood-processing firms in the Amazon. Finally, liberal export financing has been provided for Amazonian forest product producers and traders. Since 1981, export trading companies have been eligible for subsidized credits up to 100 percent of their exports in the prior year, at effective interest rates that have been considerably below the rapid rate of inflation in Brazil. Real interest rates on credits in the Amazon have been consistently negative, averaging *minus* 30 percent per year between 1981 and 1983 (Hecht 1985). This interest rate advantage is equivalent to a subsidy of up to 30 percent on the total value of forest product exports.

These industrialization incentives can contribute to an increase in local employment and income, but often do so at a heavy domestic resource cost, with lost government revenues and faster deforestation. In Ghana, as in the Ivory Coast and Indonesia, many of the mills established in response to these inducements have been small and inefficient. Recovery rates for sawn lumber and plywood have been only 70 and 67 percent, respectively, of those that are achieved elsewhere (Chapter 7). The shift to domestic processing in technically inefficient mills has the unfortunate result that considerably more logs must be harvested to meet any level of demand, disturbing much larger forest areas through selective cutting.

Furthermore, after local processing industries have come into existence with official encouragement, governments are most reluctant to reduce their supply of raw materials. This reluctance virtually assures that enough logs will continue to be harvested to feed the mills, whatever the economic or ecological reasons for reducing the harvest. In the United States, one of the main reasons why the U.S. Forest Service continues to harvest timber on lands unsuitable for commercial production, at a cost to the taxpayer exceeding $100 million annually, is to supply local mills dependent on logs from the national forests (see Chapter 8).

The Indonesian experience, discussed in Chapter 2, illustrates the fiscal costs, and the risks to the forests, that ambitious forest-based industrialization plans entail. In order to encourage local processing, in 1978 the government raised the log export tax rate to 20 percent, but exempted most sawn timber and all plywood. Mills were also exempted from income taxes for periods of five or six years. Since these tax holidays were combined with unlimited loss carryover conditions, concessionaires were frequently able to extend the holiday by declaring (unaudited) losses during the five-year holiday provision, or by simply arguing before sympathetic tax officials that the holidays were intended to apply for five years after the start of *profitable* operations.

With these incentives and the impending ban on log exports, the number of operating or planned sawmills and plymills jumped from 16 in 1977 to over 100 in 1983. By 1988, plymills will be on stream with a total effective capacity for processing 10 million cubic meters of logs per year. Sawmill capacity is expected to account for another 28.8 million cubic meters of logs. Of this total potential demand for logs, only 1.8 million cubic meters will be met from teak plantations on Java; as much as 37 million could come from the natural forests. This annual harvest level would be 50 percent greater than the maximum levels reached in the 1970s, when log exports peaked; according to Government of Indonesia long-term forestry plans, log harvests to feed the mills are expected to continue rising throughout the 1990s as well.

Because of low conversion efficiencies (2.3:1 for plymills and 1.75:1 for sawmills) and distortions in relative log and plywood prices in world markets, the jobs this harvest will create are bought at a heavy cost. For example, although a cubic meter of plywood could be exported for $250 in 1983, the export value in terms of the logs used as raw materials (the roundwood equivalent) was only $109. However, the logs themselves could be exported for $100 per cubic meter. In other words, plymills added only $9 in export earnings for every cubic meter of logs used. But, because of the export tax exemption for plywood, the government sacrificed $20 in foregone tax revenues on every cubic meter of logs diverted to plymills. At current conversion rates and projected production levels, by 1988 the revenue loss will mount to $400 million annually.

In 1983, plywood exports worth $109 at international prices per cubic meter of logs processed cost the rupiah equivalent of $133 to produce, a sacrifice of potential gains possible only because of the government's financial incentives and its forgiveness of log export taxes. The losses involved in producing sawn timber for export were more obvious. Because the average price of sawn timber exported in 1983 was only $155, a cubic meter of logs that could be exported for $100 brought only $89 if processed in local sawmills. The government actually sacrificed $20 in export taxes in order to *lose* $11 in export earnings on every cubic meter of logs sawn domestically. Economic losses, as well as wastage of the natural resources and the sacrifice of public revenues, can result when overly generous incentives permit or encourage inefficient processing operations.

Although the data with which to illustrate the effects are more complete for Indonesia, other countries' industrialization policies raise the same issues. In the Philippines, a large processing sector was created through fiscal incentives by threatened bans on log exports and by linking logging concessions to processing investments. Also, wood-processing industries have been heavily protected through differential export charges. By 1980, 209 plymills were in operation with an annual log requirement of 3.4 million cubic meters. Most were small and inefficient, with an average conversion rate of only 43 percent, far below Japan's 55 percent rate, although lower labor and transport costs offset part of this disadvantage.

In the years 1981 through 1983, production costs in Philippine mills exceeded export receipts per cubic meter of logs converted in plymills, as shown in Table 4.9, so that national rents were actually negative. Nonetheless, private interests could profit because of investment incentives and relief from log export taxes, which alone amounted to $18 per cubic meter of logs sent to domestic plymills. Although the number of mills dropped dramatically over these years through consolidation and rationalization in the plywood industry, capacity to process 3.1 million cubic meters remained in 1982.

In the Ivory Coast, plymills have been erected by timber concessionaires mainly to enable them to qualify for log export quotas, and are widely regarded as inefficient, with conversion ratios of about 40 percent.

Table 1.9. *Ivory Coast export taxes and incentives for domestic processing, selected major species*

Species	Additional domestic value-added from sawmilling (U.S. dollars per m³)	Export taxes foregone by government on sawn timber exports (U.S. dollars per m³)	Taxes foregone as percent of increased value-added
Iroko	$25.50	$52.00	204
Acajou	19.20	43.00	224
Llomba	9.24	10.00	108

Source: Chapter 7.

Because *ad valorem* export taxes on plywood are only 1 percent and 2 percent for low- and high-valued species respectively, instead of 25 percent and 45 percent for the logs themselves, domestic processing involves a considerable sacrifice of government revenues. Table 1.9 presents data for two high-valued and one low-valued species. For *iroko*, for example, more than $50 in taxes are foregone for every $25 of extra foreign exchange earnings generated when a cubic meter of log is processed into plywood (Chapter 7).

In summary, generous logging agreements that leave most of the rents from harvesting virgin forests to concessionaires, and excessive incentives to forest product industries that encourage rapid, often inefficient, investment in wood-processing capacity, combine to increase the log harvest much beyond what it would otherwise be. Poorly drafted and enforced forestry stipulations are inadequate to ensure sustainable forestry practices in the face of these powerful incentives. Forest stocks are depleted, but neither the government treasury nor the national economy benefits much from the exploitation.

Indeed, many unmeasured costs reduce the benefits even further. In all the countries studied, settlers and shifting cultivators travel the logging roads after the log harvest, completing the process of forest clearance after the commercial stems have been removed. Deforestation by shifting cultivators and timber operations are closely interlinked.

In addition, ecological losses are severe. The most dramatic were suffered in Indonesia and Sabah, during the 1982–83 drought (Chapters 2 and 3). Intense forest fires destroyed an area in East Kalimantan as large as all of Belgium – the worst forest fire ever recorded. Less is known about the extent of losses in Sabah, but damage was very extensive. Destruction in both countries was especially severe in logged-over areas, because dead trees and litter provided enough fuel to ignite remaining stems. Damage to unlogged areas was slight. Such ecological disasters, along with soil erosion and compaction, river siltation and flooding, and destruction of the habitats of indigenous peoples and wild species, are among the unpriced costs of these incentives for forest exploitation.

Policies outside the forest sector

Population growth and economic opportunities stimulate the conversion of forest land to other uses: at least a billion hectares have been deforested in the past hundred and fifty years. The reforestation of many areas in the developed countries where farming has become uneconomic demonstrates that market forces can move in either direction.

However, many governments have deliberately adopted policies that powerfully accelerate the conversion of forest lands to farming or ranching, by providing artificial incentives that lower the costs and increase the private profitability of the alternative land uses. These subsidies can become so large that they encourage activities that are intrinsically uneconomic, or push alternative land uses beyond the limits of economic rationality. When this happens, inferior and often unsustainable land uses are established because of the subsidies. Better government policies could reduce burdens on the treasury, reduce economic losses, and promote more effective resource use.

Subsidies take several forms. Governments may assume many of the costs of establishing the competing activity directly, through spending on infrastructure, grants to settlers, or budgetary losses in state-operated enterprises. Governments provide financial aid to private investors through low-interest loans and tax deductions. Also, governments may boost the profitability of agriculture or ranching activities through manipulation of farm output and input prices. The effect of all such measures is to shift the margin of relative profitability between forestry and the competing land use, encouraging more forest conversion than would otherwise take place.

In many forest-rich countries, governments actively promote and subsidize agricultural settlements in forested areas, often at very heavy cost. Multilateral development banks and bilateral aid agencies have shared these costs. In Indonesia, for example, the "transmigration" of settlers from Java to the large, sparsely populated Outer Islands is a long-standing government program. In the decade 1971–1980, approximately one million people were moved, at a cost of several thousand dollars per family. Between 1984 and 1989 the government of Indonesia plans to resettle 1 million families, about 5 million additional people, and the government's costs have risen to $10,000 per household. This is an extraordinarily high subsidy in a country where GNP per capita is only $560 per year and total annual investment per capita is only $125. The World Bank has loaned hundreds of millions of dollars to Indonesia to support the transmigration program.

In the past, many of these settlements have failed, in part because of inadequate assessment of the agricultural capabilities of the soils in the Outer Islands. Despite their low population densities, they do not offer large areas of good, unutilized agricultural land. For the most part, their tropical soils are nutrient-poor, easily leached, and erodible. Most of the nutrients are held in the biomass or the first inch or two of soil. The process of clearing the forest often impoverishes the land. Low population densities are a broad reflection of the Outer Islands' limited agricultural potential, just as Java's historically dense population is due to its deep, fertile, volcanic soils. For the most part, the better agricultural lands in the Outer Islands are already occupied, and land conflicts between Javanese transmigrants and the linguistically and culturally distinct indigenous populations have been chronic (Repetto 1986).

Partly for this reason, 80 percent of current transmigrants are to be settled in logged and unlogged primary forest. Their holdings are to consist largely of rubber, oil palm, and other commercial tree crops, supplementing subsistence crop production. This agronomic system may avoid some of the crop failures, soil depletion, and marketing gaps that accompanied efforts to introduce Javanese rice-dominated cropping patterns to ecologically different areas. But, if targets are met, it will still result in the conversion of about 3 million hectares of forest land by 1988, an area equal to 5 percent of all productive forests (Chapter 2).

Conversion of large areas of tropical forests to plantation crops can raise income and employment. Since 1950 peninsular Malaysia has converted 12 percent of its forest area to establish over a million hectares of permanent crops, such as rubber and oil palm, and become a leading exporter of these commodities. With fertilizers plantation crop yields are sustainable and provide continuous soil cover. Whether or not Indonesia's revised transmigration program will be as successful is uncertain, although it clearly promotes a rate of conversion vastly greater than that which spontaneous migration and investment would bring about in the absence of large government subsidies.

The government of Brazil has engaged in even more massive efforts to colonize its tropical forests with small farmers, despite generally unsuccessful past experiences – along the Transamazon Highway, for example.

The Northwest Development Program (POLONOROESTE) encompasses the entire State of Rondônia and part of Mato Grosso, an area where spontaneous settlement has been occurring for decades. The government undertook to demarcate plots and establish land titles. By mid-1985 the responsible agency had awarded 30,000 titles, most for 100-hectare farms, but tens of thousands of other households are awaiting titles on homesteads they have established. Settlers pay only nominal title fees for their land, can recover their relocation costs by selling timber, and become eligible for subsidized agricultural credits. Heavy government outlays on road building, agricultural development, and other infrastructure have accelerated in-migration. The budget for Polonoroeste for the period 1981 to 1986 exceeds one billion dollars, and the World Bank has concluded loans for more than $400 million to support the program. Total spending has been estimated to have reached $12,000 per settler family (Aufderheide and Rich 1985).

World Bank lending has been conditional on the creation of Indian reserves, a national park, ecological stations, and biological reserves, and restriction of agricultural settlements to suitable agricultural soils, which have been estimated to underlie a third of the region. However, these conditions have not been met. Incursions on Indian territories, rapid deforestation, and uncontrolled in-migration have taken place. In 1984, new settlers were arriving in Rondônia at the rate of 140,000 per year. Many have found that their cleared plots cannot support perennial agriculture and have abandoned them or sold them to cattle ranchers (Chapter 6). As in Indonesia, these conversions of forested areas would not take place at nearly their present rate without the heavy public expenditures that support them.

In addition to sponsoring agricultural settlements directly, many governments have provided generous indirect subsidies to other activities that encroach on the forests. In Latin America, incentives for cattle ranching are the main example: they have resulted in the conversion of enormous areas of tropical forest. Many of the ranches that were established have been uneconomic, and probably would not have been established without heavy subsidies and the hope of speculative gains in land prices. Unproductive and deteriorating pastures far from markets have not supported enough cattle to justify the costs of establishing and maintaining them. Many of these deforested lands have been sold or abandoned, while new lands are cleared for the tax benefits they offer.

In Brazil, the Amazon's cattle herd had reached almost 9 million head by 1980. At an average stocking rate of one head per hectare, conversion to pasturage for cattle ranching had accounted for 72 percent of all the forest alteration detected by Landsat monitoring up to that time (Chapter 6). Almost 30 percent of this conversion is attributable to several hundred large-scale, heavily subsidized ranches. By 1983, there were 470 cattle projects averaging 23,000 hectares each approved by the *Superintendência do Desenvolvimento da Amâzonia* (SUDAM) that had received financial assistance. On average they had converted 5,500 hectares to pasture. As indicated earlier in the discussion of the Brazilian wood-processing industry, assistance in the form of tax credits could contribute up to 75 percent of the capital requirements of a project if the parent company had other tax liabilities to offset. These capital grants have totaled more than $500 million in tax forgiveness, more than 40 percent of the total amount conferred on all Amazon investments. They have been reinforced with credits at subsidized interest rates that represented an 85 to 95 percent discount from commercial interest rates over the 1970s. Further, generous income tax holidays and depreciation allowances, combined with low overall tax rates on agricultural incomes, effectively exempted such projects from income tax liabilities, while operating losses could be written off against income from other sources.

Furthermore, after 1980, when rising credit and transport costs put a tight squeeze on smaller operations, the SUDAM-assisted projects took on even more importance. According to survey data comparing these ranches with a sample of unassisted operations, four times as much of the deforestation attributable to subsidized ranches has occurred after 1980 as that attributable to unassisted projects (see Chapter 6). The subsidized ranches have large areas of forest still uncleared and are establishing more new pastures than unassisted ranchers. One reason is that the subsidies discourage continuing outlays to combat weeds and maintain soil fertility, outlays which are usually not eligible for capital grants, and favor capital expenditures for establishing new pastures, which are eligible. A survey of SUDAM-assisted ranches estimated that 22 percent of the cleared area had already been abandoned or left to fallow by 1985 (see Chapter 6).

The overall economic worth of these large SUDAM-assisted ranching projects is highly questionable, although the many unsubsidized ranches in the Amazon, more than 50,000 in 1980, show that ranching must be economically viable in some areas and circumstances. Since initial capital costs for land clearing, pasture development, and stocking are approximately $400 per hectare, and, with only one animal per hectare, *gross* revenues are only about $60 per hectare once the operation is established, the returns must be marginal, at best.

An economic and financial evaluation of a typical 20,000-hectare cattle

Table 1.10. *Economic and financial analysis of government-assisted cattle ranches in the Brazilian Amazon*

	Net present value (U.S. dollars)	Total investment outlay (U.S. dollars)	NPV investment outlay
I. Economic analysis			
A. Base case	− 2,824,000	5,143,700	− 0.55
B. Sensitivity analysis			
1. Cattle prices assumed doubled	511,380	5,143,700	+ 0.10
2. Land prices assumed rising 5%/year more than general inflation rate	− 2,300,370	5,143,700	− 0.45
II. Financial analysis			
A. Reflecting all investor incentives: tax credits, deductions, and subsidized loans	1,875,400	753,650	+ 2.49
B. Sensitivity analysis			
1. Interest rate subsidies eliminated	849,000	753,650	1.13
2. Deductibility of losses against other taxable income eliminated	− 658,500	753,650	− 0.87

Source: Chapter 6.

ranch of the 1970s, based on sample survey data, contrasted its intrinsic economic returns with its profitability to a private investor able to take advantage of the incentives and subsidies available to investors in the Amazon during the 1970s (and continuing in large part to the present). Economic analysis of the typical cattle project found that even under optimistic assumptions, it is an extremely poor investment. The first panel of Table 1.10 shows that in the base case, which assumes a 15-year project life, after which land, cattle, and equipment are sold, an annual rise in land values 2 percent above the general rate of inflation, and a real discount rate of about 5 percent, the present value of the investment is a *loss* equal to 55 percent of total investment costs. Economic losses are US$2.8 million out of a total investment cost of $5.1 million for the typical ranch. Sensitivity analysis shows that even if land prices rose annually at 5 percentage points above the inflation rate, the typical ranch would still lose 45 percent of invested capital. Even if cattle prices were double the reported figure, the project would only be marginally viable. In other words, these ranches, which have converted millions of hectares of tropical forest, are intrinsically bad investments.

The second panel in Table 1.10 explains why these investments nonetheless went ahead. It presents their returns, not to the national economy, but to the private entrepreneur able to take advantage of all the incentives. Even though intrinsically uneconomic, the project has a present value to the private investor equal to 249 percent of his equity input, at a real discount rate of 5 percent. Sensitivity analysis shows that this present value remains positive if interest rate subsidies are removed, but turns negative if provisions for offsetting operating losses against other taxable income are also withdrawn. The implication is that government policy made it profitable to make investments that led to the conversion of large areas of tropical forest to pasturage of low productivity for livestock operations that were intrinsically uneconomic.

Not only that, but the fiscal burden of this program has been heavy as well. The government has had to provide, through one means or another, both the resources to absorb losses and those to provide large profits to private investors. For the typical ranch, the present value of the government's total financial contribution is $5.6 million (twice the cost the government would have incurred had it undertaken the investment directly). Multiplied by the 470 SUDAM-assisted ranches, the estimated total fiscal cost is over $2.5 billion.

Other countries also provide fiscal incentives for investment in activities that are potential competitors for land. In the Philippines, approved projects are eligible for (a) exemptions from duties on imported capital equipment and equivalent tax credits on equipment obtained domestically, (b) tax deductions for transportation, training, and research costs, (c) unlimited loss carry-forward against future taxable income, and (d) accelerated depreciation, and such additional benefits as eligibility for credit at concessional rates. In the Philippines, however, the few agricultural projects that have been approved for investment incentives have not drawn significantly on the area of productive forests (see Chapter 4). The scale of subsidized forest conversion for cattle ranching in the Brazilian Amazon is probably unique.

Conclusions

This report has shown that many governments of countries endowed with rich forest resources have created economic incentives that powerfully accelerate the rate of deforestation. It has identified the policies that stimulate rapid depletion of the timber resource and policies that encourage the conversion of forest land to agricultural and other uses.

Inappropriate forest revenue systems, which (a) leave enormous economic rents to timber concessionaires and other timber exploiters, (b) provide concessionaires little reason to practice sustainable long-term forestry, and (c) encourage highly selective harvesting of tropical forests with undue wastage of remaining trees, are among the forestry policies that lead to rapid depletion. Poor enforcement of forestry stipulations and underemphasis on non-timber products obtained from tropical forests reinforce these consequences.

In order to increase local employment and income, governments have overprotected domestic milling and wood-processing industries through differential taxation, investment incentives, and supply of timber from public lands at uneconomically low prices. This has encouraged technically inefficient industries to spring into existence. Once established, they also contribute to rapid deforestation. Inefficient conversion rates mean that many more logs must be processed to satisfy a given demand for wood products. Moreover, once established with government encouragement, these local industries exert a strong pressure for a continuing harvest of enough logs to use their capacity fully, whatever the case for conservation.

Generous government subsidies to competing land uses result in rapid conversion of forest lands, often to uses that are economically inferior and even non-viable without continuing government support. Large-scale agricultural settlements in forested areas, carried out at great cost to the government, have failed in several countries after deforesting huge areas. In Latin America, generous incentives made available to promote extensive cattle ranches have acted as convenient tax shelters for corporations and wealthy investors. While investors have gained, the ranches have often failed both economically and ecologically.

Although all the policies examined have been adopted in the name of development, subsequent chapters show that the issue is not between economic development and resource conservation. These policies are unsuccessful when judged only as means to promote economic development. They result in huge economic losses: wastage of resources, excessive costs, reductions in potential profits and net foreign exchange earnings, loss of badly needed government revenues, and unearned windfalls for a few favored businesses and individuals.

They also result in severe environmental losses: unnecessary destruction and depletion of valuable forest resources, displacement of indigenous peoples, degradation of soils, waters, and ecosystems, and loss of habitat for many wildlife species. Both development and environmental goals can be served by improvements in the policies described in this report. These improvements are important steps to more sustainable development of forest resources.

It is by no means too late. Although there are many countries, such as Liberia, the Ivory Coast, and the Philippines, where the rain forest has been extensively disturbed or virtually destroyed, there are other countries – Brazil and Indonesia, for example – where huge forests remain, and still other countries, such as Gabon, the Congo, and Zaire, where most forests are still untouched. In some of these countries, it is inaccessibility rather than policy that effectively restrains the timber harvest, and that inaccessibility is steadily eroding. The Transgabonnais Highway is opening up new large areas of virgin forests even while still under construction, for example. Enacting rational policies toward forest exploitation, forest-based industrialization, and conversion of forests to other land uses will conserve these important resources and forestall serious economic and fiscal losses.

Endnotes

1. A standing tree's stumpage value is its implicit market worth, estimated by subtracting from the market value of the wood products that can be derived from it all the costs of harvesting, transporting to mill, and processing.

2. Ironically, Ghana, where forestry administration has been relatively weak, operates an effectively differentiated system of specific royalties. A different royalty rate applies to each of 39 commercial species, and rates are charged per tree (rather than per cubic meter) harvested. The effect of this system is to encourage loggers to harvest a variety of species, to harvest large trees and thereby open the forest canopy for regeneration, and to utilize each stem cut as fully as possible. Unfortunately, since Ghana has almost no virgin production forests left, these beneficial effects cannot be fully realized there.

3. A loan that has a negative real interest rate is one on which the nominal rate of interest charged is less than the trend rate of inflation. With such a loan, the borrower could invest in any asset appreciating at the rate of inflation, sell it when the loan is due, repay the loan with interest, and make a profit.

References

Anderson, Dennis, and Robert Fishwick. 1984. *Fuelwood Consumption and Deforestation in Developing Countries.* World Bank Staff Working Papers No. 704. Washington, D.C.

Aufderheide, Pat, and Bruce M. Rich. 1985. Debacle in the Amazon. *Defenders,* Vol. 60, No. 2: 20–32.

Bourke, I. J. 1986. International and Trade Barriers: The Case of Forest Products. Unpublished paper. Washington, D.C.: Resources for the Future, March.

Breunig, E. F. 1985. Deforestation and Its Implications for the Rain Forests of South East Asia. In International Union for the Conservation of Nature, *The Future of Tropical Rain Forests in South East Asia*. Geneva.

Buschbacher, Robert J., Christopher Uhl, and E.A.S. Serrao. 1987. Large-Scale Development in Eastern Amazonia. In Carl F. Jordan, ed., *Amazonian Rain Forests: Ecosystem Disturbance and Recovery*. New York: Springer-Verlag.

Contreras, Armando. 1982. *Tropical Timber Processings in International Trade Development*. FAO, Forestry Department Working Paper. Rome.

FAO. 1983. *Fuelwood Supplies in Developing Countries*. Rome.

Gray, John W. 1983. *Forestry Revenue Systems in Developing Countries*. FAO, Forestry Paper No. 43. Rome.

Grainger, Alan. 1987. The Future of the Tropical Moist Forest. Unpublished paper. Washington, D.C.: Resources for the Future, May.

Guppy, Nicolas. 1984. Tropical Deforestation: A Global View. *Foreign Affairs*, Vol. 62, No. 4: 928–965.

Hartman, Richard. 1976. The Harvesting Decision When a Standing Forest Has Value. *Economic Inquiry*, Vol. 14, No. 1: 52–58.

Hecht, Suzanna. 1985. Dynamics of Deforestation in the Amazon. Paper prepared for the World Resources Institute. November.

International Institute for Environment and Development and World Resources Institute. 1986. *World Resources 1986*. New York: Basic Books.

International Institute for Environment and Development and World Resources Institute. 1987. *World Resources 1987*. New York: Basic Books.

Lanly, Jean-Paul. 1982. *Tropical Forest Resources*. FAO, Forestry Paper No. 30. Rome.

Leslie, A. J. 1987. A Second Look at the Economics of Natural Management Systems in Tropical Mixed Forests. *Unasylva*, Vol. 39, No. 1: 46–58.

Matthews, E. 1983. Global Vegetation and Land Use. *Journal of Climate and Applied Meteorology*, Vol. 22: 474–487.

Melillo, J. M., C. A. Palm, R. A. Houghton, and G. M. Woodwell. 1985. A Comparison of Two Recent Estimates of Disturbance in Tropical Forests. *Environmental Conservation*, Vol. 12, No. 1 (spring): 37–40.

Myers, Norman. 1984. *The Primary Source: Tropical Forests and Our Future*. New York: Norton.

National Research Council. 1982. *Ecological Aspects of Development in the Humid Tropics*. Washington, D.C.: National Academy Press.

Olson, J. S. 1975. Productivity of Forest Ecosystems. In National Research Council, *Productivity of World Ecosystems*. Washington, D.C.: National Academy Press.

Repetto, Robert, 1986. Soil Loss and Population Pressure on Java. *Ambio*, Vol. 15, No. 1: 14–18.

Richards, Paul. 1973. The Tropical Rain Forest. *Scientific American*, Vol. 229, December: 58–67.

U.S. Congress, Office of Technology Assessment, 1984. *Technologies to Sustain Tropical Forest Resources*. Washington, D.C.

U.S. Congress, Office of Technology Assessment. 1987. *Technologies to Maintain Biological Diversity.* Washington, D.C.

Wilson, E. O. 1987. The Arboreal Ant Fauna of Peruvian Amazon Forests: A First Assessment. *Biotropica,* Vol. 19, No. 3: 245–251.

World Resources Institute. 1985. *Tropical Forests: A Call for Action.* Report of a task force of WRI, World Bank, and UN Development Program. Washington, D.C.

2   Indonesia: public policies, resource management, and the tropical forest

MALCOLM GILLIS

The deforestation and degradation of Indonesia's tropical forest is recognized as a serious problem internationally. Herein we assess the contribution of Indonesian public policies, by design or by happenstance, to the shrinkage of Indonesia's tropical forest estate in the past two decades. At least one conclusion is clear: Indonesian deforestation would have been less rapid had government policies had more neutral effects on tropical forest land use decisions; government policies and institutions have, jointly and separately, discouraged resource conservation.

Introduction

    This chapter first details the extent and composition of the Indonesian tropical forest and identifies factors, other than government policies, important to forest destruction and/or conversion of the forest estate. The role of forestry policies in the process of deforestation or forest degradation, and the nonfiscal benefits expected from their execution, such as employment, regional development, and foreign exchange, are then examined. Further, we discuss the effects of non-forestry policies, such as tax policy, upon forest-based industry. This is followed by a consideration of the contribution of other policies, including resettlement policies, not designed as forest policies *per se* but with significant implications for the future of Indonesia's tropical forests. Finally, the chapter focuses on the degree to which Indonesian citizens have been compensated for the extraction of what must be considered now as an essentially non-renewable resource from the fragile tropical forest ecosystem, in terms of both fiscal and non-fiscal benefits. Instruments for capture of these benefits have included both forestry and non-forestry policies. Evidence suggests that these policies have been highly flawed, and that Indonesia has sold a valuable resource too cheaply, with relatively little to show for two decades of large-scale forest resource utilization.

Table 2.1. *Indonesian forest resources: Department of Forestry classification, 1985* (*million hectares*)

1. Total land area	193.6
2. Total forest area[a]	143.0
3. Elements of forest area[b]	
a. Protection forest	30.3
b. Nature conservation and tourism forest	19.0
c. Production forest (available for commercial harvesting)	64.0
i. Limited production	30.0
ii. Permanent production	34.0
d. Production forest that may be converted to non-forestry purposes	30.0
4. Area awarded to concessionaires or in process of award	65.4
a. Area under concession (holders of forest exploitation rights)	52.2
b. Areas under forestry agreements (last step prior to award of rights)	13.2

[a]Total rain forest area, 82.2 million ha; total swamp forest area, 12 million ha; total secondary forest area, 14.6 million ha; other forest area, 34.2 million ha.

[b]Because of rounding, total of these elements differs slightly from total forest area.

Sources: Forest area, 1985: Departamen Kehutanan (1985: 17); concession areas: P. T. Data Consult (1983).

Throughout this and other chapters of this book, there will be no presumption against utilization of forest endowments for commercial purposes *per se*, only a presumption against misuse of these resources, leading to avoidable wastes. Some degree of deforestation and forest degradation is viewed as not only inevitable, but advisable. Deforestation here refers to a change in forest land use from forestry to other purposes, such that it no longer functions as a forest ecosystem. Alternatively, degradation involves the depletion and/or damage of forest vegetation such that the forest area becomes suitable only for uses economically inferior to those of its undamaged state.

The Indonesian forest estate

Nearly three-quarters of the area of Indonesia is classified as forest land (Table 2.1). Tropical rain forests themselves cover more than two-fifths of the country, especially in Kalimantan (Indonesian Borneo), Sumatra, and West Irian. The vast lowland forests of Kalimantan and Sumatra, dominated by trees of the family Dipterocarpaceae, are often termed the most valuable remaining tropical forest estate in the world (Ashton 1984). The natural tropical forest covers over four-fifths of Kalimantan alone (Kuswata 1979: 129), which in turn comprises 11 percent of total national land area.

As in many countries with tropical forests, estimates of the extent and composition of forest lands vary considerably by information sources. The Indonesian Forest Department has published estimates, summarized in Table 2.1, which show forest cover to be 143 million hectares. Many observers believe that this figure may overstate forest cover in 1987 by as much as 28 million hectares. In 1981, the FAO calculated total Indonesian forest area at 157 million hectares (Table 2.2), about half the total for the nine Southeast Asian nations, including Indonesia, Burma, Thailand, Brunei, Malaysia, Philippines, Kampuchea, Laos, and Vietnam. Forest department estimates of productive forest area, at 64 million hectares, are also lower than FAO figures, but both agree that Indonesia contains 62 percent of the region's closed forest (Tables 2.1 and 2.2).

There is little disagreement, however, on the pace, if not the causes, of deforestation. Since about 1960, the nation's forests, particularly the great rain forests of Kalimantan, have been receding at rates much faster than in previous years. FAO studies have included some estimates that deforestation was 550,000 hectares annually by 1980 (Panoyouto 1983: Table 1) and would be perhaps 700,000 by the mid-1980s, more than double smaller Malaysia's level, and by far the highest level in Southeast Asia. Given Indonesia's larger forest expanse, however, the annual deforestation rate in the 1980s was but 0.5 percent per year, just below the worldwide average. But according to other recent estimates, the total loss in forest cover from 1950 to 1985 was some 39 million hectares, an average annual loss of 1.1 million hectares.

Unlike many other nations, firewood gathering has not been a particularly serious factor in overall deforestation, despite its noticeable role in Java. Published materials attribute postwar deforestation to three principal factors: shifting cultivation, population resettlement programs, and commercial logging. During the mid-1980s, shifting cultivation is believed to have resulted in about 400,000 hectares of deforestation annually, land clearing under resettlement programs (the "transmigration program") for about 200,000 to 300,000 hectares annually, and logging for between 40,000 and 80,000 (Setyono et al. 1985: Chapter 6). We will show, however, that responsibility for forest destruction in Indonesia cannot be attributed to these three factors alone. Another, rather more complex taxonomy is required for this purpose.

Perspectives on policies, forests, and deforestation

Policies. Policies affecting Indonesian forest use are here divided into two basically arbitrary groups: forest policies and non-forest policies. Forest policies include those which are intended to affect utilization and

Table 2.2. *Indonesian forest resources, FAO classification, 1980 and 1985*

	Area (million hectares)		Indonesia as percent of Southeast Asia (1985)
	1980	1985	
Total area, natural woody			
Vegetation	158.2	157.1	n.a.[a]
A. Closed forests[b]	113.9	110.9	61.8
1. Productive forests[c]	73.7	67.7	58.5
a. Undisturbed forests	38.9	33.0	56.5
b. Logged forests	34.8	34.7	68.6
2. Unproductive forests[d]	40.2	43.2	19.7
a. For physical reasons	34.7	n.a.	n.a.
b. For statutory reasons (parks, reserves)	5.4	n.a.	n.a.
B. Open forests[e]	3.0	2.8	n.a.
C. Fallows[f]	17.4	19.5	n.a.
D. Shrub formations	23.9	23.9	n.a.

[a]n.a. = not applicable.

[b]Closed forests are those that have not been recently cleared for shifting cultivation or heavily exploited. In closed forest formations, tree crowns, underlayer, and undergrowth combine to close off most of the ground from light so that continuous grass cover cannot develop.

[c]Productive forests are those from which it is both physically and legally possible to produce wood for industry.

[d]In unproductive forests, timber is exploited for statutory reasons, or because harvesting is infeasible due to difficult terrain or stand conditions.

[e]Open forest formations are marked by a continuous grass cover on the ground.

[f]Fallow refers to secondary vegetation following the clearing of forests.

Source: FAO (1981a: 40, 211–237, 277–313, 391–416).

conservation of forest materials. They are controlled primarily by the Indonesian Department of Forestry, and include such policy instruments as stumpage fees, log export taxes, concessions (forest licenses), and reforestation programs. Non-forestry policies include all those of other government agencies which, intentionally or not, have significant impact upon forest use, but which were primarily intended to further non-forest objectives. Examples include resettlement, general tax and agro-conversion policies, fuel subsidy, and even exchange rate policies.

Public policies have not been the only or even the primary cause of deforestation. Separating their role from that of other factors affecting forest utilization is a difficult task that cannot be done with any precision. Chief among other factors overtly encouraging Indonesian forest resource destruction have been endemic poverty and the effect of institutions governing both property rights and access to virgin forest stands. As in many other countries with tropical forests, limited knowledge about complex issues in forest ecology and botany is a major covert factor in deforestation in Indonesia. Ignorance about such critical matters extends well beyond the borders of Indonesia and stems largely from a lack of concerted research on such matters as tree fruiting, regeneration, and transplantation. Finally, researchers within and without Indonesia have done too little to put their conclusions in formats easily accessible to policy-makers.

If the Indonesian forest is to face an apocalypse, then, public policy is but one of the horsemen. Poverty, institutions, and ignorance round out the quartet. This chapter focuses primarily upon the first. The roles of poverty and institutions are briefly sketched, insofar as they interact with policy, while the task of illuminating the part played by ignorance is left to those with more suitable botanical and ecological training.

Poverty, institutions, and deforestation

Poverty: shifting cultivation and firewood gathering. In the postwar period, few large countries have experienced more rapid growth in national income than did Indonesia from 1967 through 1981. Over that span, annual average growth in real gross national product was 7.5 percent. Even with annual population growth rates of 2.3 percent, real per capita income for the nation's 160 million inhabitants increased more than twofold (Gillis 1984: 231–232). Although income growth rates declined sharply from 1982 through 1987, owing principally to unfavorable world market conditions for oil, timber, and other export commodities, Indonesia was less severely affected than other raw material exporting nations. Whereas Indonesia's policies toward forest utilization have had unwelcome results, the decline in both urban and rural poverty has unquestionably been largely due to deftly managed macroeconomic policies coupled with effective and flexible policies toward food crops. Nevertheless, Indonesia remains a poor country despite nearly two decades of relatively rapid growth: per capita income in 1986 was less than US$600, one-twentieth that of the United States.

Economic growth allowed substantial inroads against rural poverty after 1970 (Gillis 1984: 261) but not sufficient to curtail the two main poverty-related sources of deforestation: shifting cultivation and firewood gathering. The former is a major cause of forest destruction on sparsely settled Kalimantan; the latter remains a significant factor on populous Java, where little virgin forest remains.

Shifting cultivation, or "swidden" agriculture, has been a significant source of consumption of forest land in both Sumatra and Kalimantan for decades, but has become increasingly severe in Kalimantan since about 1970. One million families are estimated to be engaged in shifting cultivation, resulting in clearing of 400,000 hectares of forest annually (Setyono et al. 1985: Chapter 6). Some observers place this at about half of total annual deforestation. However, Soedjarwo, who as Director-General and Minister for Forests since 1966 has overseen forest exploitation, maintained in 1987 that shifting cultivators destroyed 500,000 hectares per year (Soedjarwo 1985: 1). By 1980, shifting cultivation had transformed perhaps 16 million hectares of forest land nationwide (400,000 in East Kalimantan alone) into degraded land covered mostly by the pernicious "alang-alang" grass (*Imperata cylindrica*) (Kuswata et al. 1984: 115).

The role of rural poverty and of urban-rural income differentials in promoting shifting cultivation in open access forest areas with fragile soils is well known. Not all shifting cultivation, however, is attributable to poverty, nor are primitive methods always used. In East Kalimantan some shifting cultivators commute from their urban residences in shared taxis to their fields (Brotokusomo et al. 1980: 179–190). Many shifting cultivators are not poverty-stricken, but engage in this activity to finance pilgrimages to Mecca (Mackie 1984: 67). Many migrant Kalimantan pepper growers operating on logged-over land have become prosperous, though such lands are usually rendered useless for pepper after but a decade (Kuswata et al. 1984: 115). It is not known what proportion of swidden agriculture or pepper cultivation in the former national forest was made possible by logging. Much, but by no means all, shifting cultivation takes place in the wake of extraction in logged-over forests. In such cases, the entry of swidden cultivators is merely the *coup de grace* for the forest stand, rather than the cause of deforestation (Kuswata et al. 1984: 98–102). This forest land use pattern can sometimes have catastrophic results. It is now clear that agricultural fires started by shifting cultivators were the prime cause of devastating forest fires in drought-stricken East Kalimantan and in the East Malaysian state of Sabah in 1983 (see Chapter 3 for further discussion of the fires). The fires were particularly devastating in logged-over areas (Mackie 1984: 2).

The use of non-commercial energy is often taken as one indicator of the level of poverty. Fuelwood remains Indonesia's principal source of non-commercial energy. As late as 1979, fuelwood consumption, at about 30 million tons of coal equivalent (TCE), was roughly equal to commercial energy consumption (oil, coal, natural gas) (Gillis 1980a: 3). With rising rural incomes, fuelwood consumption per capita gradually declined after 1970. Nonetheless, 1979 fuelwood consumption was still nearly 100 million cubic meters, roughly four times the volume of wood commercially harvested in that year. Deforestation related to fuelwood demand has been much more significant on Java than in the tropical rain forests of Kalimantan and Sumatra. Government programs and policies designed to combat this source of forest destruction have been ineffective; policies responsible for sustained economic growth from 1967 to 1981 had a much greater positive impact.

Institutions and deforestation. The institutional framework within which forest utilization has occurred has interacted with both poverty and public policy to yield rates of forest consumption that in all likelihood have exceeded what private owners of virgin forest land would have allowed. The institutional framework includes not only the process for assigning property rights to the forest, but their enforcement (hindered by the inability to restrict forest access) as well as the assignment of responsibilities across government departments.

The government of Indonesia owns all property rights to the natural forest through provisions established in the 1946 Constitution. The rights may be temporarily assigned for 20 to 25 years (as with timber concessions) or irrevocably transferred (as in titles to forest land awarded to transmigration families) to private parties.

Leasing of these timber concession rights exposes the forest to deforestation pressures far beyond that of logging *per se*. Relatively open forest access, conflicts in Indonesia's land law itself, and problems in lease enforcement lead to further problems. Rights assigned to timber concessionaires conflict with the rights of local people both to utilize land generally and to use all land within two kilometers of any river. Thus, by restricting access to land formerly used by local people, the concessions cause resentment and encourage excess timber harvests by local people within timber firm concession areas (Brotokusomo et al. 1980: 179–190).

Ambiguous ownership patterns, imperfect enforcement of property rights, and the inability to restrict access to the virgin forest, combined with elements of the timber concession agreements themselves, have given rise to an Indonesian version of the "tragedy of the commons." Property owned by all has again been treated as property owned by none, especially since tropical forest regeneration is a matter of at least six to seven decades.

A counterfactual example provides graphic illustration of the role of existing property rights in rapid deforestation in Indonesia. Award of all forest exploitation rights to a single profit-maximizing private monopoly firm, however unacceptable it may be for other reasons, would likely have resulted in much lower rates of forest extraction, and therefore of deforestation. This would be true even if the private firm operated with no recognition of any of the social costs of timber extraction identified later in the chapter. The reason is quite simple. From 1967 through at least 1983, Indonesian forestry policy was strongly geared to maximization of *wood production* from the tropical forest. As a result, by 1980 Indonesia exported a greater volume of tropical hardwoods than all of Africa and 'Latin America combined (Gillis 1987). The nation's share of the log export market, 41 percent, was easily large enough to affect world prices substantially. Because a monopolist maximizes profit or monopoly rent by restricting rather than expanding output, a monopolist would clearly have not followed the production maximization approach of the Forestry Department. Extraction rates would have been lower than was actually the case. Furthermore, this result would have held even if the government had appropriated nearly 100 percent of the monopolist's rent. This reasoning is demonstrated in innumerable economic texts, and need not be further belabored here (Gillis et al. 1982: Chapter 19).

The institutional framework governing forest utilization contributed to deforestation in other ways. Lacunae in the assignment of responsibilities between departments and conflicting goals among government departments were partly to blame; corruption within some of these institutions also played a significant role; the *Weltanschauung* prevailing within the Forestry Department itself was not a minor consideration.

The operations of at least six government departments directly affect the utilization of the tropical forest: the Departments of Forestry, Agriculture, Transmigration, Public Works, Energy and Mines, and the State Ministry for the Environment. The Forestry Department controls concessions policy and log harvests; the Agriculture Department controls policies affecting conversion of forest land to estate crops; the Transmigration Department identifies land sites cleared for resettlement of families from heavily populated Java and Bali to the so-called outer islands, principally Kalimantan and Sumatra; the Department of Public Works does the actual land clearing for this program; the Energy Department issues oil and mineral concessions on both forested and unforested land; and still another agency, the State Ministry for the Environment, attempts to introduce environmental considerations into the policies of the other five, but with infrequent success.

Coordination of policy toward forest utilization between the six departments has been virtually absent until recent years, and sporadic at best since then. One example, from many available, involves a 1979 plan for extensive land clearing in Sumatra for the transmigration program. This plan was formulated, and nearly implemented in the Public Works Department, without formal consultations with either the State Ministry for the Environment or the Forestry Department. More generally, the Forestry Department is not ordinarily notified when the Department of Energy and Mines issues licenses for commercial oil and mineral exploration in the forest.

The *Weltanschauung* of officials in the agencies with operations affecting the forest, particularly the Forestry Ministry itself, has had profound effects on acceleration of deforestation in recent years. The prevailing outlook of the Forestry Department from 1967 to 1983 embodied the view, aptly characterized by Manan (1974: 3), of "the forest as merely a godown that produces wood." The prime motive underlying Forestry Department policies has, as noted, been the desire to maximize production, if not value, of wood from the forest (Gillis 1987).

To illustrate, virtually all departmental pronouncements on progress in the forest sector, until 1980, referred to growth in the number of logging projects, the amount of timber sector investments, and the amount of wood harvested. Not until 1985 did public statements by the Forestry Department begin to mention environmental consequences of two decades of headlong stress on logging harvests.[1] Nor has the Forestry Department viewed as a serious matter the extensive deforestation that has resulted from the Transmigration Program or from major forest fires. Calamitous fires in the moist forest of East Kalimantan in 1983 burned an area the size of Belgium (3.6 million hectares). The initial response of the Forestry Ministry to this event exemplified the prevailing departmental view of the forest as a dispensable asset: the fire was seen as a blessing in disguise. In an interview with an international magazine, the Minister noted that "much of the area that was burned was conversion forest. So what you have is land clearing for free. The forest fire was the natural way of 'clearing the land' " (*Asiaweek* 1984).

Of the other agencies with forestry responsibilities, only the State Ministry for the Environment has explicitly recognized wider socio-economic values of virgin forest endowments. The Transmigration and Public Works Departments tend to view the tropical forest as a place for relocating transmigrants; the Department of Industry sees it as a source of feedstock for nearly 200 newly constructed and planned plywood mills; a group of officials in the Department of Agriculture has viewed forested land primarily as a barrier to successful agro-conversion projects, including cattle ranching on the island of Sulawesi.

Clash of values. Science has yet to develop measures for determining an "optimal" degree of deforestation and/or forest degradation for any given tropical forest endowment. Therefore, we lack clear criteria for judging whether or not the rate of forest conversion in Indonesia in recent years has been excessive. We may reasonably assume that the optimal rate of conversion is not zero, and that the social costs associated with deforestation and degradation (erosion, flooding, vulnerability to fire, decline in species diversity, etc.) are also not zero. It is clear that the basic institutional arrangements which would create incentives for efficient resource use are not in place. There is nevertheless a growing consensus throughout Indonesia that the rate of forest destruction has been excessive and that the problem is approaching calamitous proportions (Setyono et al. 1985: Chapter 3).

In many nations, Indonesia included, the problem is partly rooted in the way in which tropical forest resources have been viewed by the public and by public servants, including most forest specialists. The problem lies in overemphasis upon one of the several functions of tropical forests.

Virgin tropical forests have two prime functions: protection and production. The first function involves protection of soils, watersheds, climate, animal habitat, and species diversity from the effects of erosion, salinization, fire, and floods. The economic value of these services is no less tangible, but far less measurable, than that flowing from forest production. Production involves the process of transforming the natural capital of the forest into that more readily useable by humans. The natural capital of the tropical forest is found in two distinct forms: wood and non-wood resources.

In Indonesia and elsewhere, both forestry policy and those non-forestry policies affecting forest use have emphasized the productive rather than the protective function of the forest and, within the former function, wood products rather than such non-wood products as fruits, herbs, meat, oils, chemicals, and non-wood fibers. The historical stress upon wood production has led to a forestry policy focused upon wood resource extraction, rather than on maintaining economic value created by the forest's protective function or on sustaining an appropriate level of non-wood products in the forest. Policies intended to maximize short-term financial value from wood extraction are not likely to be consistent with retaining the economic value of non-wood production nor that of protective services provided by the forest.

The policy dilemma is not simply the clash of economic with broader social values; at a more elementary level we find conflict between considerations of the economic value of wood and the economic value of both non-wood products and protective services furnished by the forest. In Indonesia as elsewhere, this conflict has been consistently resolved in favor of the former. The Indonesian chapter in the short history of the search for resolution of this dilemma is studded with examples of policy and institutional failure, interacting with poverty and ignorance to yield unforeseen and largely unwanted results.

We now turn to a more detailed examination and assessment of the elements influencing forest use, forest degradation, and deforestation.

Forestry policies

Policies toward forest extraction and forest-based industry

Prior to World War II, virtually all commercial activity in Indonesian forests was confined to large teak plantations established in Java under Dutch colonial rule. While sizeable tracts of forest land had been cleared in Sumatra for rubber plantations, mechanized logging was uncommon. The natural tropical forest estate on Kalimantan remained largely undisturbed through the war. As late as 1967, little of the Kalimantan forest had been exploited, although loggers had long known of the existence of rich stands of trees, particularly in East Kalimantan, where the natural forest is dominated by trees of the family Dipterocarpaceae, which includes a number of commercially valuable species: *Shorea* spp., *Dipterocarpus* spp., and *Dryobalanops* spp. Logs from the most common of these species are commercially known as "meranti" in Indonesia, "lauan" in the Philippines.

Commercial logging was the first activity responsible for widespread tropical forest conversion in Kalimantan. Logging opened the way for more extensive shifting cultivation and for accelerated resettlement programs. Land clearing for resettlement programs and agricultural estates may constitute the prime sources of future deforestation and degradation in the tropical forest, but policies toward the wood products industry played a more significant role in the last two decades.

In the period 1960 to 1965, log harvests averaged but 2.5 million cubic meters per annum. But with the 1967 reopening of the economy to foreign investment and of the forest to both domestic and foreign investors, forest exploitation proceeded at a rapid rate, spurred by attractive income tax incentives, particularly five and six year "tax holidays." By 1970, between 7 and 10 percent of the total forest area was being utilized, and production of tropical hardwood logs had quadrupled, to 10 million cubic meters (World Bank 1970: 62). The harvest again doubled from 1970 to

Table 2.3. *Indonesia: physical volume of log harvest and export values of tropical timber (1960–87)*

Year[a]	Total log harvest[b] (million m³)	Export values (million U.S. dollars) All timber products	Logs only
1960–65 average	2.5	n.a.	n.a.
1970	10.0	110	110
1975	16.3	527	527
1976	21.4	885	885
1977	22.2	943	943
1978	24.2	1,130	1,052
1979	25.3	2,172	2,060
1980	25.2	1,672	1,428
1981	15.9	951	504
1982	13.4	899	310
1983	14.9	1,161	267
1984	16.1	1,120	178
1985	24.3	1,185	44
1986	25.0[c]	1,585	1
1987	26.0[c]	n.a.	n.a.

[a]Export values are for fiscal years, i.e., 1975 = 1975/76. The fiscal year ends on March 30.

[b]It is to be noted that different sources provide rather different estimates of the total annual harvest of logs by year, often differing by as much as 2 million cubic meters per year. The figures provided are taken from the most widely used source.

[c]Preliminary.

Sources: For 1960–85: Departamen Kehutanan, *Statistik Kehutanan Indonesia*, various issues. For 1986–87: Ministry of Finance.

1975, with most of the increase from the timber-rich province of East Kalimantan, where production of light hardwoods, primarily of family Dipterocarpaceae, peaked at 10 million cubic meters in 1978. At the zenith of Indonesia's timber boom in 1979, 25 million cubic meters of logs were taken from the tropical forest (see Table 2.3), ten times the average annual production from 1960 to 1965. Gross foreign exchange earnings from export of tropical hardwood grew from US$110 million in 1970 to US$2.1 billion in 1979 (see Table 2.3). By that year, Indonesia was the world's leading exporter of tropical logs, with 41 percent of the world market (Gillis 1987).

The national log harvest fell sharply from the 1979 peak owing to several factors discussed in ascending order of importance. First, the government doubled the log export tax, from 10 percent to 20 percent. The announced purpose for this measure was to induce holders of timber harvesting rights to move quickly into investments in domestic log processing, particularly plymills. In the short to medium term, the export tax increase slowed the rate of harvest in stands of marginal value, but had little impact on logging in the rich and dense tracts of East Kalimantan. A second factor in the log harvest decline was that the world market for tropical hardwoods weakened considerably beginning in 1980, with the onset both of world recession and large increase in U.S. residential mortgage rates after November 1979.

In 1980, further government measures contributed to sharply reducing the harvest over the next four years. The first was a May 1980 joint decree of the Ministers of Agriculture, Industry, and Trade, which prescribed a total ban on log exports by 1985 to be phased in gradually over five years. However, since the 29 plywood mills in existence in 1980 had installed capacity of but 2.5 million cubic meters of log throughput (1.1 million cubic meters of output), implementation of the 1980 joint decree drastically curtailed extraction in 1981 and 1982: log production fell to 13.4 million cubic meters by 1982 (Table 2.3), by which time installed capacity in plywood mills had risen to only 8 million cubic meters of log inputs, or about one-third of 1979 log production (P. T. Data Consult 1983: 18–37).

In a second measure in late 1980, the government imposed a reforestation deposit of US$4 per cubic meter on all logs harvested by timber concessionaires in Kalimantan and Sumatra.[2] This measure resulted in curtailment in the harvest of logs from marginal stands although this was not its purpose. The policy engendered little reforestation, but did increase government timber revenues per cubic meters of log production by 14 percent.

As a result of the factors above, timber export earnings dropped precipitously over the period 1979 to 1983 from a record US$2.17 billion in 1979 to $0.9 billion in 1982. By 1981, Indonesia's share in the world market for tropical logs had fallen to 21 percent, one-half that of 1979. And as late as 1986, overall timber exports had recovered to only two-thirds that reached 7 years earlier (Table 2.3). As new plywood mills came on stream after 1981, overall timber exports inclusive of plywood recovered to $1.2 billion by 1985, about half the 1979 peak (Table 2.3). The slowing of the timber boom in Indonesia was welcomed by various groups (including many government officials) concerned over the rapid rate of forest depletion from 1967 to 1979. A blue-ribbon team of government officials had concluded in 1979 that the area of Indonesian forests awarded or scheduled for granting to concessionaires exceeded the total area of reserve plus protected forest by 6.5 million hectares (Team Kayu 1979: 1–14). By 1983, the records of the Forestry Ministry itself indicated that the total area under concession agreements was 52.2 million hectares, not including 13.2 million hectares under Forestry Agreements (concession awards in process). Thus, the total areas conceded and in process of concession totalled 65.4 million hectares, or 1.4 million *more* hectares than in the combined area of the productive forest (see Table 2.1). And by 1985, the logged-over area of the production forest was already at least 30 million hectares, probably much more (Setyono et al. 1985: 91).

Objectives

Emphasis upon wood production in the two decades prior to 1985 embodied several objectives: generation of foreign exchange, fiscal resources and employment opportunities, promotion of industrialization through forest-based industry, and regional development.

The rapid expansion in the volume of log extraction after 1967 contributed to each of the goals, but not to the extent that the production figures might suggest. The nearly 20-fold expansion in gross foreign exchange earnings from the timber sector from 1970 to 1979 was largely illusory. Because policies were not well suited for timber rent capture, Indonesia's net foreign exchange earnings from forest harvests were significantly less than half of the gross earnings. In the early 1970s net timber export earnings (retained value to Indonesia) were as low as 25 percent of the true value of gross timber export earnings,[3] and remained as low as 50 percent by 1978 (Gillis 1981: 96). Only after major increases in export taxes and reforestation fees from 1978 to 1980 did net timber export earnings exceed 50 percent of gross values.

Until 1978, fiscal resources from the timber sector were a small share of the value of log production. Taxes from the timber sector never exceeded 20 percent of reported log export values until 1978; this percentage was consistently half that recorded by the East Malaysian State of Sabah (Gillis 1981: Chapter 5). Tax collections from the timber sector have never exceeded 5 percent of total government revenues, and have typically been less than 3 percent (Table 2.4). Throughout much of the past two decades and particularly before 1980, timber tax collections have fallen far short of levels that would have been obtained if timber tax systems of Malaysia and much of Africa had applied in Indonesia.

Job creation has consistently been a major objective of forest policy in Indonesia (Departamen Kehutanan 1985: 20), but has not been well fulfilled. Employment opportunities created first by logging and then by plymills have been limited. In some logging districts in East Kalimantan, only 12 percent of total jobs provided by logging camps were taken by local people in 1977–78; jobs created indirectly by logging activity were

Table 2.4. *Indonesia: tax and royalties collected from the timber sector as a percent of total central government tax collections*

Fiscal year	(1) Taxes and royalties on timber sector (billion rupiah)	(2) Total tax collections (billion rupiah)	(3) Timber as percent of total
1971–72	12.4	428	2.8
1972–73	21.3	590	3.6
1973–74	44.3	967	4.6
1974–75	40.2	1,753	2.3
1975–76	33.3	2,241	1.5
1976–77	54.3	2,906	1.9
1977–78	77.8	3,535	2.2
1978–79	158.1	4,266	3.6
1979–80	307.8	6,696	4.6
1980–81	341.8	10,227	3.3
1981–82	221.7	12,213	1.8
1982–83	164.5	12,418	1.3

Sources: Table 2.14 (column 1); Department of Finance (column 2).

not numerous either. For example, locally-purchased food accounted for only 5 percent of camp spending (Kuswata et al. 1984: 118).

Employment in the forest sector did expand rapidly from 1967 to 1980, but declined precipitously in 1980 as quantitative restrictions on log exports were implemented. By 1982, the total number of Indonesians employed in the entire forest-based sector was but one percent of the labor force. The surge in plywood plant construction after 1980 did allow employment in the sector to double in 1983. Nevertheless, from 1970 through 1982 there was no year in which the value of payrolls for Indonesians in the sector exceeded the estimated value of timber rent accruing to private firms in the industry (see Table 2.21). And we will see that at the peak of the timber boom in 1979, timber rents accruing to loggers exceeded payrolls for employees in the sector by about US$1 billion, an amount then equal to 10 percent of total government revenue, including oil revenue.

Forest policy also sought to use the wood-producing potential of the forest to secure regional development goals (particularly in Kalimantan). Benefits to timber provinces were substantial for a short period, and were probably greater initially than the localized costs associated with expansion in forest-based activity in the affected regions. As with the central government, the principal benefit to the regions was fiscal in nature. Provincial governments receive 70 percent of timber royalties and license fees. These fiscal flows to provinces peaked at US$43 million in 1979, when they were as much as 45 percent of provincial budgets in East and Central Kalimantan (Gillis 1980b: 26–28). Royalties temporarily allowed governments in both provinces to spend about three times the national average. Other benefits to timber-producing provinces were relatively limited. Although development benefits from fiscal resources were not trivial, the timber industry made but a limited contribution to the livelihood of local people. For example, while timber firms made large investments in logging roads and housing, the roads were essentially access roads specialized to logging projects, and the housing units were either mobile or temporary.

Rapid growth in the timber regions also involved social costs; East and Central Kalimantan in particular both suffered from a "boom-town" syndrome. Prior to the onset of logging activity in 1968 the then-isolated province of East Kalimantan (with the country's richest stands) had about one-half million inhabitants. By 1981, immigration had swelled population to 1 million. For the period 1967 to 1981, this implies a 4.9 percent annual rate of population growth, roughly 2.15 times faster than in the country as a whole for the same period. But as log production began to decline precipitously in 1980, both because of extraordinarily high previous rates of extraction and because of government-imposed cutbacks in log exports, job opportunities were sharply curtailed.

The type of "boom-town" process discussed above carries with it a potentially sizeable social cost. In the early stages of rapid immigration of people attracted by timber employment, the "boom town" or "boom region" (East Kalimantan) encounters difficulties in financing additional requirements for social infrastructure (roads, water supply, hospitals, schools, etc.) since they generally precede the flow of boom-related taxes. The quality of life in the affected region temporarily declines until revenues provided later in the boom period begin to allow sufficient infrastructure construction. Later, as forest resources are depleted and production diminishes, large numbers of the population are left jobless and poorly adapted to the economic conditions of the post-boom period. Further, the social capital built up during the boom becomes redundant and underutilized (relative to post-boom population). Operating and maintaining this infrastructure becomes a financial burden upon the region or locality.

A final policy goal became prominent only after 1978: promotion of forest-based industrialization. Strong incentives for such investments were initially established in that year, when the Ministry of Finance doubled taxes on log exports (but exempted most sawn timber and plywood).

These incentives were reinforced in 1980 with the enactment of a phased-in ban on log exports. By 1982 forest-based industrialization had become the overriding forest sector policy issue. Progress in the sector was no longer identified merely with growth in log production and export, but also with the new capacity in plywood manufacture. The number of operating mills in fact rose rapidly, along with log intake capacity, which doubled from 1980 to 1982 and was scheduled to double again by 1987 (to 15 million cubic meters per year).

Thus, forestry policies under the control of the Forest Department have been strongly focussed upon the value of wood products and the economic and social value of other agricultural uses of virgin forest lands. This narrow policy focus has become the principal barrier to reform of policies impinging on tropical forest utilization, and it is the principal underlying reason for policy-induced deforestation.

Forestry policy instruments

The Indonesian Forestry Department is responsible for a wide variety of functions germane to the future of the tropical forest. These include not only the establishment of royalties (stumpage fees) and area licenses, reforestation, felling regulations, and period of concession, but also mapping and inventory of forest uses. We will be concerned only with those policies that impinge directly or indirectly on extraction decisions.

Royalties and license fees. Two types of royalties are collected from timber concessionaires operating on government-owned forest stands. The first is the ordinary royalty, *Iuran Hak Pengusahan* (IHH). The IHH is imposed at a uniform *ad valorem* rate of 6 percent of posted prices (check prices) drawn up every quarter for all timber species by the Department of Trade. The rate is uniform for all logs regardless of species, primarily to simplify royalty administration (see Table 2.5). The timber license fee is collected annually from concessionaires at a rate of 1,000 rupiah (Rp.) per hectare of concession area. (This amount equals US$0.66 at the October 1986 exchange rate.)

Both royalties and license fees were increased after 1979. This is one reason why the government's share in timber rents was, in the mid-1980s, significantly higher than a decade earlier. But we shall see that instruments other than royalties under the Forestry Department's control have been primarily responsible for increasing the government's share in rents after 1978.

The royalty structure also has implications for forest depletion. In Indonesia and most other countries with tropical forest endowments (ex-

Table 2.5. *Indonesia: structure of taxes and royalties on timber sector as of June 1985 (at June 1985 and October 1986 exchange rates)*

Type of tax charge[a]	Rate and base
Timber royalty	6% of posted export prices (all species)[b]
Timber license fee	Rp. 1,000 per hectare[b] (US$0.89 in 1985, $0.66 in 1986)
Timber export tax	Logs: 20% of posted export prices[c]
Additional timber royalty	Rp. 1,686 per m³ (US$1.50 in 1985, $1.06 in 1986)[d]
IPEDA (property tax)	20% of value of all timber royalties[b]
Grading and scaling fees	Rp. 250 per m³ (US$0.22 in 1985, $0.16 in 1986)[e]
Replanting deposit	US$4 per m³ on logs extracted by concession holders in Kalimantan and Sumatra (refundable when company produces evidence of adequate replanting)[f]
Tax on expatriate forest workers	US$100 per month for each foreign employee[g]

[a]Table does not include a fee paid for use of harbors ($1.00 per m³) and local taxes.
[b]Decree #451/KPTS/Um/7/1979 of July 1979, Minister of Agriculture.
[c]Decree #10A/KMK/06/1978 (February 1978) and #157/KMK/06/1978 (April 1978), Minister of Finance.
[d]Decree #368/KMK/.011/1979 (August 1979), Minister of Finance.
[e]Decree #457/KPTs/EKKU/1979, Minister of Agriculture.
[f]Joint Decree of Ministers of Agriculture, Industry, and Trade.
[g]Decree #55/1974, President of the Republic of Indonesia.

cept Thailand), timber royalties are assessed on the basis of timber removals, rather than on the basis of the stock of merchantable stems in the stand. Where selective cutting[4] is the prescribed method of harvest (as in Indonesia), then such a uniform *ad valorem* royalty (as well as an *ad valorem* export tax) can result in more serious depletion of tropical forest resources than would occur under other royalty regimes of equal revenue yield or under perfectly administered income taxes (see Appendix A). In particular, under a uniform (flat rate) *ad valorem* royalty, forest quality suffers greater damage than would be the case under a system of finely differentiated and constantly updated specific royalties. The latter type of system would place the highest specific (i.e., volume-based) charges on the most valuable stems, and progressively lower charges on less valuable stems. An ideal royalty system requires that the Forestry Department base the charges upon the structure of differential rents accruing to different species in varying merchantable sizes, locations, and times. This might possibly be feasible in the more homogeneous temperate forests of North

America, but would involve onerous informational requirements in the Indonesian tropical forest, with its extreme variegation of species. The attempt would require establishment and updating of prices for 150 commercial species under changing market conditions[5] and detailed information on logging costs by location as well as on the composition and density of stems, stand by remote stand.

Most countries with tropical forest resources lack both the information and administrative resources required to operate such a system. In such circumstances, a uniform *ad valorem* system of royalties may represent the least objectionable alternative method for imposing royalties.

Appendix A supports the notion that *ad valorem* royalties and export taxes in Indonesia involve more significant depletion of tropical forest resources than would a perfectly administered, and therefore unattainable, ideal form of royalty (or income tax). Yet the remedy is not necessarily abandonment of the *ad valorem* royalty, as long as the Indonesian version of selective logging is continued. The flat-rate *ad valorem* royalty is fairly effective in capturing rents and is relatively simple to administer provided the tax collectors stay abreast of changing world market conditions in timber.

Indonesia's royalty system suffers from another problem that has little to do with its *ad valorem* structure. When royalties are based on actual log harvests from a selectively logged forest stand rather than the potential yield from mature trees in the stand, the logger has no incentive to harvest defective and oversized stems of commercial (and lesser known) species. Indeed, harvesting such stems (which do have commercial value once sawn) is actively discouraged, since the concessionaire is charged full royalties on defective trees (Ross 1984: 54). Progressive improvements in yields from the forest require that defective trees be cut down. Harvest of defective and oversized trees not only reduces competition for light for immature stems, but increases availability of nutrients by reducing competition for them and adds more nutrients from the non-useable residues of harvested defective and oversized trees (Ross 1984: 53–54).

Reforestation and reforestation fees. Reforestation activities fall into two categories. On government-owned forest land (over 90 percent of the total forest area) the efforts are called, naturally enough, "reforestation" programs; on privately owned land (primarily in Java), they are called "regreening" programs. Our focus is largely upon the former.

From 1946 through 1983, cumulative reforestation programs had covered 2.3 million hectares, about one-twelfth the area deforested since independence. From 1974 to 1983 alone, recorded deforestation totalled about 4.95 million hectares; reforestation and regreening programs covered but 2.68 million hectares (FAO 1981a: 107). For the period 1983–1988, deforestation is projected at about 0.6 million hectares per year, while 1985 plans called for reforestation programs to reach 0.9 million hectares per year (Departamen Kehutanan 1985: 39, 108–109). Thus, over the period 1974 to 1988 alone, the increase in the deforested area will have exceeded the total area where reforestation has begun by about 0.77 million hectares. That being the case, Indonesian forest cover may be expected to decline only slightly after 1988, assuming 100 percent success ratios in all reforestation/regreening efforts. Experience, however, suggests that reforestation efforts will be considerably less than 100 percent successful: survival rates for reforestation plantings were but 72 percent in Sumatra and only 54 percent in Kalimantan in the 1970s (FAO 1981a: 225). Even given 100 percent success ratios, these figures may not be very meaningful in terms of implications for the virgin forest. Deforestation since 1968 has been concentrated in Kalimantan, but only 1 percent of the area reforested before March 1983 was located there. The position of forest resources and forest cover remains tenuous, reforestation programs notwithstanding.

The government enacted reforestation deposit fees in late 1980. At $4 per cubic meter of tropical timber extracted in Kalimantan and Sumatra, these deposits account for almost as much revenue as collected from timber royalties in each region. The deposit may be refunded to extractive firms when they present verification of adequate replanting programs, but thus far there is no record of such refunds. For all practical purposes, the reforestation deposit is merely another tax, and it could not encourage replanting even at higher rates. This could not be otherwise, given both the costs of reforestation ($1,000 per hectare) and the duration of timber concessions granted by the government.

Size and duration of concession. The average size of timber concessions and the duration of concessions affect production capacities and incentives for conservation of natural forest resources.

Most Indonesian timber concessions are quite large relative to those in other tropical nations. According to one authoritative source, average concession size is about 98,000 hectares (nearly 1,000 square kilometers). Concessions of such size cannot be easily policed, whether by forest officials (to insure adherence to regulations by firms) or by the concessionaires (to prevent illegal harvest of logs by poachers).

The duration of timber concessions in the Indonesian tropical forest is ordinarily 20 years; very few concessions run as long as 25 years. Concessionaires, thus, face very weak incentives to take steps to safeguard the forest's productivity or to reduce logging damage, which can be substantial. As much as 40 percent of stems remaining after logging may not survive until the next harvest, in stands where loggers entered only once (Appendix A provides similar estimates).[6] There is ample scope for reducing damage to stands from logging: Whitmore, a leading authority on tropical forests, maintains that felling trees in the direction that causes least damage, plus well-planned skidding, reduces damage to the residual stand by up to half (Whitmore 1984: 271). However, Indonesian policies provide extremely weak incentives for adopting those less damaging options.

Due to the long growing cycles of commercial tropical hardwoods, a second harvest in a given stand should be delayed for at least 25 to 35 years after the initial harvest. Given the present duration of concessions in Indonesia, concessionaires have no financial interest in maintaining long-term forest productivity;[7] in fact, they have every incentive to re-enter a stand within 5 to 10 years, as many have done.[8] Repeated logging within such short periods inevitably damages the new growing stock of seedlings and saplings (see Appendix A). As a partial remedy various specialists have proposed extending the concession length to 70 years.[9] Indeed, under the Indonesian Selective Logging System prescribed by the Forestry Department, a minimum rotation of 35 years is envisaged for second generation dipterocarps (Setyono et al. 1985: 108).

Harvesting methods and allowable cut. The prescribed method for harvest of tropical timber in Indonesia is said to be based on the notion of "sustained yield," developed to apply to temperate forests. As commonly interpreted, maximum sustainable biological yield for a forest is obtained when the volume of timber harvested in a given period is equivalent to the growth of the timber stock in that period. Maximization of biological yield requires that mature trees be cut when their incremental growth falls below the average growth rate. Sustained yield is a dubious notion when applied to tropical forests, given the very long (70 or more years) growing cycles of commercial species, the great diversity in species, and logging damage to the residual stand.

Holders of timber exploitation licenses in Indonesia are allowed to use only one method of harvesting logs from tropical forests: the Indonesian Selective Logging System (ISLS), one variant of a group of selective – as opposed to uniform – cutting systems (Appendix A).[10] ISLS embodies a 35-year harvesting cycle and forms the basis for the determination of the Annual Allowable Cut (AAC) for each concessionaire by the Forestry

Department. The ISLS prescribes the selection of large stems (those with diameter of 50 cm. or more) of so-called "primary," or commercially desirable, species.

Limiting the harvest to large stems of desirable species means that, with each entry, only a small proportion of the standing stock is removed. Between harvests, the forest is left to develop as it will, with the expectation that subsequent harvest yields (nominally, after 35 years) will be at least the same as in the initial harvest (Ross 1984: 50). Thus, the assumption underlying the ISLS is that natural regeneration will allow "sustained yield" of wood products from the forest from one cutting cycle to another.

Quite apart from the fact that concessionaires have not always adhered to a 35-year cutting cycle, the Indonesian selective logging system is inconsistent with any sensible concept of resource conservation in a tropical forest. As noted by Spears, ISLS fosters continuing low yields in the production forest (Spears 1983). As few as three to five trees are harvested per hectare, or about 1 percent of the 310 to 325 trees (of diameter greater than 10 cm.) per hectare typical of East Kalimantan forests. In some very rich concessions, however, as many as 20 trees per hectare have been extracted on initial entry (6 percent of the standing stock of trees) (Kuswata et al. 1984: 119).

Low yields from harvested stands imply greater present and future pressures on unexploited forest areas, and therefore unnecessarily rapid depletion of virgin forest resources. Low-intensity cutting, or "high-grading," of the forest leaves many commerical trees uncut. Not only may these die by the next cutting period (35 years in Indonesia nominally), but low-intensity cutting may also harm immature dipterocarps because it may not open the forest canopy enough to secure the best emergence of dipterocarp saplings and other immature trees. Low yields of the Indonesian productive forest estate under concession licenses are evident in comparing natural growth rates with apparent productivity: natural growth is estimated at 1 to 2 cubic meters per hectare annually and productivity but 0.5 cubic meter per hectare annually (Setyono et al. 1985: 117). And high-grading will likely reduce future natural growth rates. If some form of selection, rather than uniform, system must be used, then a switch from the Indonesian Selective Logging System (also known as the "Zero Treatment Option") to a selection system involving *some* silvicultural management and treatment would increase yields in the production forest. This would reduce both pressure on the remaining forest estate and needless deforestation. However, marginal reform of the cutting system may not provide very large benefits, even should new rules be adequately enforced, until such time as a sound ecological basis for any selection system in the tropical forest has been clearly established (Ashton 1980: 43–54).

Policies toward non-wood forest products. Non-wood products of the Indonesian tropical forest include rattan, charcoal, tengakawang,[11] damar, lopal, bird's nest, resin, sandalwood, sago, honey, natural silk, fruit, sizeable quantities of meat, and an immense variety of products of significant present and likely future use in pharmaceutical and cosmetic compounds (Ashton 1984: 7–8). These commodities are not only important to the rural economy as a source of livelihood,[12] but several have potentially sizeable export markets. Except for rattan and charcoal, non-wood forest products can generally be harvested by local families from the forest without extraction of wood stems. Moreover, in an undisturbed or carefully managed natural forest, the harvest of these products provides a stream of local income in perpetuity, whereas log harvests in the tropical forest furnish income in the current period and again, at best, perhaps 30 years hence. It is not at all clear that the discounted present value of annual income (in perpetuity) per hectare from non-wood forest products must be less than the discounted present value of log extraction per hectare. While further research is required to test this last proposition, it is not wildly implausible.

It is nevertheless clear that the availability of non-wood forest products increases the economic value of natural stands of tropical forest. The greater the utilization of non-wood products, the more valuable becomes the tropical forest in its natural state, and the less is the incentive to extract wood products only. Therefore, policies that promote – or at least do not hinder – efficient utilization of non-wood forest products are consistent with maintenance of both the protective *and* productive function of the forest. Policies in place throughout Southeast Asia, however, have the opposite effects.

Exports of non-wood forest products have in fact expanded significantly since the early 1970s, particularly after 1979 (see Table 2.6). By 1982, such exports had reached $120 million, exceeding export values for copper, aluminum, tea, pepper, tobacco, and several other traditional exports for which Indonesia is well known. Also significant, the share of non-wood products in total forest product export value increased from only 2.9 percent in 1973 to 11.2 percent in 1981 and 13.3 percent in 1982, and Indonesia has just begun to tap the markets for these items. Although the forests of Indonesia are celebrated among botanists for their wealth of non-timber resources, exploitation and management of these assets has been decidedly suboptimal (Ashton 1984: 7).

Table 2.6. *Indonesia: exports of non-wood products and wood products, 1973–82*

Year	Value of non-wood product exports (million U.S. dollars)	Value of wood product exports (million U.S. dollars)[a]	Non-wood export values as percent of export values of logs plus process products
1973	17.0	583.4	2.9
1974	24.9	725.5	3.4
1975	21.6	527.0	4.1
1976	34.7	885.0	3.9
1977	48.3	943.2	5.1
1978	58.6	1,130.6	5.2
1979	114.0	2,172.3	5.2
1980	125.6	1,672.1	7.5
1981	106.0	951.8	11.2
1982	120.0	899.4	13.3
1983	127.0	1,161.1	11.0
1984	n.a.	1,120.2	n.a.
1985	n.a.	1,185.6	n.a.

[a]Estimate: Rattan exports alone were $82 million in 1982.
Source: Value of non-wood exports: Departamen Kehutanan (1984: 96–99).

There are diverse reasons for the neglect of opportunities for expanding production and export of non-wood forest products. Perhaps the main reason is the overriding view within government (the Forestry Department in particular) that the principal value of the forest lies in its productive function, specifically in its wood-producing capacity. When the tropical forest is widely seen as essentially a "godown that produces trees," it is not surprising that a small proportion of official attention – not to mention share of the Forestry Department budget – goes toward increased utilization of non-wood forest products. This was true even as late as 1985: in the Draft Long-Term Forestry Plan, non-wood forest products receive scant attention (four minor citations in 88 pages); virtually the entire document focuses upon wood utilization. Export performance and prospects for non-wood products are nowhere mentioned in the draft. Forestry Department budgetary resources have been heavily concentrated upon utilization of the forest's wood-producing capabilities. The Department has estimated its total financial resources in the years from 1983 to 1988 to be about US$2.95 billion.[13] If programs directed toward improving identification and utilization of non-wood products had received the same budgetary share as, say, the proportionate value of non-wood to wood exports in 1982, then about $384 million would have been spent upon these activities over the period 1983 to 1988. That is about

US$75 million per year, roughly 62 percent of the recent annual value of non-wood forest exports. But projected spendng in programs and projects directed toward non-wood products was less than $3 million per year over the period.

A larger budget share devoted to research and support for expanded production and improved marketing of non-wood products is particularly needed to finance exploration programs to discover and classify non-wood products of potential value in medicinal and cosmetic products. Especially promising is one species of *Voacanga*, a tree whose latex yields compounds active in treatment of cardiac disease. Systematic exploration of the forest would also help clarify the nature of the world's stake in the genetic richness of the Indonesian tropical forest, where not more than 50 percent of the flora is known to science (Kuswata 1979: 131–137). There are no exploration programs under way in the tropical forests of Kalimantan and Sumatra, and none are contemplated in the current draft long-term forestry plan. Assistance in quality control and marketing for non-wood products could also have sizeable benefits since high variability in quality has resulted in many non-wood forest products being exported at uneconomically low prices (Ashton 1984: 7).

Non-forestry policies toward forest-based industry

Non-forestry policies include those with significant effects on decision-making in the forest sector, but for which responsibility is vested in government departments other than the Forestry Department.

Tax policy

The Department of Finance formulates and implements policies pertaining to generally applicable taxes. Three general tax policy instruments have significant implications both for government revenues from timber and for decision-making in extraction of wood and non-wood products of the tropical forest: income tax incentives, export taxes, and rural property taxes (IPEDA).

Income tax incentives. From the reopening of the tropical forest to investment in 1967 until December 31, 1983, Indonesia offered generous income tax incentives to logging investments, principally including a five-to-six year income tax holiday for timber sector investments. As the tax rate normally applicable to logging investment incomes was 45 percent, the holidays nearly doubled the effective after-tax return on eligible investments, for a period *longer* than the holiday granted. The tax holidays nominally were confined to five years (or at most six). In practice, most timber companies—foreign and domestic—paid no income taxes at all from 1967 through 1983.

There were two basic reasons why five-year tax holidays could become holidays of 10 to 15 years. First, companies were allowed unlimited carry-over of losses incurred during the tax holiday period. While few firms incurred actual losses, most were still able to report large losses for the initial years of operation. This was easy to do: since exempt firms (both domestic and foreign) were not audited during the tax holiday period, there was nothing to prevent them from reporting losses in each year of their first five to six years of operation, although we will later see that logging was an extremely profitable enterprise from 1967 to 1981. High reported losses during the tax holiday period were merely applied to any positive profits generated and reported in the post-holiday period. In this way, many firms were exempt from income taxes for a decade or more.

Second, several timber firms managed to stretch the duration of five-year income tax exemptions by more fraudulent means, with or without the connivance of the income tax administration. One method particularly favored by domestic firms and those from nearby Asian countries was to persuade tax officials that the tax holiday legislation as written meant something else entirely. Whereas the law required that the tax holiday period began at the inception of operations, some firms successfully contended that the holiday begin only in the first year of *profitable* operations. Since many firms were able to report losses until, say, 1980, their holidays were deemed to begin in that year. Other stratagems employed to escape timber profit taxes included gross overstatement of debt on firms' capital structures and other, even less savory, practices. As late as 1983, one-third of all timber processing firms reported debt-to-equity (D/E) ratios of much greater than three to one (P. T. Data Consult 1983: 18–20) when the true underlying D/E ratios did not likely exceed 3:2.[14] Since for income tax purposes interest is deductible from income and dividends are not, the inflated D/E ratios resulted in large government income tax revenue losses. In 1983 alone, tax losses from this source of evasion alone were about US$10 million in the timber sector.

The principal implication of the above is that from 1967 until 1983, Indonesia collected insignificant amounts of income tax revenues from logging and processing firms, most of which were domestically owned. More effective income taxation would not have reduced the rate of exploitation of the tropical forest, but would have clearly raised the benefits of timber exploitation to the owner of the forest resource. Tax revenues foregone because of the availability of and abuses of tax holidays, and other forms of tax evasion in the timber sector, were probably on the order of US$2 billion from 1967 to 1983, conservatively estimated.

Such abuse of tax holidays was a major factor in the government's decision to end all tax holidays in 1983. It was concluded that tax holidays were not only ineffective in stimulating investment (particularly in natural resource industries) but that the privilege would inevitably be abused. The abolition of tax holidays and all other income tax incentives was a prominent feature of a comprehensive tax reform of 1983 (Gillis 1986). Future earnings from new investments in timber – and other sectors – will be taxed at a maximum rate of 35 percent, and withholding taxes will apply to profits remitted by the small number of foreign firms remaining in the Indonesian timber sector.

Export taxes. Indonesia has imposed *ad valorem* export taxes upon a variety of commodities since 1965: tin, copper, rubber, palm oil, and of course timber. Several of these taxes, including that on rubber, were reduced in 1976. Others were abolished in 1979; log exports remain taxable.

From 1965 to 1978, log exports were subject to a 10 percent tax on "check prices" established each quarter. The check price system is a form of posted price system adopted in the 1960s to prevent undervaluation of exports subject to export taxes. However, the system was highly defective from 1965 to 1972; check prices were often as much as 50 percent below actual f.o.b. prices for timber. Check prices on timber were drastically increased in 1972 and until 1981 remained reasonably close to actual f.o.b. values. As a result, the log export tax was the most significant instrument in the capture of logging rents throughout the 1970s.

In 1978, the government doubled the log export tax to 20 percent. This measure was undertaken not primarily for revenue reasons, but to strengthen incentives for domestic processing of wood products. Sawn timber exports were initially taxed at a 5 percent rate in 1978, but were later exempted (except for roughly sawn timber of a thickness in excess of 5 cm.). Plywood exports have always been exempt from export tax.

The configuration of export tax rates of 20 percent on logs and 0 percent on sawn timber and plywood would be expected to significantly increase the rate of exploitation of Indonesia's forests, since taken by itself it furnished strong incentives for substantial investments in sawmills and plymills. Indeed, from 1977 to 1983, the number of operating and planned plymills increased from 16 to 182 (P. T. Data Consult 1983: 18–30).

Incentives for investment in timber processing were particularly attractive before 1985 because the export tax structure by itself created high rates of effective protection to the processing industry (see Appendix B for an elucidation of the concept of effective protection as applied to export industries). For example, the average f.o.b. value of logs in 1983 was just under US$100 per cubic meter, while the f.o.b. value of plywood was about $250 per cubic meter. But Indonesian plywood mills have been relatively inefficient in recovering wood value from logs made into plywood, achieving recovery rates of but 43.5 percent (Departamen Kehutanan 1985: 22), versus 50 percent elsewhere in Asia and up to 55 percent in Japan. Because of low rates of recovery, the value, in roundwood equivalent, of plywood exports was but $109 (see Table 2.16). This means that each cubic meter of logs, with an export value of $100 in 1983, had a value as plywood of $109, i.e., additional domestic value-added created by processing was $9 or 9 percent. Moreover, by processing the log into plywood, the exporter saved $20 in export taxes (20 percent of $100). Taxes saved by the firm were lost by the government, in amounts more than *twice* the additional domestic value-added created by processing activity. In fact, taxes saved by plywood firms were 222 percent of the value of additional domestic valued-added created in plywood processing. This is the relevant measure of the protection afforded on domestic value added to plywood mills by the Indonesian export tax structure. That is to say, the export tax structure allowed Indonesian plywood firms to have labor and capital costs that were almost 2.2 times higher than plywood mills buying Indonesian logs in Korea, Japan, and Taiwan, and still remain profitable. For each dollar of additional domestic value-added in plywood milling made possible by the export tax structure, the government gave up $2.20 in export tax revenues. The Indonesian government became in effect a heavy investor in the domestic plywood industry without receiving equity shares for the investment.

The situation in sawmilling was even worse. In all years from 1981 to 1985, the roundwood equivalent value of sawn timber was *less* than the value of logs. In 1983 in particular, the f.o.b. export value of an average cubic meter of sawn timber product was $155.00 (see Table 2.16). The f.o.b. value, in roundwood equivalent, of the 1.8 cubic meters of logs required to make one cubic meter of sawn timber was $85. Therefore, domestic value-added for sawn timber – measured at world prices – was *minus* $15. Here, sawn-timber exporters saved $20 in taxes for each cubic meter of log input in order to obtain a *negative* $15.00 in domestic value-added. Sawmilling was therefore socially wasteful but privately profitable in Indonesia.

Effective protection to sawmills and plywood mills became infinitely high in 1985, when *all* log exports were banned. Effective protection rates under the export tax structure from 1979 to 1984, before the log export ban was fully effectuated, had already embodied strong incentives for rapid depletion of forest resources.

The connection between accelerated depletion of these resources and strong incentives for domestic processing is not an obvious one. We have seen how government tax and export policies provided irresistible signals for investment in plywood mills and for their subsequent inefficient operation. These mills have a claim on Indonesian log inputs that no foreign mill could ever exercise. During periods of very low world prices for tropical hardwood products, these mills will not have to suspend operations because of their heavy indirect subsidization by government policy. They will likely be *unable* to suspend operations, for two reasons. First, laying off workers is very difficult in Indonesia, because of policies enforced by the Ministry of Manpower. Second, governments may force the mills to operate even when faced with large losses rather than cope with the political problems associated with over 100,000 jobless plywood workers.

In addition, high rates of effective protection to sawmills and plymills give managers little incentive to curb operational inefficiency, such as the very low recovery rates of Indonesian mills. Indonesian plywood mills require 15 percent more log feedstock per cubic meter of plywood output than mills elsewhere in the region. Thus, for every cubic meter of Indonesian plywood, 15 percent more stems must be cut relative to countries that import Indonesian logs for processing.

In sum, the non-neutralities implicit in the export tax structure on wood products have created sizeable economic inefficiencies by skewing incentives.

Property taxes. The forestry sector has been an important source of rural property tax (IPEDA) revenues since the early seventies. In recent years the sector's share in total IPEDA collections has varied between 15 percent and 22 percent. Since 1979, IPEDA liabilities of concessionaires have been defined as one-fifth the amount of timber royalties paid by the firms. The tax is therefore not a true property tax on concession areas, but an *ad valorem* tax on the harvest that magnifies (by 1.2 times) any effects of the *ad valorem* royalty on timber. Inasmuch as the present *ad valorem* royalty has some slight but unavoidable effects in increasing forest damage from harvest, the IPEDA contributes marginally to these damages.

Policies for forest-based industrialization

We have seen that the overriding thrust of public policies toward the forest sector since 1978 has been promotion of forest-based industrialization first through tax policy instruments then by a progressive tightening of log export quotas beginning in 1980, culminating in a total ban on log exports in 1985. There are significant differences between the short and long term effects of restrictions on log export.

In the first three years of the restrictions, the effect was clearly that of slowing both the rate of log harvest and of deforestation, since, as noted earlier, capacity for processing logs into sawn timber and plywood was sufficient to handle only one-fourth of log production at 1979 levels (see Table 2.7). But according to the most recent Forestry Department Long-Term Plan, plywood mills with installed capacity for nearly 20 million cubic meters of log intake will be on stream by 1988. If *effective* capacity is approximately half installed capacity, then the forest will be expected to supply 10 million cubic meters of logs for the mills by that year. Other wood products (sawn timber, moulding, etc.) for domestic use will necessitate a projected harvest of another 28.8 million cubic meters of logs. Thus, according to the Forestry Department, total log demands will reach 38.8 million cubic meters by 1988. Only 4.6 percent of total log production is expected to come from tree plantations on Java (Departamen Kehutanan 1985: 22). The natural forest is therefore to provide 37 million cubic meters of logs by 1988 (Table 2.8), 2.7 times the log harvest in 1982.

Even further strains will be placed upon the natural forest estate if production plans for 1993 and 1998 are realized. By 1993, current plans call for the total harvest from the natural forest to be more than three times 1982 levels. By 1998, the natural forest is expected to produce 45.7 million cubic meters, or 3.4 times the 1982 harvest.

The respite in pressure upon the natural forest estate provided by the log export restrictions in 1980–84 has therefore been very temporary. Industrialization policies were to have spawned 137 plywood mills by 1988, thereby creating heavy demands for steady growth in log harvests. The protection to domestic timber-producing industries provided by government policies insures that millions more logs will be harvested than would be needed by efficient mills to produce planned plywood output. By 1988, perhaps as many as 150,000 steady state jobs will exist in plywood mills in Indonesia. Based on past patterns of volumes taken per hectare, log production at levels projected for the period 1988–93 will require logging on about new 800,000 hectares per year. Therefore, between 1988 and 1998 alone, 8 million new hectares will be logged in order to support, at most, 150,000 new jobs. That is, each job will require the

Table 2.7. Indonesia: production and export of tropical timber, 1973–83 (thousands of cubic meters in roundwood equivalent)

| | (1) | (2) Exports | | | | (3) |
Year	Log production by timber concession holders	(2A) Logs	(2B) Sawn timber[a]	(2C) Plywood[b]	(2D) Total exports	Export of tropical timber as percent of total production
1973	26,297.0	19,095	608	—	19,703	74.9
1974	23,280.0	17,728	637	—	18,365	78.9
1975	16,296.0	13,511	738	1	14,250	87.4
1976	21,427.0	17,877	1,159	19	19,055	88.9
1977	22,234.8	19,212	1,069	26	20,307	91.3
1978	24,742.9	19,444	1,367	161	20,972	84.8
1979	25,313.6	18,205	2,311	269	20,785	82.1
1980	25,190.4	14,583	2,165	564	17,312	68.7
1981	15,954.4	6,391	2,273	1,777	10,981	68.8
1982	13,376.5	3,103	2,200	2,834	8,137	59.1
1983	14,996.0	2,988	2,517	4,162	9,617	65.2
1985	24,277.0	0	6,469	4,983	11,452	59.5
1986	25,000.0	0	6,440	5,300	11,740	47.0
1987	26,000.0	0	6,500	6,000	12,500	48.1

[a]Converted to log roundwood equivalent using Department of Forestry figure of 1.82 m³ of sawn timber to 1.0 m³ of logs (recovery of 54.6%).
[b]Converted to log equivalent using Department of Forestry figure of 2.3 m³ of plywood to 1.0 m³ of logs (recovery of 43.5%).
Sources: Columns 2A, 2B, and 2C: Departamen Kehutanan (1984, Tables 2.7, 2.9, and 2.11). Column 1, 1977–82: Departamen Kehutanan (1984); 1973–76: International Monetary Fund, Recent Economic Development (various issues, 1973–79); 1983–87: 1984–87 Ministry of Finance, Kuala Lumpur.

Table 2.8. *Indonesia: present and projected log requirements from the natural forest, 1982–98 (million cubic meters)*

Year	Requirement
1982	13.4
1988	37.0[a]
1993	41.0[a]
1998	45.7[a]

[a]Projected.
Source: Departamen Kehutanan (1985: 43).

utilization of 53.3 hectares of forest (which will all be damaged). The economic and environmental implications of these projected demands on the natural forest estate have only begun to be addressed.

Regulations on entry terms for foreign investment. The timber boom in Indonesia began with substantial foreign participation. Such large multinationals as Weyerhauser, Georgia Pacific, and Unilever had been awarded large concessions in Kalimantan by 1970. Then, in that year, regulations were changed to limit foreign involvement in concession awards to Indonesian-foreign joint venture companies. In 1975, the regulations were again altered to limit concession awards to domestic firms only.

By 1983, foreign firms held but nine of the existing 521 forest concessions. All of the large Western multinationals had by then almost fully divested of their holdings in Indonesia (Gillis 1987); the remaining foreign firms were principally enterprises from Japan, Korea, and Malaysia.

Entry terms and utilization rates. Entry terms may affect resource conservation to the extent that foreign firms differ from domestic firms in utilization rates per hectare harvested. Low utilization rates result in more hectares being cut each year, and therefore greater damages to the forest estate. Larger foreign firms would ordinarily be expected to apply more capital-intensive cutting methods, and thus to have higher utilization rates than domestic loggers. There is some evidence that this was in fact the case. Large foreign firms tended to have all-weather operations based in permanent logging roads. They also employed the most modern logging equipment, including mechanical skidders, powerful loaders, and very large logging trucks. The concessionaires thus were able to remove most of the volume of harvestable trees. Harvestable trees constituted about 19.2 percent of all trees per hectare in a typical stand structure for East Kalimantan (Long and Johnson 1981: 80). Moreover, yields from capital-intensive all-weather operations were appreciably higher than for smaller dirt-road loggers. The all-weather logger typically removed about 84 percent of the harvestable stems from a stand while the smaller, dirt-road loggers usually removed only 47 percent (Schoening 1978: 163).

It therefore appears that post-1970 policies designed to reduce logging participation by foreign firms, however justifiable these policies may have been on other grounds of national policy, reduced utilization rates and resulted in more hectares being harvested annually, thus increasing the extent of forest damage.

Entry terms and plantations. It is doubtful that the Indonesian variant of so-called "sustained-yield" management of virgin tropical forest estates has had the desired results (Setyono et al. 1985: 129–131), especially since concessions are of short duration (20 years). While silvicultural treatments in virgin stands (weeding around small trees, enrichment planting, etc.) may enhance growth rates and stocking levels, available evidence does not suggest that such methods will be successful, short of repeated, and prohibitively expensive, maintenance treatments (Long and Johnson 1981: 82–83).

Plantation forestry might appear to be supportive of goals of forest conservation. Forest plantations require a much smaller land base for the same production as that possible from a much larger area of natural forest; that volume can be achieved in less time than in natural stands. Long and Johnson estimate that only 15 to 20 percent of a concession needs to be converted to plantations to sustain yield (Long and Johnson 1981: 83).

Tropical plantations of fast-growing forest trees therefore furnish an alternative to the repeated relogging of the natural forest. Large foreign firms are the most experienced in plantation undertakings in logged-over forest areas, and they possess the requisite capital and technology.[15] However, foreign firms are no longer allowed to receive concession agreements, nor do current land tenure policies allow these firms to share with domestic firms property rights in tree plantation projects. Without secure property rights, foreign firms will not invest in plantation projects with long gestation periods. Dozens of proposals for plantation projects, both for hardwoods and tree crops, have run afoul of these limitations.

The *Draft Long-Term Forestry Plan* makes no mention of new government forest plantation projects, which in any case would require 25 years or more before the first harvest. Without government or private investment in plantations in logged-over areas, the natural forest will continue to be exploited at a rate several times faster than would otherwise be the case. To be sure, government teak plantations in Java now cover 1.5 million hectares, but production from these long-established estates is forecast at less than 4 percent of total log production through 1999 (Departamen Kehutanan 1985: 43).

Other policies affecting forest utilization

A different set of policies designed to achieve non-forest-related policy objectives nevertheless has significant implications for deforestation and forest degradation. These policies are controlled by agencies other than the Forest Department. Transmigration policy is, naturally enough, the responsibility of the Transmigration Department. Exchange rate policy is crafted by several agencies under the jurisdiction of the State Minister for the Economy, Finance, and Industry, as is kerosene subsidy policy. National parks policy is formulated jointly by several agencies, but coordinated by the State Minister for the Environment.

Of these, transmigration policy represents the most serious direct threat to the future of the natural tropical forest. Exchange rate policy has on occasion had a negative indirect impact on forest conservation. Subsidies on kerosene consumption, originally justified partly as retardants to deforestation, have been extremely expensive and largely ineffective. The recently revitalized program of expanding national parks and forest reserves stands virtually alone in a prevailing mix of policies that have consistently placed the commercial value of wood and agriculture above all other economic and social values of the forest estate.

Transmigration policy and programs

Since the beginning of the century, transmigration has been a prime objective of government policy, first under the Dutch Colonial Administration and then (after 1946) under the independent Republic of Indonesia. The prime purpose of transmigration programs has been to encourage a shift in population away from densely populated islands of Java, Bali, and Madura to the thinly populated Outer Islands, to relieve population pressures on critical watersheds, and to open up new agricultural areas. In the past 80 years, at least 2.5 million people have been relocated under these programs. Government-supported programs, as opposed to "spontaneous transmigration" (unassisted migration of households in response to economic opportunity in the Outer Islands) took a growing share of the national budget from 1970 to 1985. By 1984, transmigration program outlays were 2.9 percent of total government expenditures. By comparison, total outlays on health and family planning programs together were 2.5 percent of total government spending. Transmigration development costs are estimated at about $2,667 per hectare, or about $10,000 per family for a typical tract of 3.75 hectares (wetland tracts average about 3 hectares, dryland tracts, about 5 hectares) (Ross 1984: 57; Ross 1982: 94).

The current five-year development plan contemplates resettlement of 1 million families from 1983 to 1988; 80 percent of settlements will fall within logged and unlogged primary forest. One authoritative source has estimated that clearing the forest land targeted before 1983 for transmigration use will result in cutting down 35 million cubic meters of large commercial logs annually, 23 million cubic meters of which would likely be wasted (Ross 1982: 17). Unfortunately, few of the proposed settlement tracts were in the large area (3.5 million hectares) burned by the great forest fires of East Kalimantan in 1982–83 (see Chapter 3).

The ultimate objective of the transmigration program is to relocate 15 million more people by the century's end. Reaching this objective will eventually require clearing an estimated 360,000 hectares of forest land per year (Ross 1982: 59), or about 12 million more hectares before the year 2000.

Annual losses in forested area due to the transmigration program are considerable and threaten to surpass all other sources of deforestation combined. At present, there is no way to ascertain whether the benefits to Indonesia – short or long term – of the transmigration program are greater or smaller than the social costs of attendant deforestation. Existing evidence is not encouraging. The program has been bedeviled, particularly in recent years, by natural limitations and management problems. Primary forests provide 80 percent of transmigration sites. The lush forest growth gives the appearance of great soil fertility, but this is not generally so in Kalimantan or Sumatra. The forest ecosystem is extremely efficient in recycling nutrients, maintaining appropriate soil and ground temperature, and retaining moisture. Once cleared mechanically for transmigration sites, the generally poor quality of the original soil deteriorates further, as nutrients are leached away from the exposed and compacted soils. Few cleared forest sites have proven able to sustain agricultural yields sufficient to support households using dryland agricultural methods essential for the nearly two-thirds of the transmigration sites planned before 1986.

Management problems in the transmigration program have been legion. To illustrate, until 1971, the Department of Transmigration had never consulted the Directorate of Communicable Diseases in the Department of Health to identify diseases endemic to transmigration areas and those likely to be carried by transmigrants. Several published reports by Secrett (1986: 80–81) and Whitten (1986: 4–6) identified serious problems in selection of transmigration sites and in provision of basic infrastructure, including roads and potable water. Poor soils, as well as managerial lapses in the operation of the program, often impel transmigrants to clear more of the neighboring primary forest, thus establishing shifting cultivation in areas where none was present before (Setyono et al. 1985: 64–65).

Further, additional deforestation often follows even on the heels of agricultural success at official transmigration sites. Success attracts both new spontaneous transmigrants, usually relatives of the relocated households, and displaced transmigrants ("shifted cultivators") whose officially-sponsored programs have failed. Such newcomers have little experience in farming forests on a sustained basis and tend to resort to slash-and-burn methods unsuited to the poor forest soils of Kalimantan and Sumatra (Secrett 1986: 77–85).

For these and other, more political reasons, the Indonesian Transmigration Program has attracted very substantial worldwide attention, nearly all of which has ranged from moderately critical to hypercritical.

The record does strongly indicate that the economic and ecological arguments employed thus far to justify the Transmigration Program require reassessment given the high costs of the program, relative to any demonstrated benefits. The program clearly did not achieve target relocation levels in the 1970s. Present plans to expand the program, if implemented without major reforms, will compound decades of overemphasis on the wood and agricultural potential of the forest, to the continued detriment of its non-wood productive potential and of its protective function.

Exchange rate policy

While a nation's exchange rate policy affects decisions across the entire economy, the focus here is upon implications for foreign trade. An overvalued exchange rate, as defined below, decreases the attractiveness of exporting; the Indonesian rupiah has been significantly overvalued at various times since 1972.

Inasmuch as more than 90 percent of the harvest of tropical logs has been exported (first as logs and later as processed products), overvaluation might have been expected to induce a reduction in log harvests, and therefore to slow the pace of deforestation. Thus, overvaluation, generally regarded as inimical to the economy as a whole,[16] could conceivably be seen as a counterweight to other policies that encouraged excessive rates of forest utilization.

Unfortunately this outcome is not likely. On balance, overvaluation of the Indonesian rupiah has probably played at least a small role in promoting deforestation. Exports of Indonesian logs and wood products have been little discouraged by adverse exchange rate policy, given the existence of substantial timber "rents" as identified in the final section of this chapter. At the same time, incentives for export of non-wood forest products have likely been decreased by exchange rate overvaluation. Overvaluation means that exporters receive fewer rupiah per dollar's worth of exports than would be the case without overvaluation. Since the primary present and potential market for non-wood forest products is overseas, overvaluation reduces incentives for gathering those products from the natural forest relative to incentives to utilize the wood-producing capacities of the natural forest. Exchange rate overvaluation therefore reinforces incentives to cut down rather than conserve forest resources. In addition, overvaluation diminishes the profitability of such traditional agricultural exports as rubber, pepper, and palm oil, exacerbating rural poverty and therefore the problem of shifting cultivation.

Assessing the effects of exchange rate policy upon deforestation must begin with examination of movements in the nominal exchange rate: the rate quoted on any given day to buyers and sellers of foreign exchange. Since 1970, the exchange rate regime has been one of fixed rates, with periodic devaluations (four since 1970) intended to maintain competitiveness for Indonesia's exports of non-oil products. In August 1971, the nominal rupiah/dollar rate was changed from 378 to 415. This rate was maintained until November 1978, when it was again raised to Rp. 625 = $1. The rupiah was then allowed to depreciate slowly against the dollar, rising to 694 in January 1983. On March 30, 1983, the nominal rate was adjusted formally to 970 per dollar. Further depreciation in 1984 and 1985 brought the June 1985 nominal rate to 1,118 rupiah to the dollar, and the most recent devaluation (in September 1986) raised the rupiah/dollar rate to 1,600 to 1.

Incentives to export and to import ultimately depend on the real effective rate of exchange, not on the nominal rate. The real rate of exchange can be defined in several ways, but one common method is to adjust a nation's nominal exchange rate for the difference in inflation between that nation and its major trading partners. Given any fixed nominal ex-

Table 2.9. *Indonesia: indices of the real effective exchange rate, wood product export values, and non-wood product export values, 1971–83 (1971–73 = 100)*

Year (midpoint)	Index of the real effective exchange rate	Index of wood export values	Index of non-wood export values
1971	100[a]	52	39
1972	90	70	97
1973	102	177	165
1974	125	221	232
1975	140	152	213
1976	155	238	329
1977	157	291	464
1978	122[a]	307	562
1979	110	557	1,104
1980	115	571	1,210
1981	124	279	1,026
1982	131	259	774
1983	98[b]	334	n.a.

[a]Devaluation years: 1971—August 21; 1978—November 15; 1983—March 30.
[b]Estimation from IMF data.
Source: Effective exchange rates: Kincaid (1984: 62ff.).

change rate, high rates of inflation in Indonesia relative to its major trading partners cause the real exchange rate to rise, indicating a decline in export profitability and competitiveness. This is commonly known as increasing exchange rate overvaluation. With an overvalued exchange rate, the probability of exporting is less than would be the case without overvaluation.

The concept of the "real exchange rate" can be further refined by allowing for the effects of changes in export taxes, so as to calculate the real "effective" exchange rate. Increases in export taxes (as on log exports) raise the real effective exchange rate, thereby decreasing export competitiveness. A decrease in export tax rates has the opposite effect.

Table 2.9 displays the evolution of real effective exchange rates and of wood product and non-wood product exports over the period 1971–1984. It can readily be seen that between the devaluations of 1971, 1978, and 1983, the exchange rate moved strongly toward overvaluation. There was a particularly long period, from 1972 to 1977, when the degree of rupiah overvaluation increased steadily. Overvaluation then decreased sharply after the 1978 devaluation, but reemerged in 1980.

To the extent that log and wood product exports responded to movements in the real exchange rate, exchange rate policy throughout most of

1972–84 would have tended to reduce both the log harvest and non-wood exports. The latter is likely, the former doubtful. The presence of rents, together with supply constraints, resulted in export supply elasticities much lower for wood products than for non-wood products. Two types of supply constraints for wood and wood products held during the period. The first involved capacity limitations in the wood products industry itself. From 1967 to 1973, they constrained logging operations as the industry was gearing up for massive extraction activities in Kalimantan. Capacity constraints were also important in plywood manufacture, as the number of mills rose slowly from less than five in 1977 to over 130 in 1984. Capacity constraints are insignificant for non-wood forest products, which arise essentially from gathering activities that can be easily varied according to prices received by producers. A second supply constraint on the wood-products industry, not affecting non-wood products, was the enactment of log export quotas beginning in 1980.

Available empirical evidence also suggests that timber exports were much less responsive to movements in the real exchange rate than were exports based on gathering activities. While there is little direct evidence regarding the supply responsiveness of non-wood products to real exchange rates, price incentives have been shown to be significant for such gathering activities as rubber collection. For the 1970s, Gillis and Dapice found short and long-run supply elasticities of close to 0.2 (Gillis and Dapice, in press). Since gathering of tropical non-wood forest products is even less dependent on capacity constraints than rubber, we may expect supply elasticities in these activities to be even higher than for rubber.

Table 2.9 supports the view that non-wood products have responded more to real exchange rate movements than have wood products. There was sharp growth in non-wood forest product exports following both the August 1971 and November 1978 devaluations. There are, therefore, ample reasons for believing that long periods of deep overvaluation, especially in 1972–1978 and 1980–1983, resulted in a decline in the forest's value as a source of non-wood products relative to its value in producing wood. This, together with the likely increase in rural poverty and thus in shifting cultivation due to overvaluation, would encourage more forest destruction than a more neutral exchange rate policy. While the empirical basis for asserting that overvaluation affected non-wood forest product exports to a much greater extent than wood products is not definitive, it is more than plausible that this effect has been present.

Subsidies and parks

We have seen that some forestry policies and a few non-forestry policies have, jointly and separately, contributed to rapid rates of deforestation in Indonesia. Moreover, many forestry policies designed to encourage reforestation have had little positive impact. Vested interests benefiting from Indonesian forestry policy since 1970 have proven too strong to permit major shifts in these policies. Not surprisingly, environmentally conscious policy-makers have turned to other measures that, while primarily geared to non-forestry objectives, could help curtail the pace of deforestation. Two particularly noteworthy examples are the subsidization of kerosene consumption, aimed primarily at deforestation in Java, and a program for expansion of national parks and wildlife refuge areas in Sumatra and Kalimantan. The first has been a spectacular failure as a conservation measure; the second has reasonable prospects for some limited success.

The kerosene subsidy. Although fuelwood scarcity has been a persistent problem on Java since the beginning of the century (Setyono et al. 1985: Chapter 2), deforestation and erosion arising from fuelwood demands became increasingly serious in the 1960s on this densely populated and mountainous island where 61 percent of the nation's 162 million inhabitants resided by 1985.[17]

The incidence of poverty has also been higher in Java than on the larger islands of Sumatra and Kalimantan. Although per capita income for the nation as a whole was estimated at $560 for 1983 (World Bank 1985: Appendix Table 1), material standards of living have been consistently higher in Sumatra and Kalimantan than in Java since 1950. The combination of rural poverty and population pressure on Java has induced substantial timber harvesting for fuelwood as well as increased cultivation of such root crops as cassava on slopes formerly not subject to wood harvests or agriculture. Serious erosion has resulted, particularly in east and central Java, to the extent that rivers that were formerly perennial are now annual, due to heavy siltation. The 24 percent of Java still under forest cover (largely secondary forest and forest plantations) remains at risk.

Domestic kerosene consumption was heavily subsidized in Indonesia throughout most of the 1970s and through the early 1980s. By 1979–80, the economic cost of this subsidy was estimated at US$1.6 billion, or just over 4 percent of gross domestic product (GDP).[18] The justification for this expensive subsidy program was twofold: income redistribution and curtailing deforestation. The subsidy, however, was effective in neither case (Gillis 1980a: 62–69).

At first glance, a kerosene consumption subsidy, however expensive, might seem to be an efficacious measure for forest conservation and erosion control, to the extent that erosion is traceable to harvest of timber for fuelwood on mountain slopes and to the extent that the subsidy reduces this activity. The subsidy was in fact justified largely on environmental grounds, a reasonable rationale if kerosene and fuelwood were good substitutes in rural areas. Unfortunately, this was not the case in Indonesia in the period 1970–1985. Kerosene in rural households was used primarily for lighting and rarely for cooking. Fuelwood has been used exclusively by households for cooking; 85 percent of rural Java households still used firewood for cooking in 1980 (Biro Pusat Statistik 1981: Table 4.1.3). Inasmuch as fuelwood is not ordinarily used for lighting, the kerosene subsidy could not be justified as a conservation measure, given its limited impact in encouraging households to switch from fuelwood to kerosene for cooking. Even if subsidized kerosene prices had served to reduce fuelwood demand for cooking purposes, the program would have protected only a relatively small area of forest at a very high price.[19] As a forest conservation program, the kerosene subsidy in 1979 involved a cost of at least US$77,000 for each hectare protected from cutting. Inasmuch as the most expensive and most complete replanting program then involved costs of not more than US$1,000 per hectare (Gillis 1984: Table IV–1), the subsidy was a grossly inefficient means for curtailing deforestation.

Moreover, far more effective methods for reducing firewood demand existed at the time, at costs of but a small fraction of annual kerosene subsidy costs, which climbed to $2.2 billion in 1980. One promising policy option was that of subsidizing use of liquified petroleum gas (LPG) to induce its substitution for firewood in rural areas. Indonesia in those years had substantial excess capacity in LPG production, and even during the years of oil boom sold LPG abroad at a heavy discount. Indeed, all of the approximately 12 million rural households on Java could have been provided a free LPG cookstove in 1980 at a one-time total cost of about US$650 million. Since the kerosene subsidy cost in excess of US$5 billion in the three years prior to 1981, its replacement by the LPG option would have saved Indonesia over $4 billion in those years alone, and would also have provided greater protection to forested areas. Put another way, the net saving from substituting one forest conservation policy for another could have financed replanting 4 million hectares of land over a three-year period.

National park programs and policies. Forest and other nature conservation values have been most evident in recent years in policies governing the establishment of National Nature Reserves, wildlife refuges, and park lands. In the seven years following the 1978 appointment of the first Minister for the Environment, sizeable new areas were added to these lands. Nature reserve areas expanded by four and a half times, and that for wildlife refuges and tourist parks tripled (Departamen Kehutanan 1985: 121) to a total of nearly 19 million hectares (Table 2.1). Most significantly for the tropical forest estate, by 1982 a full 20 percent of the area of Kalimantan and 27 percent of Sumatra were designated as Nature Conservation forests. One seasoned foreign observer has termed these measures as generous, perhaps too generous given the eventual needs for agriculture.[20]

Impetus for these initiatives has arisen primarily from the Minister for the Environment, and from non-profit organizations inside and outside Indonesia, rather than from the Forestry Department. The costs of financing increments to Nature Reserves and Parks, however, have been borne almost exclusively by Indonesia's government budget.

Timber rents and rent capture

Deforestation and forest degradation, as we have seen, proceeded at a rapid rate from 1967 to 1979, slowed briefly from 1980 to 1984, and, according to the present long-term forestry plan, is to again accelerate from 1985 until the end of the century. A major issue, given these developments, is the extent to which tangible benefits flowing from utilization of the natural forest counterbalance the tangible, not to mention the intangible, losses arising from consumption of forest resources.

We have noted that employment, regional development, and foreign exchange benefits from resource exploitation have not been sizeable by any standard. Large efficiency losses in forest-based industrialization were also depicted. Information from these sections may now be combined to assess the extent to which the resource owner – the government – succeeded in appropriating a respectable share of the rents available from tropical timber. This requires examination of the efficiency of fiscal policy (including forest fees and taxes) in capturing rents for the government. It will be evident that the Indonesian experience in capturing timber rents evidently has not been an especially good one, particularly in comparison with Sabah (Chapter 3). However, the record has been one of slow and marginal improvement since 1970, though recent increases in the rent share captured by the government are largely a by-product of environmental and industrialization policies, rather than the result of a conscious attempt to increase the owner's share.

Rent in tropical timber

Resource rents arise from the scarcity value of resources provided by nature, such as copper, ore, crude oil, and natural tropical forests. More generally, resource rent indicated surplus value available for appropriation either by the owner of the property right to the resource, by the owners of resource extraction enterprises, or by the government when it is not the resource owner. The policy importance of the existence of resource rents is that their full capture by the government, either as resource owner or through its taxing powers, will affect neither the production not the capital invested in its extraction. By definition, a rent is a return to a factor of production beyond that required to attract that factor to a given economic activity.

In principle, rent can be determined in any natural resource activity by deducting from gross extractive income the cost of labor, materials, and capital inputs (including costs of paying a "normal" return to capital). In forest economics, the concept of "stumpage value" is very close to that of economic rent. Higher quality or more accessible stands will command a higher stumpage value. This is essentially the same thing as saying that better quality stands will receive higher rents, given differential resource rents.

Rents in natural forest projects may also include so-called monopoly rents, arising from imperfections in competition in the resource industry. In tropical timber, rents consist primarily of straightforward resource rents. While monopoly and oligopoly rents arise in tropical forest operations when the supply of timber concessions is limited and when firms do not bid for them, the structure of the international forest products industry is far less oligopolistic than that of the world oil or hard minerals industry; hence monopoly (oligopoly) rents are far less important in tropical timber.

Rents are clearly present in tropical timber, although it is much easier to describe them than for owners of timber rights to capture them. Thus, government timber policies not oriented toward rent capture fail to protect the owners' interests. The government as owner, however, cannot realistically hope to capture exactly 100 percent of rents from tropical timber harvest, because in practice rents are difficult to determine for any given resource endowment. Rather, a sensible approach to rent capture is to first recognize that rents are present, then to devise administerable instruments that enable the government-as-owner to appropriate as large a rent share as is practical (Bucovetsky and Gillis 1978: Chapter 4).

Estimates of timber rent in Indonesia

While there are numerous examples of estimated stumpage values and rents for particular temperate softwood forests, there are few published estimates of rents in tropical timber. Coincidentally, one of the most detailed of these studies is one done by Ruzicka for Indonesian tropical timber stands for the 1973–1979 period.

In one set of calculations, Ruzicka estimated rents for logs produced by large firms in the timber-rich province of East Kalimantan, the source (then and now) of over half of Indonesian timber exports. Rents averaged $45 per cubic meter over four quarters in 1973–74. Potential rents per hectare available under full utilization of the commercial standing stock were estimated at between US$1,873 and $2,257. Rents actually generated under harvesting methods in use in East Kalimantan at the time were placed at between US$1,479 and $1,642 per hectare (Ruzicka 1979: 49–56).

During that period, timber enterprises appropriated about 75 percent of the actual rents generated in East Kalimantan forest exploitation. This was a large proportion for rent capture for enterprises in natural resource extraction; it corresponded to about 62.5 percent of the gross export value of timber (Gillis 1983).

Governments attempt to increase the proportion of rent captured for the domestic economy in a variety of ways, but taxation has proven the most effective. This section examines rent capture by the government in the more recent period 1979–83.

The estimates are somewhat more crudely based than Ruzicka's for 1973–77, as his data base was considerably more detailed, derived as it was from internal company data. Nevertheless, the estimates developed herein are based on information as recent and comprehensive as is available to researchers without access to recent concessionaire records.

The data from which the rent and rent capture estimates are drawn are presented in Tables 2.10 through 2.20.

Tables 2.10 through 2.13 show collections of government timber revenues by type of levy for all recent years for which data were available. Table 2.10 presents timber royalties, Table 2.11 shows IPEDA (property tax) collections, Table 2.12 timber export taxes, and Table 2.13 revenues from replanting deposits. Table 2.14 combines all government revenues from timber. The table indicates that overall timber-source revenues grew much more slowly than production or export until 1978, the year log export taxes were doubled. This is verified in Table 2.15, which portrays timber taxes as a percent of total reported timber exports. Timber taxes and royalties never exceeded 20 percent of reported export value until 1978. Thereafter the percentage increased sharply until 1982–83, when it fell to 27.6 percent. The behavior of these ratios after 1978 was due primarily to variation in revenues from export tax on logs. As the proportion of logs in total exports declined due to stepwise reductions in log

Table 2.10. *Indonesia: timber royalties and license fees, 1969–70 through 1982–83*

Fiscal year	Regional government share (billion rupiah)	Central government share (billion rupiah)	Total, all levels of government[a] (billion rupiah)	Total, all timber royalties (million U.S. dollars)[b]
1969–70	1.0	0.4	1.4	3.7
1970–71	3.1	1.3	4.4	11.6
1971–72	3.7	1.6	5.3	12.8
1972–73	6.7	2.9	9.6	25.5
1973–74	10.2	4.4	14.6	35.2
1974–75	8.1	3.5	11.6	28.0
1975–76	6.8	2.9	9.7	23.4
1976–77	11.9	5.1	17.0	40.7
1977–78	22.7	9.7	32.4	78.1
1978–79	26.0	11.2	37.2	49.7
1979–80	35.7	15.3	51.0	81.6
1980–81	72.8	36.7	109.5	173.8
1981–82	39.9	19.1	59.0	93.7
1982–83	40.7	18.9	59.6	91.7

[a]Includes "additional" royalty of US$1.50 per m^3.
[b]Converted at prevailing rates of exchange for each year.
Sources: For years prior to 1979, Directorate-General of Domestic Monetary Affairs. For years after 1979, Departamen Kehutanan (1984).

Table 2.11. *Indonesia: property tax collections (IPEDA) on the forestry sector, 1971–72 through 1984–85*

Year	IPEDA on forestry (billion rupiah)
1971–72	0.072
1972–73	0.144
1973–74	0.366
1974–75	0.373
1975–76	1.555
1976–77	4.073
1977–78	7.931
1978–79	9.120
1979–80	11.600
1980–81	13.384
1981–82	20.005
1982–83	15.201
1983–84	18.032
1984–85	19.744

Source: Departamen Keuangan, *Nota Keuangan*, various issues.

Table 2.12. *Indonesia: timber export taxes on logs and sawn timber, 1971–72 through 1984–85*

Year	Export tax collections[a]	
	Billion rupiah	Million U.S. dollars
1971–72	6.9	17.3
1972–73	10.4	25.1
1973–74	29.1	70.1
1974–75	27.5	66.3
1975–76	20.8	50.1
1976–77	31.7	76.4
1977–78	35.6	85.8
1978–79	102.7	197.5
1979–80	223.4	357.4
1980–81	196.4	311.8
1981–82	84.9	134.8
1982–83	42.4	65.2
1983–84	52.5	53.4
1984–85	37.6	35.6

[a]Timber export tax rates were as follows: 1971 through January 1978: 10 percent *ad valorem*, logs only; January 1978 to present: 20 percent *ad valorem* on logs, 5 percent on roughly sawn timber (thickness more than 5 mm).
Source: Ministry of Finance, Government of Indonesia (personal communication).

Table 2.13. *Revenues from replanting deposit ($4.00 per cubic meter of logs harvested in Kalimantan and Sumatra)*

Year	Revenues (billion rupiah)
1980–81	7.2
1981–82	31.7
1982–83	28.8
1983–84	17.2[a]

[a]Estimate based on official figures for log exports.
Source: Departamen Kehutanan (1984).

export quotas from 1980 to 1984 (and their ultimate ban in 1985), so did the ratio of taxes to export values.

Table 2.16 presents, for recent years, potential rents per cubic meter of roundwood equivalent. These computations begin with f.o.b. prices for exported logs, sawnwood, and plywood. Most sawmill and plymill output

Table 2.14. *Indonesia: total identifiable tax collections on forestry sector, 1971–72 through 1982–83 (billion rupiah)*

	(1) Total export tax, royalty, and license fee tax collections (all levels of government)				(2) Revenues from replanting deposit	(3) Total corporate tax collections on timber companies	(4) Total taxes on timber sector
Year	Export tax	Royalty and license fees	IPEDA	Subtotal			
1971–72	6.9	5.3	0.1	12.3	—	0.1	12.4
1972–73	10.4	10.6	0.1	21.1	—	0.2	21.3
1973–74	29.1	14.6	0.4	44.1	—	0.2	44.3
1974–75	27.5	11.6	0.4	39.5	—	0.7	40.2
1975–76	20.8	9.7	1.6	32.1	—	1.3	33.3
1976–77	31.7	17.0	4.1	52.8	—	1.5	54.3
1977–78	35.6	32.4	7.9	75.9	—	1.9	77.8
1978–79	102.7	37.2	9.1	149.0	—	9.1	158.1
1979–80	223.4	51.0	11.6	286.0	—	21.8	307.8
1980–81	196.4	109.5	13.4	319.3	7.2	15.3	341.8
1981–82	84.9	59.0	20.0	163.9	31.7	17.0	221.7
1982–83	42.4	59.6	15.2	117.2	28.8	18.5	164.5
1983–84	—	—	—	—	17.2	—	—

Sources: Column 1: Tables 2.10 through 2.12; Column 2: Departamen Kehutanan (1984); Column 3: Directorate-General of Monetary Affairs, Departamen Keuangan.

Table 2.15. *Indonesia: total identifiable taxes on forestry sector as percent of total recorded timber export values (logs, veneer, plywood)*

Year	(1) Total taxes on timber sector[a] (billion rupiah)	(2) Total reported timber exports (billion rupiah)	(3) Exports in million U.S. dollars	(4) Taxes as percent of total reported exports
1971–72	12.4	68.0	170	18.2
1972–73	21.3	114.1	275	18.7
1973–74	44.3	298.8	720	14.8
1974–75	40.2	255.2	615	15.8
1975–76	33.3	218.7	527	15.2
1976–77	54.3	367.3	885	14.8
1977–78	77.8	391.4	943	19.9
1978–79	158.1	587.6	1,130	26.9
1979–80	307.8	1,353.2	2,172	22.7
1980–81	341.8	1,048.3	1,672	32.6
1981–82	221.7	601.0	951	36.8
1982–83	164.5	594.2	899	27.6
1983–84	—	1,055.3	1,161	—

[a]From Table 2.14.

is vertically integrated back to the logging stage. For that reason, and others indicated later, *product* prices (other than logs) must be converted into roundwood equivalents to provide meaningful estimates (Column 2). Potential rent per unit is then derived by subtracting appropriate costs, including logging and transport cost for logs, plus additional milling costs for sawn timber and plywood. Cost estimates were taken from a number of sources detailed in the notes to Table 2.16, and were checked against those reported for West Africa (FAO 1981b: 67–76, 120–149), for Malaysia, and by private consulting firms in Indonesia (P. T. Data Consult 1983: 38–49). It was necessary to use the same cost figures for all producers, a simplification that will underestimate rents for some firms and overestimate rents for others.

The pattern of potential rent that emerged in all activities is at first glance surprising. Except for 1979, per unit rents in logging were consistently higher than when the same logs were used for sawn timber or plywood (Column 5). Also, after 1979 potential rents obtainable from timber used in plywood manufacture were consistently negative. Column 5 is a jarring illustration of the process of rent destruction. Recent high rates of effective protection to plywood and increased utilization of logs in

Indonesian plywood mills in 1980–83 rather than for export meant that logging rents were used to subsidize inefficient investments in forest-based industry. As long as plywood recovery rates remain low, most of the 140-odd plymills will swallow timber rents as surely as black holes do nearby celestial bodies.

Table 2.17 brings together information from the preceding seven tables to obtain an estimate of hypothetical rent per cubic meter and the share captured by government. This table illustrates the share of rent that would have been captured by government *if* all timber royalties and timber taxes (other than income taxes) had been collected at 100 percent efficiency. Collection efficiency was not near 100 percent due to administrative problems and corruption. Table 2.17 thus displays policy intentions in rent appropriation, and it indicates that the policy thrust was to increase government shares in timber rent from logs over the period, and to capture lower shares of rent on sawn timber and plywood, at least until 1981. However, the replanting deposit, a by-product of an environmental measure, doubled the government share of rent in sawn timber in 1981, and nearly doubled it again in 1982.

Table 2.17, however, presents a picture only of the effects of government tax policy in capturing hypothetical rents per unit of production from the wood-based industry as it was actually constructed from 1979 to 1982. The table does not encompass all government policies affecting the aggregate magnitude of potential log rents that would have been available had log exports not been sharply curtailed by policies. That is to say, the growing governmental share in per unit rents depicted in Table 2.17 was a result not only of more aggressive tax policy but of marked shrinkage of aggregate rents. Shrinkage of rent available for capture by the government was due largely to strong government incentives for investments in sawmills and plywood mills. As a result, substantial amounts of rent were destroyed in the process of encouraging forest-based industrialization. From 1979 to 1982, the government's share in *potential* rents (actual rent plus destroyed rents) from exploiting tropical forest resources actually declined. This may be seen in Tables 2.18 through 2.20.

Table 2.18 presents production information used to derive estimates of the volume of logs harvested but not utilized in sawnwood and plywood production. Column 4 thus indicates the total amount of logs available for export in log form in each year. Table 2.19 draws on information in tables 2.18 and 2.17 to estimate aggregate tropical timber rents given actual utilization of logs (as logs for sawnwood and for plywood) in the years 1979–1982. Column J shows estimated aggregate rent generated in sales

Table 2.16. *Indonesia: hypothetical rent for tropical timber in roundwood equivalent, 1979–83 (U.S. dollars)*[a]

	(1) Average f.o.b. value per m^3 of product	(2) Average f.o.b. value per m^3 of roundwood equivalent	(3) Logging and transport costs (to port) per m^3 of roundwood equivalent	(4) Milling costs per m^3 of roundwood equivalent	(5) Potential rent per m^3 of roundwood equivalent: Col. 2 − (Col. 3 + Col. 4)
1979					
Logs	85.21	85.21	29.84	n.a.[b]	55.37
Sawnwood	183.42	100.78	29.84	10.36	60.58
Plywood	271.11	117.87	29.84	56.48	31.55
1980					
Logs	106.93	106.93	34.24	n.a.	72.69
Sawnwood	214.55	117.24	34.24	11.92	71.08
Plywood	227.49	98.90	34.24	64.93	− 0.27
1981					
Logs	95.84	95.84	37.93	n.a.	57.91
Sawnwood	175.31	96.32	37.93	13.21	45.18
Plywood	195.23	84.88	37.93	71.99	− 25.04
1982					
Logs	100.59	100.59	41.00	n.a.	59.59
Sawnwood	150.78	82.84	41.00	14.28	27.56
Plywood	229.12	99.61	41.00	77.82	− 19.21
1983					
Logs	99.10	99.10	45.92	n.a.	53.18
Sawnwood	154.70	85.00	45.92	16.00	23.08
Plywood	250.63	108.97	45.92	87.16	− 24.11

a All timber products are valued at opportunity costs (world prices). Conversion factors are: *sawnwood*, 1.82 m³ of logs for each m³ of sawnwood; *plywood*, 2.3 m³ of logs for each m³ of plywood. Source: Departamen Kehutanan (1984). Note that other sources report higher rates for log recovery in Indonesia. Source for export values: Departamen Kehutanan (1984). Source for cost data: Based on information in Hunter (1984: 109–110) and corroborated by cost data in P. T. Data Consult (1983: 9). Sawmill cost data were corroborated with 1979 costs reported for a mill with annual capacity of 5,000 m³, using mixed dipterocarp timber in Indonesia and Malaysia (FAO 1981b). Hunter's figure for logging plus transport cost to mills for 1982 was $35.00 (average for East Kalimantan, source of 21 percent of Indonesia's timber exports). This figure does not, apparently, include harbor fees and associated export costs of approximately $2.00 per m³. Further, an estimate by Buenaflor for logging costs, cited in Repetto et al. (1987: 35), indicates such costs at $44.60 per m³ in 1980. In view of the discrepancies in these estimates, a figure of $41.00 per m³ was employed for logging/transport costs in 1982.

For other years, costs were assumed to rise as follows: one-half of costs rose *pari passu* with world inflation; one-half rose *pari passu* with domestic inflation (wholesale price index for manufactures). Note that the years 1979–81 were all years in which the rupiah's nominal value was fixed at 630 per $1.00. Thus, costs were estimated as follows, per unit of product: logging and log transport costs were US$29.84 per m³ of logs in 1979, $34.24 in 1980, $37.93 in 1981, $41.00 in 1982, and $45.92 in 1983; milling costs for sawnwood were $18.86 per m³ in 1979, $21.69 in 1980, $24.05 in 1981, $26.00 in 1982, and $29.12 in 1983; milling costs for plywood were $129.91 per m³ in 1979, $149.34 in 1980, $165.58 in 1981, $179.00 in 1982, and $200.48 in 1983.

Because the comparisons in the table—and subsequent tables—are expressed in roundwood equivalents, milling costs must also be converted to their roundwood equivalents—i.e., we need to know the costs required to transform a m³ of log to sawn timber or plywood product. Standard Indonesian conversion ratios were used (1.82 for sawn timber and 2.3 for plywood), and results are presented in the following paragraph. The use of one conversion ratio for all species and all firms of course masks differences not only in recovery from different species but also in the relative efficiency of different firms. However, more detailed information is unavailable.

Expressed in roundwood equivalents, the logging and log transport costs were US$29.84 in 1979, $34.24 in 1980, $37.93 in 1981, $41.00 in 1982, and $45.92 in 1983. Milling costs, roundwood equivalent, for sawnwood were $10.36 per m³ in 1979, $11.92 in 1980, $13.21 in 1981, $14.28 in 1982, and $16.00 in 1983. Milling costs, roundwood equivalent, for plywood were $56.48 per m³ in 1979, $64.93 in 1980, $71.99 in 1981, $77.82 in 1982, and $87.16 in 1983.

b Not applicable.

Table 2.17. *Hypothetical timber rent capture by government: Kalimantan and Sumatra, 1979–82, exclusive of income taxes (U.S. dollars)*

| | Government revenues from timber per m³ of roundwood equivalent | | | | | | |
| | (1) | (2) | (3) | (4) | (5) | (6) | (7) |
	Potential rent[a]	Timber royalty (6 percent of f.o.b. log value at export point)[b]	Timber export tax (20 percent for logs)	Replanting fee ($4.00/m³)[c]	Other levies, IPEDA, the "additional" royalty	Total government revenue (Cols. 2 + 3 + 4 + 5)	Government timber revenue as percent of potential rent (Col. 6 ÷ Col. 1)
1979							
Logs	55.37	5.11	17.04	n.a.	2.52	24.67	44.6
Sawnwood	60.58	5.11	n.a.	n.a.	2.52	7.63	12.6
Plywood	31.55	5.11	n.a.	n.a.	2.52	7.63	24.2
1980							
Logs	72.69	6.41	21.39	n.a.	2.78	30.58	42.1
Sawnwood	71.08	6.41	n.a.	n.a.	2.78	9.19	12.9
Plywood	− 0.27	6.41	n.a.	n.a.	2.78	9.19	
1981							
Logs	57.91	5.75	19.17	4.00	2.65	31.57	54.5
Sawnwood	45.18	5.75	n.a.	4.00	2.65	12.40	27.5
Plywood	− 25.04	5.75	n.a.	4.00	2.65	12.40	
1982							
Logs	59.59	6.03	20.12	4.00	2.71	32.86	55.1
Sawnwood	27.56	6.03	n.a.	4.00	2.71	12.74	46.2
Plywood	− 19.21	6.03	n.a.	4.00	2.71	12.74	

[a] Derived from Table 2.16 (potential rent per m³).

[b] Timber royalty structure was changed in July 1979. Since then, a 6 percent *ad valorem* rate has applied to the f.o.b. value of all timber produced by concession holders. The 6 percent royalty applies to the "check price" (posted price), which since 1972 has ordinarily been 5–9 percent below actual export values. The estimates in this table apply to values in roundwood equivalent at the check price, and thus understate rents by perhaps 5 to 10 percent per year.

[c] Replanting fee enacted in 1981; imposed on log production in Kalimantan and Sumatra.

Table 2.18. *Indonesia: total production of logs, sawnwood, and plywood (million cubic meters), 1977–82, in roundwood equivalents*

Year	(1) Total log harvest	(2) Sawnwood production	(3) Plywood production	(4) Logs not utilized in production of sawnwood and plywood: Cols. 1 − (2 + 3)
1977	22.235	3.51	0.28	18.45
1978	24.234	3.50	0.47	20.26
1979	25.314	3.40	0.69	21.24
1980	25.190	3.41	1.01	20.77
1981	15.954	3.51	1.55	10.89
1982	13.377	3.75	2.20	7.43

Sources: Log harvest: Table 2.3. Sawnwood and plywood production: Takeuchi (1983).

of sawn timber, plywood, and logging in each year. Not surprisingly, aggregate rents declined after 1980 as the restrictive effects of log export quotas commenced.

The data in Tables 2.18 and 2.19 enable an estimate of potential timber rents available in the absence of strong incentives for sawmill and plywood investments: that is, we estimate what total volume of rent would have been available for capture by the government had all logs harvested from 1979 through 1982 been exported as logs. We do this in Table 2.20, where yearly figures for hypothetical rent per cubic meter of logs are multiplied by annual log production. Table 2.20 calculations embody the assumption that greater availability of Indonesian logs on world markets from 1979 through 1982 would have depressed world wood product prices no more than by supplying more plywood and sawn timber in response to incentives for forest-based industrialization. This implies that all logs harvested in Indonesia from 1979 to 1982 could have been exported at per unit f.o.b. prices actually prevailing for logs in those years. This assumption accords with those used by other analysts for the same period.[21]

Table 2.20 thus indicates that total potential timber rent in the absence of domestic processing of the tropical logs would have been marginally higher than actual rents in 1980, 23 percent higher in 1981, and 58 percent higher in 1982. The government share of these potential rents is depicted in Table 2.21. Counting potential rents dissipated in plywood manufacture, the government share in rents did not increase after 1979, but declined slightly from nearly 35 percent to about 33 percent. Furthermore, the restrictive effects of government policies for encouraging for-

Table 2.19. *Indonesia: rents in tropical timber—logs, sawnwood, and plywood, 1979–82*

	Logs		
	(A) Logs not utilized in sawnwood and plywood (million m³)[a]	(B) Rent per m³ (U.S. dollars)[b]	(C) Aggregate rent in logging (Cols. A × B) (million U.S. dollars)
1979	21.24	55.37	1,176.1
1980	20.77	72.69	1,509.8
1981	10.89	57.91	630.6
1982	7.43	59.59	442.8

	Sawnwood		
	(D) Production in roundwood equivalent (million m³)	(E) Rent per m³ of roundwood equivalent (U.S. dollars)[b]	(F) Aggregate rent in sawnwood (Cols. D × E) (million U.S. dollars)
1979	3.50	60.58	212.0
1980	3.41	71.08	242.4
1981	3.51	45.18	158.6
1982	3.75	27.56	103.3

	Plywood			
	(G) Production in roundwood equivalent (million m³)	(H) Rent per m³ of roundwood equivalent (U.S. dollars)[b]	(I) Aggregate rent in plywood (Cols. G × H)	(J) Total rent: (Cols. C + F + I) (million U.S. dollars)
1979	0.47	31.55	14.8	1,402.9
1980	1.01	− 0.27	− 0.3	1,751.9
1981	1.55	− 25.04	− 38.8	750.4
1982	2.20	− 19.21	− 42.3	503.8

[a]Logs produced but not utilized in sawnwood or plywood manufacture, from Table 2.18, Column 4.

[b]Columns B, E, and H from Table 2.17, Column 1 (hypothetical rent capture per cubic meter).

est-based industrialization, coupled with world recession, caused total potential rents to drop by half from 1979 to 1982. The government therefore captured a relatively constant share of declining aggregate rents in the timber industry.

Findings of substantial rent destruction through imposed forest-based

Table 2.20. *Potential timber rents in the absence of domestic processing, 1979–82*

	Hypothetical rent per m³, logs[a] (U.S. dollars)	Total log production[b] (millions m³)	Total potential rent, logs (Col. 1 × Col. 2) (million U.S. dollars)
1979	55.37	25,314	1,401.6
1980	72.69	25,190	1,831.1
1981	57.91	15,954	923.9
1982	59.59	13,377	797.1

[a]From Table 2.17.
[b]From Table 2.18.

Table 2.21. *Government share in timber rents: no domestic processing of timber*

	(1) Total rents available[a] (million U.S. dollars)	(2) Government tax collection[b] (million U.S. dollars)	(3) Government share, percent (Col. 2 ÷ Col. 3)
1979	1,401.6	488.6	34.9
1980	1,831.1	542.5	29.6
1981	923.9	351.9	38.1
1982	797.1	261.1	32.7

[a]From Table 2.20.
[b]From Table 2.14, converted to U.S. dollars at prevailing exchange rates.

industrialization in Indonesia is supported by other researchers. Fitzgerald (1986) found that, in constant dollars, for every dollar gained in value of plywood exports, more than four were lost in log exports. He estimates that forced investments of between US$1.0 billion and $2.1 billion in plywood factories caused a net drain on the nation's resources of $956 million, as plywood was exported at prices below production costs. Under these policies, domestic timber processing produces net social costs, not the benefits intended by the policy. In effect, potential rents were dissipated in the high costs of investing in and operating inefficient plywood and sawnwood mills established merely to retain timber concessions. Under these policies, Fitzgerald notes Indonesians paid $956 million to be plywood exporters instead of log exporters, and lost value-added that would have resulted from additional exports of more than 10 million cubic meters of logs per year.

In sum, government policies intended to capture timber rents were not particularly successful. Policies intended to create additional domestic value-added in timber processing destroyed sizeable amounts of rent. These conclusions offer scant hope that the costs of deforestation in Indonesian tropical forests may have been offset by material benefits resulting from forest resource exploitation.

Appendix A: Selective cutting, *ad valorem* royalties, and damage to forest quality

This appendix portrays the separate and joint effects of selective cutting policies on the one hand and royalty structures on the other in the context of tropical timber endowments.

Selective cutting systems, or selection systems, are best characterized in contrast with "uniform cutting systems." Under any of several forms of selective cutting systems, *only* trees above a certain diameter size (e.g., 50 cm. d.b.h.) may be cut, but only those with desirable traits (merchantability, uniformity) are taken; low-quality trees above minimum size are left. Under uniform cutting systems,[22] *all* trees above a certain diameter are cut or poisoned, depending on whether they are to be utilized. Many countries, including Indonesia and the Philippines, prescribe some form of selective cutting; many, including Indonesia, utilize *ad valorem* fees as the basis for the timber royalty system. Both selective cutting[23] and flat-rate *ad valorem* royalties separately lead to significant *additional* depletion of tropical forest resources above and beyond the volume of wood removed from the forest. As practiced in most countries, including Indonesia, combining the two factors yields results that are more damaging than the sum of their results.

Selective cutting

Except in Thailand and under some variants of the Malayan Uniform System, enterprises logging in tropical forests are ordinarily charged royalties and taxes on the basis of timber *removals*, without reference to the stock of merchantable and legal-sized stems in the stand. This contributes to unavoidable features of selective cutting and unintended depletion of tropical forest resources.

First, logging operations under selective cutting involve removal only of the best stems of prime species; stems *not* taken are often badly damaged by felling, skidding, and other associated extractive activities. To use an oversimplified example, consider a forest stand initially worth $10 million from which $5 million of stems are taken on first entry. Due to the value of the stems damaged during logging, the worth of the remaining stand may be only $3 million. The damage to the residual stand on public lands constitutes a social cost for which the logger is not charged.

In reality, these damages can be very substantial. Several studies of logging in insular Southeast Asia summarized by Kuswata (1980: 120) suggest that selective cutting as usually practiced can severely damage as much as half of the standing trees exceeding 10 cm. diameter at breast height. Ashton (1980:48) cites the effects of road-building, yarding, and other operations upon regeneration. Thaib and Karnasudirja (1981) report markedly different stand damage in Indonesia depending on the logging method. Damage to residual stands from tractor logging varied from 4 percent to 51 percent; under high-lead logging (using long cables) damage was 72 percent. Post-felling operations (ground skidding and handling) are responsible for most of the damage. The stock of trees for logging in the next cutting cycle (30 to 35 years later) may be reduced substantially by these extractive activities; at a minimum, the commercial value of the logged-over stand may be diminished by an amount well in excess of the value of removed stems. Furthermore, the low-intensity felling in most selective cutting systems (two to five trees per hectare), may be insufficient to open the forest canopy to provide enough light for the best emergence of immature dipterocarps in the residual stand. In addition, extractive activities may require that as much as 30 percent of the logged-over area be devoted to roads, skid trails, and log yards, all of which form bare areas. Besides increasing erosion (a separate social cost) and soil compaction, this tends to disturb forest water courses, causing many old and young trees to die in the ponds that may form as a result.

The Philippine experience with selective cutting is similar to that of Indonesia. Concessionaires who began operating in the 1950s were, by 1986, just reaching the end of the first 35-year cutting cycle provided under that system. It is now clear that the quantity and the quality of the second cycle harvest will be much below the earlier optimistic estimates of the government. This has been attributed, as in Indonesia, to residual stem damage and premature reentry into stands, as well as to shifting cultivation, fire, and illegal harvests.

A second type of deterioration of stand quality arises not from damage to the unlogged standing stock but from the effects of certain kinds of forest fees (particularly some types of royalties) in inducing "high-grading" of tropical forest stands. Tax-induced "high-grading" or "cream-skimming" is of two distinct types (Gillis 1980b): (a) that within a particular timber stand (high-grading at the intensive margin); (b) high-grading as between different timber stands of different qualities (high-grading at extensive margins).

Only the first type of high-grading directly reduces forest quality, and it is the only type discussed here. Tax-induced high-grading exacerbates the degree of selective cutting within stands. This is because virtually all the most common systems of forest fees and taxes encourage operators to remove only the most valuable stems of so-called "primary" species while bypassing less profitable, "secondary" species as well as very large less merchantable stems of primary species. Some forms of forest fees and taxes induce particularly marked high-grading activity. Uniform specific royalties involve the most serious degree of high-grading. These are royalties expressed as a flat currency (e.g., a dollar) amount per physical unit (e.g., cubic meter) of timber extracted. Flat rate *ad valorem* royalties and export taxes (those expressed as the same percent of value of timber extracted) involve less high-grading than uniform specific royalties that produce the *same* government revenue (royalty yield). Finally, among output-based taxes (not profits taxes), a finely differentiated system of non-uniform *ad valorem* royalties involves the least degree of high-grading. Such royalties are differentiated according to species *and* grade, with higher royalty rates on the best grades of the most valuable species and lower rates royalties on lower grades of desirable species.[24] Granted that different forms of forest fees and taxes do induce different degrees of high-grading, it remains to be shown that high-grading involves social costs. There is at least one way that high-grading may lead to social costs in the form of a decline in the quality of the standing stock; this pertains to the effects of relogging.

It might be thought at first glance that high-grading in tropical forests is of little consequence, since stems bypassed because of tax-induced high-grading can always be later harvested. However, some analysts report that although the Indonesian Selective Logging System requires a 35-year cutting cycle, relogging of previously harvested hectares within five years is frequently practiced by Indonesian timber concessionaires.[25] Concessionaires presumably return to previously logged-over stands for two reasons: to harvest stems that previously fell short of minimum diameter requirements on first logging, and to take stems of lower quality bypassed earlier due to tax-induced high-grading. From the firm's point of view this would ordinarily be a sensible procedure: take the best quality (lower-cost) stems first, returning some years later both for those left behind in the first cut and for those that have reached minimum diameter in the intervening years. Given that in insular Southeast Asia, and particularly in Indonesia, firms seek an average of 35–60 cubic meters per hectare before entering a stand in the first place (compared to 20 to 80 cubic meters per acre in Asia as a whole) and given that, for example, East Kalimantan meranti stems may be expected to expand only about 1 cm. in diameter per year, it is difficult to see how any previously logged-over tract would be attractive enough to permit relogging within 5 or even 10 years of natural growth based solely on the previously immature standing stock. At 1 cm. diameter growth per year but a few trees would attain the minimum after only five years. But natural growth *plus* the presence of previously bypassed stems (owing to high-grading) might easily furnish sufficient volume to justify relogging within a 5–10 year period.

However, this relogging – made more attractive as a consequence of previous high-grading – results in further damage to the residual trees and soils and retardation of the forest recovery process. Thus, with high-grading and subsequent relogging, the forest is twice damaged under selective cutting of the Indonesian type: once on initial logging, with as much as 50 percent of the larger standing trees damaged, and again on relogging, with perhaps even greater damage.[26]

Ad valorem *royalties and high-grading*

"High-grading," or "cream-skimming," is widely viewed as a significant problem in the tropical forest (Setyono et al. 1985: 119–20). While transport costs, customary species selection, and other factors influence grade and species selection, taxes and fees used to collect revenues from forest activity are also significant factors leading to high-grading in tropical forestry. Tax-induced high-grading arises when timber taxes and royalties are imposed only with short-term government *revenue* objectives in mind, without regard to their effects on decision-making by loggers. High-grading contributes to a decline in the value of the forest as an asset, and encourages needless waste in timber extraction.

Forest fees and taxes imposed without due attention to these effects, combined with selective cutting harvest methods, encourage firms to cut only the most valuable timber, while both less attractive stems of valuable species and secondary species are left or damaged in the process. Any system of forest taxes or royalties that is presently administerable in the context of tropical forests in Indonesia involves some degree of high-grading. Therefore, the issue is what degree of high-grading the government (as landowner and holder of timber rights) is willing to accept. Sensible forest policy would seem to call for greater reliance on taxes and fees that involve less high-grading for a given amount of revenue and de-emphasis on taxes and fees known to have the most deleterious impact on harvesting methods.

We employ a heuristic exercise to illustrate the potential impact of different types of forest fees and taxes on harvesting decisions in tropical forests. This exercise considers effects of high-grading in a hypothetical stand, with incidence of high-value primary species conforming to patterns found in the natural rain forest in East Kalimantan in Indonesia. The exercise focuses upon the effects of different *forms* of tax on high-grading, rather than the effects of different *levels* of tax collections from a *given* stand.

Consider four types of taxes and/or charges commonly employed at different times in Indonesia and in the forest sectors of nations in the Asia-Pacific region: (1) income taxes; (2) differentiated royalties; (3) *ad valorem* royalties; (4) specific uniform royalties. Expressed in terms of effects on high-grading, these revenue devices stand in the relation as indicated in Inequality 2.1. Here again the focus is upon high-grading within given stands.

$$T_Y < T_D < T_A < T_S \qquad (2.1)$$

where

T_Y = high-grading effects of flat-rate income taxes
T_D = high-grading effects of finely differentiated royalties
T_A = high-grading effects of flat-rate *ad valorem* royalties
T_S = high-grading effects of uniform specific royalties

Note that Inequality 2.1 clearly applies when rates of the four alternative forms of tax are set so as to yield the same amount of revenue. This is summarized in Equation 2.2 below:

$$R_{Ya} = R_{Db} = R_{Ac} = R_{Sd} \qquad (2.2)$$

where

R_{Ya} = income tax revenues from a flat-rate income tax set at rate a
R_{Db} = royalty revenue from a differentiated royalty with rates set to generate revenues as close as possible to those available under the income tax set at rate a and to diminish high-grading to the maximum extent consistent with revenue level R_{Ya}
R_{Ac} = royalty revenues from a flat-rate royalty set to raise revenues as close as possible to R_{Ya}
R_{Sd} = royalty revenues from a uniform specific royalty with rates high enough to approximate R_{Ya}

The general principles portrayed in Inequality 2.1 and Equation 2.2 can be best illustrated by comparing the effects of various forms of *revenue equivalent* taxes and royalties on high-grading in a hypothetical stand having properties as indicated in Table 2.22.

Table 2.22. *Hypothetical tropical timber stand*

(1) Species and grade	(2) F.o.b. unit value (U.S. dollars per m³)	(3) Standing stock (thousand m³)	(4) Variable cost of extraction and transport (U.S. dollars per m³)
Alpha 1	100	5	40
Alpha 2	90	10	40
Alpha 3	75	15	40
Alpha 4	60	10	40
Beta 1	90	10	40
Beta 2	75	10	40
Beta 3	55	5	40
Gamma 1	70	15	40
Gamma 2	50	10	40
Delta 1	50	10	40
Total		100	

A. Total sales revenue = $7.1 million (Column 2 × Column 3).
B. Total variable cost = $4 million ($40 × 100,000).
C. Fixed cost = $1 million.
D. Surplus rent = $2.1 million [A − (B + C)].

The hypothetical stand is of relatively high density containing 100,000 cubic meters of standing stock of stems of over 50 cm. in diameter. There are 65,000 cubic meters of desirable species (Alpha and Beta, both of varying grades) and the remainder is in "secondary" species Gamma and Delta. Costs are divided into variable and fixed components. In the examples that follow variable costs are defined as those directly associated with the harvesting of an additional stem (labor, fuel, blades, etc.). Fixed costs are those associated with the exploitation of the stand as a whole (e.g., cost of moving equipment to the site, wear and tear on capital goods). Average *variable* costs are assumed to be constant at $40 per cubic meter. Fixed costs are set at $1 million per stand. Costs as stated include a 10 percent "normal" rate of return on capital.

To keep the illustration simple, we assume that the stand can be fully harvested within one year's time, thereby avoiding the need for discounting.

Absent any royalties or taxes *all* commercial stems in the stand will be harvested. Without taxes, incremental costs of extraction are, by construction, in all cases less than the unit values for all grades and species, *and* net revenues are sufficiently high to cover fixed costs as well. The total surplus after covering all costs, including normal returns on capital, is $2.1 million per stand.

Table 2.23. *Summary of differential implications of various forms of revenue equivalent taxes and royalties for high-grading, government revenues, and division of rents*

	Specific uniform royalty	*Ad valorem* royalty	Differentiated royalty	Flat-rate income tax
Royalty rate or tax rate (per m³)	$21	22.4%	variable[a]	65%
Quantity harvested (thousand m³)	65	80	100	100
Revenue to firm (sales value) in million U.S. dollars	5.225	6.100	7.100	7.100
Gross rent (million U.S. dollars)	1.625	1.900	2.100	2.100
Royalty or tax collections (million U.S. dollars)	1.365	1.366	1.365	1.365
Surplus (rent) left to firm (million U.S. dollars)	0.260	0.534	0.735	0.735
Extent of high-grading by-passed because of taxes (thousand m³)	35	20	0	0

[a]Involves high rates of royalty per m³ on the most desirable species, with lower rates on less desirable (lower-valued) species as follows: Alpha 1, U.S.$53; Alpha 2, $43; Alpha 3, $24; Alpha 4, $14; Beta 1, $17; Beta 2, $0; Beta 3, $0; Gamma 1, $0; Gamma 2, $0; and Delta 1, $0.

The effects of different *forms* of revenue collection on tropical forestry may be summarized as in Table 2.23. *Each* of the four forms of tax and/or royalty involves approximately equivalent amounts of revenue for the government, but the differentiated royalty and the income tax involve no high-grading at all, while high-grading is present under both the specific uniform royalty *and* the *ad valorem* royalty. The specific royalty is the worst: out of 100 cubic meters available from the stand, only 65 will be cut. The *ad valorem* royalty involves less high-grading than the specific uniform royalty, but substantially more than either the differentiated royalty *or* the income tax.

Furthermore, the surplus left over after all costs (*including* taxes) is much higher under the differentiated royalty and the income tax. This surplus need not all be left to the firm; the pattern of harvest under the differentiated royalty or the income tax would allow even *higher* government revenues than available under either the uniform specific royalty or the uniform *ad valorem* royalty of 22.4 percent.

A clear conclusion emerges from the foregoing discussion: flat-rate income taxes and differentiated royalties are far superior to either a uniform specific royalty or an *ad valorem* flat-rate royalty, in terms of minimizing the decline in the asset value of the forest associated with high-grading. However, both the income tax and a finely tuned system of differentiated royalties involve very substantial information requirements and large quantities of what are now very scarce administrative resources if they are to be operated with any success at all. The discussion therefore suggests that given problems of access to requisite information, and given limited resources in tax administration in nearly all tropical countries, income taxes should not be employed to *fully replace* royalties in forest revenue tax systems. Instead, income taxes may feasibly be employed as *one* component of any administrable system of forest taxation designed to raise revenues and reduce high-grading in tropical timber. The example also indicates that the *ad valorem* royalty has significant advantages over the specific royalty, in terms of reducing high-grading, and is more practical to administer than the differentiated royalty, making it a more advisable selection than either of the other two forms of royalty. A differentiated royalty structure that can be feasibly administered would not be discriminating enough to avoid high-grading and therefore has only limited efficiency advantages over a flat-rate *ad valorem* royalty.

Appendix B: Effective protection, log export taxes, and domestic processing of tropical forest products

The theory and practice of analysis of effective protection was developed by economists in the early 1960s to portray the effects of import duties on incentives to invest in import-substituting industries.[27] This appendix shows how effective protection measures may be also used to evaluate effects of export taxes.

The concept focuses upon protection given to domestic value-added in producing a given product, not upon protection to the product itself, which is meaningless. Value-added is the margin by which the sales price of a product exceeds the cost of inputs purchased from other firms, and those imported from abroad. Within this margin, the firm must pay wages, rent, and interest on borrowed funds and extract any profits. The greater the margin, the greater can be the firm's cost of production, or the higher its potential profit. Domestic value-added (measured in domestic prices) can be increased through import duties; tariffs on competing finished products can be raised and/or those on imported inputs reduced. Effective protection calculations measure this dual effect of the tariff structure; it shows the extent to which the tariff structure allows domestic value-added for a given product to exceed value-added for the same product being exported to the world market by foreign producers. In a sense, it measures the degree to which domestic production may be shielded from international competition. The simplest expression of effective protection is as in Equation 2.3:

$$ERP = \frac{VA_d}{VA_w} - 1 \qquad (2.3)$$

where

ERP = effective rate of protection
VA_d = value-added at domestic prices
VA_w = value-added at world prices

The numerator (VA_d) is nothing more than the domestic unit price of the protected good (P_d) minus the domestic costs of material inputs per unit of output (C_d). The denominator is defined as the difference between the world price (exclusive of tariffs) of competing imports (P_w) and the cost of inputs, also valued at world prices. Therefore, Equation 2.3 can be refined, as in Equation 2.4:

$$ERP = \frac{P_d - C_d}{P_w - C_w} - 1 \qquad (2.4)$$

Now, the domestic price (P_d) cannot exceed the world price (P_w) plus the amount of the tariff on competing imports (t_o). Similarly, domestic costs of inputs (C_d) cannot exceed the cost of inputs at world prices (C_w) plus any tariff on imported inputs (t_i). Thus we may further refine the numerator to obtain Equation 2.5:

$$ERP = \frac{P_w (1 + t_o) - C_w (1 + t_i)}{P_w - C_w} - 1 \qquad (2.5)$$

This may also be expressed as

$$ERP = \frac{(P_w) (t_o) - (C_w) (t_i)}{P_w - C_w} \qquad (2.6)$$

Equation 2.6 may be used to find protection provided by import tariffs to an import-substituting domestic industry. Suppose that a $100 pair of shoes, valued at world prices, would require $50 of material inputs, such as leather and rubber, also valued at world prices. Thus, value-added is $50 at world prices. If the government imposes a 50 percent tariff on imported shoes and a 10 percent import duty on imported inputs, then the protection to domestic value-added in the shoe industry is:

$$ERP = \frac{100(0.50) - 50(0.10)}{100 - 50} = 90 \text{ percent} \tag{2.7}$$

In this case, nominal protection, or the protection directly associated with the tariff on imported shoes, is but 50 percent. But effective protection, the protection furnished to domestic value-added by the import duty structure, is much higher, at 90 percent.

The concept of effective protection may also be applied to export taxes. Suppose that an export tax is applied to a product that is also an input into a domestic industry, and further suppose that when the input is used domestically, it is taxed at a lower rate than that of foreign competitors that also use the input. In this case also, value-added in the domestic industry can exceed that of competing industries abroad. Thus, export taxes may protect domestic value-added in the export-substituting industry from foreign competition.

Consider the Indonesian export tax structure on wood products from 1978 to 1985. Plywood, the final good, was subject to an export tax of zero percent. Logs, the most important input in plywood manufacture, were subject to a 20 percent export tax. The world price of plywood per cubic meter was $250.70. But since 2.3 cubic meters of logs are required to produce one cubic meter of plywood in Indonesia, the world price of plywood must be expressed in its log equivalent, or $109. Thus, using 1981 data, we have

P_w = $109 (world price of plywood, in roundwood equivalent)
C_w = $100 (world price of logs)
t_o = 0 percent (export tax on plywood)
t_i = 20 percent (export tax on logs)

It is to be noted that in the context of the export tax structure, the 20 percent export tax on logs must be seen as a subsidy (negative tax) on logs not exported but used in domestic plymills.

Using the formula, it is evident that effective protection to Indonesian plymills was

$$ERP = \frac{109(0.0) - 100(-0.20)}{109 - 100}$$

$$ERP = 2.22 = 222 \text{ percent} \tag{2.8}$$

Therefore, even though the export tax on logs was but 20 percent, domestic plymills enjoy a margin of 222 percent over value-added at world prices: the log export tax artificially allows value-added in Indonesian plymills to be 2.2 times as high as in competing mills elsewhere. Efficient Indonesian plymills may therefore earn very high profits. On the other hand, very high value-added afforded by protection may allow plymill owners to accommodate inefficiencies and costs substantially above those of foreign competitors.

Chapter 2 indicates that the efficiency losses arising from protection to Indonesian plymills were extremely high in 1978–85. This is a natural consequence of the absence of competitive pressure inevitable in the presence of very high rates of effective protection.

Endnotes

1. The statement reads: "During the first plan, our attention seemed to be focussed on earning badly needed foreign exchange so that activities which promoted other functions of the forest received inadequate attention." See Departamen Kehutanan (1985: 7).

2. The reforestation fee is in principle refundable to the concessionaire following presentation of proof of establishment of a workable reforestation plan.

3. Underreporting of timber export values was particularly severe and widespread prior to 1972, before the government updated and enforced a posted price system for timber exports (Gillis 1984).

4. In 1978, one foreign firm reported the following, regarding the implications of selective cutting in Indonesia. "A typical stand of virgin timber in East Kalimantan would consist of 77 m^3 per hectare of all species of merchantable size (50 cm. diameter at breast height). Under direct road logging, only 47 percent of the timber of that size would be harvested; all weather logging by large firms might extract 85 percent" (Schoening 1978).

5. A rain forest may contain as many as 3,000 species that will reach commercial size, but only 150 of these may be currently considered to be of commercial value (Setyono et al. 1985). In a typical stand in the relatively homogeneous forests of East Kalimantan, there may be as many as 210 different dipterocarp species (Kartawinato et al. 1984: 115).

6. Some forestry experts maintain that prudence requires an assumption that 60 percent of residual stems are liable to logging damage on first entry alone. This estimate includes stems damaged from felling of trees harvested (crown and bark damage) and trees that die later because of compaction of soil (from heavy equipment) or from severing of surface roots. Ross (1984: 144) and Whitmore (1984: 270) report 55 percent of trees over 10 cm. (the basal stock) were damaged.

7. Indonesia allows trees to be harvested only when they exceed 50 cm. diameter at breast height (d.b.h.). Loggers prefer stems with d.b.h. of 60 cm. The mean growth rate for Indonesian commercial species is 1 cm. d.b.h. per annum (Ross 1984: 144).

8. Kuswata (1980) reports that it has been common for firms with concessions to return to logged-over areas after five years. Ashton (1980: 45) notes examples of re-entry into harvested stands in Southeast Asia, euphemistically called "cleaning operations." See also Appendix A for further evidence.

9. Ross (1984: 53) and Whitmore (1984: 100–101) also stress that need for much longer concession agreements.

10. The Indonesian acronym for the Selective Logging System is TPI, for *Tebang Pilih Indonesia*.

11. Tengakawang is a particularly valuable nut produced by some species of *Shorea*. It is a source of oil that can be used as a substitute for cocoa fat. See Tontra (1981: 29–32).

12. Forest products, including non-wood forest products, are a particularly important source of income for government-sponsored transmigrants in East Kalimantan. There families receive lands averaging only about 1.12 hectares per household. To obtain additional income for subsistence, as few as 38 percent of the families at some sites and 68 percent at others collected forest products (Kartawinato et al. 1984: 118).

13. This amount consists of Rp. 2,546,064 million in rupiah funds in the state budget, converted to Rp. 1,200 per dollar and US$868 million in foreign aid.

14. One-sixth of such firms reported D/E ratios in excess of 5:1.

15. Ross estimates that costs involved in establishing forest plantations may be about $2,500 per hectare until harvest. If so, to supply all of Indonesia's requirements for industrial logs from forest plantations would cost $12.5 billion, or half the level of total export earnings in 1982 (Ross 1984: 57).

16. For a state-of-the-art discussion of the implications of overvaluation for developing nations see Dornbusch (in press).

17. Population density on Java averages 748 persons per square kilometer, a density of 12 times higher than in Sumatra, and 35 times that of the island of Kalimantan, site of most of the remaining tropical forest (Biro Pusat Statistik 1985: Table I.1.2 and II.1.7).

18. A full history of the kerosene subsidy may be found in Gillis 1980a: 50–70.

19. A 50 percent increase in the price of kerosene in 1979 would, at most, have resulted in harvesting of an additional 2 million cubic meters of fuelwood of Java annually, on about 20 thousand hectares of land (Gillis 1980a: 59).

20. Westoby notes, however, that the manpower expansion and proposed training and education targets to enable effective parks administration provide little assurance that earmarking of larger areas for parks and reserves will have the intended effects (Westoby 1985).

21. Fitzgerald notes that had Indonesia continued to export at 1979–80 levels, it would have increased supplies of logs to the world by about 6 percent and would have decreased plywood supplies by about 4 percent (Fitzgerald 1986: 22). Since higher availability of Indonesian plywood would tend to reduce the prices that Indonesia would have received for its logs to be made into plywood, it does not seem unreasonable to assume that logs harvested in Indonesia from 1979 to 1982 could have been sold at prices received for logs by Indonesia over the same period. Prices used in the estimates of rent in Table 2.20 reflect actual prices received in 1979–82. The prices Indonesia received for logs did decline precipitously after introduction of the log export ban in 1980, falling from a peak of US$193 per m³

to US$135 in 1983 (World Bank 1986: 4 of summary). Although this was a period of world recession, the withdrawal of more than 10 million cubic meters of Indonesian logs from world markets in 1981–83, relative to 1980, was insufficient to prop up world log export prices. Thus far, Indonesia's market power has proven insufficient to materially affect either world log prices or world plywood prices.

22. Under the uniform system, the logging process itself provides the silvicultural treatment that stimulates the growth of younger trees in cut-over stands. Underlying this system is the view that the trees that have dominated the forest are the fastest growing, and will therefore emerge from the residual stand of immature saplings to gain dominance in the forest (Setyono et al. 1985: 84).

23. Whitmore (1984: 271) argues, for example, that sensible *versions* of selective cutting can reduce damages to logged-over stands by 50 percent or more. He presents comparisons of several types of cutting methods, ranging from monocycle methods (Malayan Uniform System) to various polycyclic systems (Malayan Selective Management System). See Thaib and Karnasudirja (1981: 61–65).

24. Examples of *uniform specific royalty* systems by *species* include those used in the Philippines since 1960 and that used by Sarawak. No major producing country employs a royalty system that involves the same absolute royalty for *all* species. The Philippine royalty is uniform for all grades of lauan (30 pesos per cubic meter). In Sarawak, all hill species meranti pays $5.53 per cubic meter. Examples of *ad valorem* royalties and export taxes include Indonesia's 6 percent timber royalty (since 1979) on all species and 20 percent export tax on logs, and Sabah's timber royalty, which rises from 44 percent at low f.o.b. prices to percent at higher prices. There have been few examples of finely differentiated systems of specific royalties, although the system used in Indonesia from 1970 to 1979 best exemplified this approach. Under this system, generally higher royalties are imposed on desirable species and generally higher royalties on higher grades of all species.

25. Thaib and Karnasudirja (1981: 63) and Kuswata (1980) also report significant non-compliance with the system on the part of concessionaires. Whitten reports that the enforcement of harvest regulations by the Forestry Department is so lax that genuine second-cycle logging (relogging after 35 years) is close to a myth (Whitten 1986: 8).

26. Some experts employ even higher figures for damages to standing stock associated with logging and relogging. First, in one widely publicized report, it was stated that "in the Dipterocarp Rain Forests of Tropical Malaysia, only 10 percent of the trees with diameter greater than 10 cm. were harvested, 55 percent were severely damaged and only 35 percent were undamaged" (U.S. Interagency Task Force: 18). Second, Ashton (1980: 48) concludes that regeneration is effectively eliminated in forests which have experienced more than one mechanized felling within a decade.

27. The discussion of effective protection provided by export taxes in this appendix is based on a modification of the discussion of effective protection furnished by import taxes found in Gillis et al. (1982: 435–436).

References

Ashton, Peter. 1980. The Biological and Ecological Bases for the Utilization of Dipterocarps. *Bio-Indonesia*, Vol. 7: 43–54.

Ashton, Peter. 1984. Aide Memoire on the State of Rain Forest Research and Its Application in Indonesia. Cambridge, Mass.: Arnold Arboretum of Harvard University, July 20.

Asiaweek. 1984. Fire in the Earth. July 13.

Biro Pusat Statistik. 1981. *Statistik Indonesia 1980–81.* Jakarta: Biro Pusat Statistik.

Biro Pusat Statistik. 1985. *Statistik Indonesia 1984–85.* Jakarta: Biro Pusat Statistik.

Boyce, Stephen G., ed. 1979. *Biological and Sociological Bases for a Rational Use of Forest Resources for Energy and Organics.* Asheville, N.C.: U.S. Department of Agriculture, Forest Service.

Brotokusomo, Mohammed, C.J.P. Colfer, and A. P. Vayda. 1980. Interaction Between People and Forest in East Kalimantan. *Impact of Science on Society,* Vol. 30, No. 3 (July–September): 179–190.

Bucovetsky, Meyer, and Malcolm Gillis. 1978. The Design of Mineral Tax Policy. In *Taxation and Mining: Non-Fuel Minerals in Bolivia and Other Countries,* ed. Malcolm Gillis et al. Cambridge, Mass.: Ballinger.

Departamen Kehutanan. 1984. *Statistik Kehutanan Indonesia, 1982–83.* Jakarta.

Departamen Kehutanan. 1985. *Draft Long-Term Forestry Plan.* Jakarta.

DiCastri, F. 1984. *Ecology in Practice.* Dublin: Tycoly.

Dornbusch, Rudiger. In press. Overvaluation and Trade Deficits. Chapter 2 in *A Policy Manual for the OPEN Economy,* ed. Rudiger Dornbusch.

Fitzgerald, Bruce. 1986. An Analysis of Indonesian Trade Policies: Countertrade, Downstream Processing, Import Restrictions and the Deletion Program. Washington, D.C.: World Bank, CPD Discussion Paper No. 1986–22, July.

FAO. 1981a. Tropical Forest Assessment Project. *Forest Resource of Tropical Asia.* Rome: FAO.

FAO. 1981b. *Small and Medium Sawmills in Developing Countries.* Rome: FAO.

Gillis, Malcolm, et al., eds. 1978. *Taxation and Mining: Non-fuel Minerals in Bolivia and Other Countries.* Cambridge, Mass.: Ballinger.

Gillis, Malcolm. 1980a. Energy Demand in Indonesia: Projections and Policies Development. Discussion Paper No. 92. Cambridge, Mass.: Harvard Institute for International Development, April.

Gillis, Malcolm. 1980b. Fiscal and Financial Issues in Tropical Hardwood Concessions. Development Discussion Paper No. 110. Cambridge, Mass.: Harvard Institute for International Development.

Gillis, Malcolm. 1981. Foreign Investment in the Forest-Based Sector of the Asia-Pacific Region. Prepared for FAO, Rome.

Gillis, Malcolm. 1983. Taxes, Returns and Risks in Petroleum Contracts in Developing Countries. Development Discussion Paper No. 142. Cambridge, Mass.: Harvard Institute for International Development, January.

Gillis, Malcolm. 1984. Environmental and Resource Management Issues in the Tropical Forest Sector of Indonesia. Development Discussion Paper No. 1971. Cambridge, Mass.: Harvard Institute for International Development, May 30.

Gillis, Malcolm. 1986. The Micro and Macro Economics of Tax Reform. *Journal of Development Economics,* No. 142.

Gillis, Malcolm. 1987. Multinational Enterprises and Environmental and

Resource Management Issues in the Indonesian Tropical Forest Sector. In *Multinational Corporations, Environment, and the Third World: Business Matters*, ed. Charles Pearson. Durham, N.C.: Duke University Press.

Gillis, Malcolm, and David Dapice. In press. External Adjustments and Growth: Indonesia Since 1965. Chapter 6 in *A Policy Manual for the OPEN Economy*, ed. Rudiger Dornbusch.

Gillis, Malcolm, Dwight Perkins, Michael Roemer, and Donald Snodgrass. 1982. *Economics of Development*, 1st ed. New York: W. W. Norton.

Hunter, Lachlan. 1984. Tropical Forest Plantations and Natural Stand Management: A National Lesson from East Kalimantan. *Bulletin of Indonesian Economic Studies*, Vol. 20, No. 3: 109–110.

Kincaid, G. R. 1984. A Test of Exchange Rate Adjustments in Indonesia. *IMF Staff Papers*, Vol. 30 (February): 62–78.

Kuswata, Kartawinata. 1979. An Overview of the Environmental Consequences of Tree Removal from the Forest in Indonesia. In *Biological and Sociological Bases for a Rational Use of Forest Resources for Energy and Organics*, ed. Stephen G. Boyce. Asheville, N.C.: U.S. Department of Agriculture, Forest Service.

Kuswata, Kartawinata. 1980. East Kalimantan: A Comment. *Bulletin of Indonesian Economic Studies*, Vol. 16, No. 2 (November): 120–121.

Kuswata, Kartawinata, Soedarsono Riswan, and Andrew P. Vayda. 1984. The Impact of Man on a Tropical Forest in Indonesia. *Ambio*, Vol. 10, Nos. 2–3: 115.

Kuswata, Kartawinata, and Andrew Vayda. 1984. Forest Conversion in East Kalimantan. In *Ecology in Practice*, ed. F. DiCastri. Dublin: Tycoly.

Long, Alan, and Norman Johnson. 1981. Forest Plantations in Kalimantan, Indonesia. In *International Symposium on Tropical Forests*, ed. Francois Mergen. New Haven: Yale School of Forestry.

Mackie, Cynthia. 1984. The Lessons Behind East Kalimantan Forest Fire. *Borneo Research Bulletin*, Vol. 16, No. 2: 67.

Manan, Sjatii. 1974. Ecology vs. Economics. Unpublished ms. Institute Pertanian Bugor, Indonesia.

Panoyouto, Theodore. 1983. Renewable Resource Management for Agriculture and Rural Development: Research and Policy Issues. Bangkok: Agricultural Development Council, November.

P. T. Data Consult. 1983. *A Comprehensive Report on the Indonesian Plywood Industry*. Jakarta, April 1.

Repetto, Robert, Michael Wells, Christine Beer, and Fabrizio Rossini. 1987. Natural Resource Accounting for Indonesia. Unpublished ms. Washington, D.C.: World Resources Institute, May.

Ross, M. S. 1982. *The South-Sea Log Market in Relation to the Indonesian Transmigration Program*. Jakarta: East Kalimantan Transmigration Area Development Project.

Ross, M. S. 1984. Forestry in Land Use Policy for Indonesia. Ph.D. dissertation, University of Oxford.

Ruzicka, I. 1979. Rent Appropriation in Indonesian Logging: East Kalimantan, 1972–73 to 1976–77. *Bulletin of Indonesian Economic Studies*, Vol. 15, No. 2 (July): 49–56.

Schoening, Jack. 1978. *Weyerhauser's Experience in Southeast Asia.* Madison: University of Wisconsin.

Secrett, Charles. 1986. The Environmental Impact of Transmigration. *The Ecologist,* Vol. 16, Nos. 2/3: 77–85.

Setyono, Sastrosumarto, Herman Haerum, Atar Sibero, and M. S. Ross. 1985. A Review of Issues Affecting the Sustainable Development of Indonesia's Forest Land, Vol. 2. Unpublished ms. Jakarta, November 30.

Soedjarwo. 1985. Forest Devastation – Very Serious. *Indonesian Observer,* May 25: 1.

Spears, John. 1983. The Role of Afforestation as a Sustainable Land Use and Strategy Option. In *Let There Be Forest.* Agricultural University of the Netherlands.

Takeuchi, Kenji. 1983. Export Prospects for Forest Products in Indonesia, 1983–1990. World Bank Working Paper No. 1983–1. Washington, D.C.: World Bank, January.

Team Kayu. 1979. *Laporan.* Jakarta: Ministries of Agriculture, Finance, and Planning, August.

Thaib, Jurnali, and Saparman Karnasudirja. 1981. Review of the Application of Selective Cutting Outside Java. *Indonesian Agricultural Research and Development Journal,* Vol. 3, No. 3: 61–66.

Tontra, I.G.M. 1981. Tengakawang. *Indonesian Agricultural Research and Development Journal,* Vol. 3, No. 2: 29–32.

U.S. Interagency Task Force on Tropical Forests. 1980. *The World's Tropical Forests.* Washington, D.C.: U.S. Government Printing Office.

Westoby, Jack C. 1985. Indonesia's Long-term Forestry Plan: Comment. Unpublished ms.

Whitmore, T. C. 1984. *Tropical Rain Forests of the Far East.* Oxford: Clarendon Press.

Whitten, Anthony J. 1986. Indonesia's Long-term Forestry Plan: Comment. Unpublished ms.

Whitten, Anthony J. 1986. Indonesia's Transmigration Program with Special Reference to Sumatra. Paper presented at Environmental Studies Association Workshop on Role of Applied Ecology in Economic Development, August.

World Bank. 1970. *The Indonesian Economy 1970–72. Vol. I: The Main Report.* Washington, D.C.: IBRD, November 27.

World Bank. 1985. *World Development Report.* Washington, D.C.: World Bank.

World Bank. 1986. *Price Prospects for Major Primary Commodities.* Washington, D.C.: World Bank, November.

3   Malaysia: public policies and the tropical forest

MALCOLM GILLIS

There can be no meaningful discussion of the implications of "Malaysian" policies impinging upon Malaysia's natural forest estate. Subnational governments in Malaysia possess great autonomy; in forestry policy autonomy has been defined with particular clarity. We may, however, speak of Sabah policies, Sarawak policies, and to a very limited extent, Peninsular Malaysian policies affecting the forest sector.

Accordingly, this chapter is divided into five distinct sections. The first sketches the salient features of forest endowments, forest utilization, and forest policies in Malaysia as a whole. The second section focuses upon forest exploitation and public policies in the state of Sabah, where timber harvest and deforestation rates have been the most rapid in recent years. The third section examines forest issues and forest policies in Sarawak state, and the fourth deals with these questions in the 12 states of Peninsular Malaysia. The final section focuses upon the implications of national-level non-forestry policies on deforestation in all three regions.

Malaysia: east and west

Malaysia is composed of three distinct and geographically separate regions: Peninsular Malaysia, containing 12 states, and the East Malaysian states of Sabah and Sarawak in the northern portion of the island of Borneo. Peninsular Malaysia, also known as West Malaysia, contains 40 percent of the nation's total land area (Table 3.1), but 81 percent of the population, 14.9 million people in 1983. Malaysia is a relatively wealthy country: 1985 GNP per capita, at US$2,000, was nearly four times that of neighboring Indonesia (World Bank 1987: 202).

Most of the factors responsible for deforestation in Indonesia since 1950 have also been operative in Malaysia: poverty, institutions, and public policies. However, the relative roles of these factors have been quite different in the two countries. Generalizations about the role of public

Table 3.1. *Malaysia: summary of forest endowments and land use, 1981 (million hectares)*

Nationwide (14 states)	
Total land area	33.2
Total forest area	20.4
Area in rubber plantations	2.0
Area in oil palm plantations	1.0
By region	
Peninsular Malaysia (12 states)	
Total land area	13.2
Total forest area	6.3
Sabah	
Total land area	7.4
Total forest area	4.7
Sarawak	
Total land area	12.5
Total forest area	9.4

Sources: Land area: Nor (1983: 164); Peninsular Malaysia forest area: Thang (1984: 34–35); Sabah forest area: U.S. Department of State (1985); Sabah forest area: Nor (1983: 164).

policy in Malaysian deforestation are made especially difficult both by the unusual institutional structure governing Malaysia's forest utilization and by differing forest endowments in the country's three distinct regions.

In 1981, forests covered about 60 percent of Malaysia's total land area (Table 3.1), with the largest remaining forest estate in Sarawak. Roughly 30 percent of the nation's area had been converted to agriculture by 1981; rubber and oil palm plantations alone occupy almost one-tenth of total land area (Table 3.1).

Total deforestation in Malaysia has been running at about 250,000 hectares per year (Table 3.2), less than .1 percent of total land area, and about 1 percent of total forest area. Therefore, in recent years, the rate of annual deforestation as a percent of forest area has been about one-fifth the median rate reported in Simon and Kahn (1984: 160–161) for 48 tropical countries. This is, however, not particularly good news. First, deforestation rates in Sabah and Sarawak have accelerated since the 1970s (Table 3.2). Second, large areas of Peninsular Malaysia's closed forest had already been converted to agricultural estates in the seven decades prior to 1980: forest area shrank by 30 percent over that period.

Logging has played a significant role in Malaysian deforestation. Malaysia has consistently accounted for about 20 percent of world tropical log

Table 3.2. *Malaysia: total annual deforestation due to all causes, 1976–85*

	Annual deforestation rate (thousand hectares)		Deforestation as percent of:			
			Land area		Forest area in 1981	
	1976–80	1980–85	1976–80	1980–85	1976–80	1980–85
Total Malaysia	230	255	.07	.08	.11	1.25
Sabah	60	76	.08	.1	.13	.17
Sarawak	80	89	.07	.07	.09	.1
Peninsular Malaysia	90	90	.07	.07	.14	.14

Sources: Table 3.1 and FAO (1981: 289–290, 308–309, 326–327).

harvests and an even larger share of world trade in tropical hardwoods. For logs alone, the Malaysian share of world trade was nearly 40 percent from 1977 to 1981, though log exports from Peninsular Malaysia were progressively reduced after 1972, and virtually ceased after 1978.

The timber sector has been a major source of foreign exchange but a much less significant source of employment in Malaysia. In the period 1981–84 the total annual value of wood and wood product exports surpassed $4 billion every year, exceeded only by revenues from exports of petroleum and oil palm. Table 3.3 indicates that while wood products (sawn timber, plywood, and veneer) dominated exports from Peninsular Malaysia, log exports were the majority of timber exports from both Sabah and Sarawak.

Rarely does timber sector employment exceed 1 percent of the labor force, even in countries where the sector is a major source of export earnings. An exception is Canada, where this figure has been about 2 percent in the 1980s. Labor utilization in the Malaysian timber sector conforms to the general pattern: in the late 1970s total employment in the logging and wood products industry was estimated at 0.8 percent of the labor force (Gillis 1981: 86). The rate for Sabah and Sarawak, however, is nearly 10 times as high, but these two states have less than 20 percent of Malaysian population.

Regional commonalities and regional differences in policies toward the forest sector

There are several commonalities and differences in forest endowments, forest utilization, and policies between the three principal timber regions. There are four commonalities. Malaysia's overall deforestation rate is relatively low by world standards, and these rates are

Table 3.3. *Value of Malaysian exports of timber products, 1981–84*
(million Malaysian dollars)[a]

	1981	1982	1983	1984
Peninsular Malaysia	1,262	1,192	1,282	1,054
Logs	21	2	11	5
Sawn timber	779	747	824	647
Plywood	283	253	287	172
Veneer	32	32	25	19
Molding	144	127	136	209
Other (chipboard, etc.)	2	31	0	2
Sabah	1,812	2,353	2,110	1,927
Logs	1,642	2,092	1,692	1,528
Sawn timber	135	227	349	319
Plywood	18	15	15	20
Veneer	18	19	53	60
Sarawak	975	1,430	1,243	1,389
Logs	812	1,261	1,093	1,227
Sawn timber	85	101	85	68
Plywood	7	6	6	17
Veneer	0	0	5	8
Molding	59	52	54	64
Other	12	10	0	5
Total	4,049	4,975	4,636	4,369

[a]Exchange rates: 1981, US$1 = M$2.24; 1982, US$1 = M$2.31; 1983, US$1 = M$2.34; 1984, US$1 = M$2.35; 1985, US$1 = M$2.42.
Source: Timber Trade Review, Vol. 12, No. 4 (1983); Vol. 14, Nos. 2 and 3 (1985).

similar across the three regions. Also Table 3.3 shows that while Sabah accounts for about 45 percent of forest product exports, all three regions have a major stake in world tropical hardwood trade. In addition, in all regions, but less so in Sarawak, policies provide strong incentives for forest-based industrialization, and forestry policy has generally placed heavy emphasis upon the wood-producing potential, rather than non-wood productive potential or the protective function, of the forest estate. Finally, state governments in all three regions, particularly Sabah, have derived very significant tax benefits from forest exploitation.

Regional differences are more striking than the commonalities; the most significant lies in the legal and institutional framework governing rain forest utilization. Malaysia may be viewed as one nation united by a common language (Bahassa Malay), but separated by geography and hampered by the lack of central government control over forest utilization. Sharp distinctions in forest policy exist between Sabah and Sarawak in East Malaysia, and Peninsular Malaysia in the West. States in the three regions are semi-autonomous in many matters, but in forestry the states of Sabah and Sarawak have virtually complete independence. Three entirely separate Forest Departments oversee forest use in the three regions. In fact, the annual report of the Malaysian Director-General of Forests is concerned with Peninsular Malaysia alone. In addition, neither Sabah nor Sarawak yet adheres to the much discussed National Forestry Policy, adopted in 1978. One result of assigning separate forest responsibility, and a major reason it persists, is that both foreign exchange revenues and forest production royalties accrue entirely to Sabah and Sarawak state governments. Neither the residents of West Malaysia nor the national government derive any measurable benefits from forest exploitation in East Malaysia.

Nor has the adoption of the National Forestry Plan harmonized forestry policy in Peninsular Malaysia. The 12 states retain significant autonomy in forest sector policy-making. Indeed, the National Forestry Department functions to *advise* state governments in Peninsular Malaysia in developing timber-based industries, managing forest lands, and recruiting foresters. Also, all federal foresters are seconded to, paid by, and responsible to the state governments (Callahan and Buckman 1981: 33). All policy decisions relating to forest concession licenses, timber royalties, and annual cut are the domain of the state governors. Therefore, even in Peninsular Malaysia, responsibility for policies toward forest endowments is balkanized among several states.

Regional differences in forest endowments further complicate assessing the role of policy in Malaysian deforestation. Sabah contains some of the world's richest rain forest stands. In both Sabah and Peninsular Malaysia, the family Dipterocarpaceae, containing over 350 species of trees, is dominant and found throughout. Mangrove forests are extensive in coastal zones in both Sabah and Peninsular Malaysia. Peat swamps are important only in Sarawak, where they have long been exploited for valuable *ramin* trees (*Gonystylus bancanus*).

Another major difference between regions involves the relative roles of factors responsible for deforestation. In Peninsular Malaysia, agro-conversion has long been the leading cause of deforestation, with shifting cultivation a distant third, after logging. This may partly be due to the lower incidence of rural poverty in Peninsular Malaysia, which in turn may result from the region's 70-year history of headlong forest conversion to such permanent tree crops as rubber and oil palm. In Sabah and Sarawak, shifting cultivation is generally identified as the leading cause of deforestation, well ahead of logging. As in Indonesia, however, shifting cultivators gained access to a large but undocumented portion of deforested area initially through road building and felling by loggers.

Table 3.4. *Volume of Malaysian exports of timber products, 1981–84*
(thousand cubic meters)

	1981	1982	1983	1984
Peninsular Malaysia	3,163	3,175	3,183	2,461
Logs	232	273	132	59
Sawn timber	2,254	2,244	2,323	1,832
Plywood	442	443	524	325
Veneer	92	86	65	43
Molding	143	127	139	202
Other	1	2	0	0
Sabah	9,098	10,485	10,893	9,012
Logs	8,698	9,827	9,475	7,632
Sawn timber	386	642	936	854
Plywood	14	16	18	31
Veneer	0	0	464	495
Sarawak	7,096	9,393	9,357	9,156
Logs	6,923	9,200	9,171	8,982
Sawn timber	163	184	153	130
Plywood	10	9	8	15
Veneer	0	0	25	29
Total	19,357	23,053	23,433	20,629

Source: Timber Trade Review, Vol. 12, No. 4 (1983); Vol. 14, Nos. 2 and 3 (1985).

Regions also differ in the policy emphasis they place on forest-based industrialization. Peninsular Malaysia was earliest and most ambitious in promoting domestic processing of tropical logs through log export quotas and other policy devices. By 1984, only 2.4 percent of the region's timber export volume was in log form. Corresponding figures for Sabah and Sarawak were 79 and 98 percent respectively (Table 3.4).

Finally, regions have differed markedly in the degree of their success in capturing timber rents: Sabah has clearly performed best relative not only to the rest of Malaysia, but to Indonesia as well. This is due both to Sabah's aggressive use of timber taxes and royalties and its less vigorous pursuit of forest-based industrialization relative to Peninsular Malaysia. Sarawak did poorly in rent capture, particularly prior to 1980, largely because of low timber taxes and royalties.

Sabah

Forest endowments and forest use
By the end of 1980, forest land covered five-eighths of total Sabah land area. The state contained nearly 2 million hectares of undisturbed

Table 3.5. *Sabah: forest estate, estimated, end 1980*
(thousand hectares)

A. Productive forest	3,200
1. Undisturbed	1,920
2. Logged	1,280
B. Unproductive forest	1,797
1. Unproductive for physical reasons	1,622
2. Parks, reserves	175
C. Total, forest land	4,997
Composition of total forest land	
Mixed lowland and hill dipterocarp forests	2,781
Montane dipterocarp forest	831
Reach and swamp forest	190
Other forest	845

Source: FAO (1981: 35, 297–305).

forest (Table 3.5), dominated by rich stands of trees of the Dipterocarpaceae family, principally *Shorea* spp. and *Parashorea* spp. Potentially recoverable yields of commercial species have been significantly higher in Sabah than in Indonesia, and much higher than in Africa: in eastern Sabah, potential yields are as high as 140 cubic meters per hectare. Actual yields run about half that for large loggers and one fourth that for small logging companies (U.S. Department of State 1985).

According to some estimates, only 2.5 million hectares of the total forest area was available for commercial harvesting in 1985, as the logged-over forest area increased from about 1.3 million hectares in 1980 (Table 3.6) to about 1.5 million in 1985. By 1980, all of Sabah's productive forest was committed to exploitation (*Draft Sabah Regional Planning Study* 1979).

Shifting cultivation has been the immediate cause of most recent deforestation in Sabah. Deforestation due to logging plus conversion of forest land to permanent agriculture (primarily cocoa) accounted for 40 percent of estimated deforestation in 1976–80 and 45 percent in 1980–85 according to FAO estimates (Table 3.6). Therefore, shifting cultivation appears to have caused more than half of annual deforestation; the total area affected by shifting cultivation was nearly three times that of the logged-over forest by 1980 (Table 3.6). Sabah's story, like Indonesia's, is not quite that simple. A mix of institutions (including property rights and land use laws), rural poverty, and logging has furnished both opportunities and incentives for shifting cultivation on formerly forested areas. The state government owns all property rights to virgin timber stands. But by pre-independence laws, any Sabah native may obtain title to forest

Table 3.6. *Sabah: utilization of the tropical forest estate: deforestation due to shifting cultivation, area utilized for logging, end 1985 (thousand hectares)*

I. Logged-over forest (stock)	1,280
II. Unlogged natural forest under logging license (stock)	2,029
A. Under regular concession (21-year lease)	869
B. Under special license (10-year lease)	782
C. Under Form I license (one-year lease)	77
D. Licenses pending	301
III. Area affected by shifting cultivation (stock) (1970–80 only)	3,650
IV. Annual deforestation rate due to shifting cultivation (flow)	
A. 1975–80	36
B. 1980–85	42
V. Annual deforestation rate due to logging and conversion of land to permanent agriculture	
A. 1976–80	24
B. 1980–85	34
VI. Total annual deforestation rate due to all sources (IV + V)	
A. 1976–80	60
B. 1980–85	76

Source: FAO (1981: 301–309).

land by clearing and working it. Combined with logging activities this law provides the opportunity for natives to enter logged-over areas to clear and claim them (Callahan and Buckman 1981: 5). Rural poverty provides the incentive to do so. In 1982, 51.2 percent of the population of Sabah fell below the poverty line as defined by the government, compared to 37.7 percent for all of Malaysia. Sabah's rural areas had an even higher poverty incidence, at 59 percent (Segal 1983: 46).

In what follows, we focus on the role of government policy in encouraging deforestation and in capturing benefits from forest exploitation. We do not systematically discuss the interactions between rural poverty – and policies that may perpetuate it – and forest damage.

Historical sketch of commercial activity in the timber sector

The history of the timber sector in Sabah consists of four distinct periods: (1) the era of the British Borneo Timber Company's timber monopoly (1919–1952), (2) the postwar entry of other foreign timber concessionaires (1952–1966), (3) the early transition years following the creation of the Sabah Foundation (1966–78), and (4) the years of full control over forests by the Sabah Foundation (1979 onwards). Over the past 66 years, then, the forest estate has passed from monopoly to oligopoly ownership and back to monopoly, with the present monopolist being a domestic foundation instead of a foreign firm.

The British Borneo Timber Company behaved as would any self-respecting monopolist: Sabah's highly desirable hardwoods were extracted at low levels in order to keep prices high. With the demise of the BBT monopoly in 1952, logging expanded quickly, with the entry of three other large foreign firms and eight local companies (Lee 1982: 4). By the end of the 1950s, virtually all logging was mechanized. The industry expanded throughout the 1960s, as heavier and more powerful machinery enabled harvests in new areas (Fox 1968: 326–346).

Creation of the Sabah Foundation in 1966 signaled a new era in forest use. By 1970, the state government had granted to the Foundation a 100-year license to 855,000 hectares (3,300 square miles) of forest, to be developed on behalf of all citizens of the state. The Foundation's role was gradually redefined over the next decade. By 1979 it had become a statutory body of the state government (Hepburn 1982: 400–402), exercising virtually full control over the region's richer timber stands and undertaking multifarious economic and social objectives. It is not only a development agency, but a dispenser of endowment income for poverty relief from the earnings from its 855,000 hectares, making annual cash payments to all citizens over age 21, much like the Permanent Fund of the U.S. state of Alaska. These payments peaked at US$46.50 per capita in 1981, but were temporarily suspended in 1983 due to low world prices for log exports (Davies and Lauriat 1980: 23–26).

During the first three periods (1919–1977) timber enterprises, mainly British, secured traditional long-term concessions. By the early 1960s these concession tracts amounted to about one-fourth of the entire state's area. By the end of 1979, 10 of these two dozen agreements remained in force, including seven granted prior to Malaysia's independence from Britain (Chong 1980). All were scheduled to expire before 1986, although the government reserves the right to grant three-year extensions. In addition, two major 25-year concessions were granted in 1979, involving 280,000 acres, with the Sabah Foundation as a joint venture partner with Kuwaiti investors (*Draft Sabah Regional Planning Study* 1979: 17–23). Since 1970, the Foundation has undertaken several joint ventures in many fields, including three that combine logging and processing and one in reforestation. The three joint ventures in logging and processing include one with the Japanese Yuasa Company (Sinora Sdn. Bhd), one with the American-based Weyerhauser (Pacific Hardwoods Sdn. Bhd), and another with a Philippine company (Sabah Melale Sdn. Bhd). In general, these agreements involve a 51 percent equity share for the Foundation, with an agreement life of 12–14 years.

In Sabah, as in Indonesia, firms from neighboring less-developed countries (LDCs) began to play a significant role in the forest sector after 1970. Aside from the aforementioned joint venture in logging and processing with the Philippine-based Melale, the Sabah Foundation has an agreement with the Birla Group of India and Malaysia's Fibre Chemicals Company to undertake a pulp and paper mill.

By 1983–84, the bloom had faded from the rose for the Sabah Foundation and, temporarily, Sabah generally. Total value of timber exports, which had tripled from 1971 to 1976 and doubled between 1977 and 1982, fell to M$1.9 billion in 1984. Large projects in tropical timber, including the joint venture with Weyerhauser, were cancelled in 1983. Declining world prices for oil and timber coupled with massive spending by the Foundation and the government had, in the words of a new Chief Minister elected in 1985, "brought Sabah near bankruptcy" (Berthelsen 1985) with foreign debts per capita at $3,000, far higher than more well-known debtors like Brazil or Mexico.

Benefits from timber exploitation

Nationally, value-added in production of primary commodities stood at about 28 percent of gross domestic product by 1982. In Sabah, value-added in this sector was almost half of gross regional product (GRP) (Segal 1983: 54). Two newer extractive industries, oil and copper, came on stream in 1975, but neither provides large benefits to the Sabah economy. Most of the tax revenues flow to the national government. The timber sector – primarily logging – has furnished about one-third of GRP since 1970. Timber sector employment, at 35,000 in the late 1970s (Chong 1980: 42) was then about 7 percent of Sabah's labor force, much higher than in Peninsular Malaysia or Indonesia.

The most significant contributions of timber sector to the regional economy have been export earnings and tax revenues. Since 1977, logs and (to a much lesser extent) sawn timber have consistently accounted for about 41 percent of Sabah's total exports. Total timber export value rose from 420 million Malaysian dollars in 1971 (US$175 million) to M$2,250 million in 1982 (US$974 million) but fell to less than US$820 million in 1984 (Table 3.7). Until 1979, virtually all exports were in log form; since 1979, sawn timber exports have been between 7.5 and 10 percent of total timber exports.

The share of the timber sector in overall government revenues has greatly exceeded its share in Sabah's gross regional product (GRP) or in

Table 3.7. *Sabah: export values for logs, sawn timber, and plywood, 1971–84 (million Malaysian dollars)*

Year	Log exports	Sawn timber, plywood, and other processed products	Total value of timber exports
1971	419.0	0.7	419.7
1972	409.2	1.0	410.2
1973	806.9	1.7	808.6
1974	885.3	0.5	885.8
1975	597.1	0.7	597.8
1976	1,210.9	1.5	1,212.4
1977	1,262.2	6.2	1,268.4
1978	1,422.9	6.5	1,429.4
1979	2,050.9	45.3	2,096.2
1980	1,783.0	78.0	1,861.0
1981	1,642.0	170.0	1,812.0
1982	2,092.1	226.9	2,319.0
1983	1,692.0	418.0	2,110.0
1984	1,528.0	399.0	1,927.0

Source: Timber Trade Review, various issues, Vols. 2–14.

exports. From 1970 to 1979, timber taxes and royalties consistently accounted for between half and two-thirds of total government revenues (U.S. Department of State 1985: 16). Since 1979, timber's share has been as low as 61 percent (1985) and as high as 71 percent (1980). Timber royalties have been the most important fiscal levy on the sector since 1970 (Table 3.8), providing upwards of 90 percent of government forest revenues in the mid-1970s and between 80 and 85 percent since 1978.

Apparently, Sabah was consistently more aggressive than Indonesia or Sarawak in using taxes to capture timber rents, at least through 1982 (Table 3.10). Sabah taxes and royalties totalled but 26 percent of export values in 1974 to 1976, but increased to about 60 percent in 1980, nearly double that of Indonesia in 1980. In both Sabah and Indonesia, however, the ratio of timber taxes to export values fell precipitously after 1981, to 36 percent in Sabah and 28 percent in Indonesia. Timber sector taxes had sharply declined, given stagnant world markets and a diminishing share of high-valued logs in total exports (Table 3.9).

Sabah, like most of its Asian neighbors, has tended to treat its forest endowments as a non-renewable asset, to be mined in the same fashion as a copper or oil deposit. Foreign exchange, tax, and employment benefits flowing from the mining of timber have been substantial in Sabah; but so has the rate of deforestation arising from logging and associated shifting

Table 3.8. *Sabah: government revenue from the forest sector, 1973–85 (million Malaysian dollars)*

Year	(1) Total government revenue from forest sector (Cols. 2 + 3 + 4 + 5 + 6)	(2) Timber royalty and timber export duty	(3) Timber development charge	(4) Timber extraction charge	(5) Timber cess	(6) Miscellaneous forestry revenue	(7) Forest revenues in U.S. dollars (millions) (Col. 1 times prevailing exchange rate)
1973	183.9	167.3	n.a.	6.5	8.4	2.7	85.5
1974	240.1	225.2	n.a.	7.0	7.2	1.7	111.7
1975	151.7	142.6	n.a.	1.7	6.3	1.6	70.6
1976	326.6	300.9	n.a.	9.3	14.2	2.6	151.9
1977	497.0	349.9	75.5	22.3	46.7	2.4	231.2
1978	510.3	433.4	23.3	13.8	48.7	2.8	237.3
1979	1,110.0	964.6	20.0	17.2	63.0	3.3	511.6
1980	1,099.0	n.v.	n.v.	n.v.	n.v.	n.v.	488.4
1981	780.0	n.v.	n.v.	n.v.	n.v.	n.v.	342.1
1982	830.0	n.v.	n.v.	n.v.	n.v.	n.v.	360.9
1983	800.0	n.v.	n.v.	n.v.	n.v.	n.v.	346.3
1984	701.0	n.v.	n.v.	n.v.	n.v.	n.v.	298.3
1985	650.0	n.v.	n.v.	n.v.	n.v.	n.v.	275.4

n.a. = not applicable or not in existence.

n.v. = not available.

Sources: 1973–79, Sabah Forest Service (1979); 1980–83, Segal (1983: 56–59); 1984–85, U.S. Department of State (1985: 16).

Table 3.9. *Sabah: total value of forest sector taxes as percent of export value, 1974–85*

Year	(1) Total forest sector taxes and royalties (million Malaysian dollars)	(2) Total timber export value[a] (million Malaysian dollars)	(3) Total forest sector taxes and royalties as percent of total export values (Col. 1 ÷ Col. 2)
1974	240.1	885.8	27.1
1975	151.7	597.8	25.3
1976	326.6	1,212.4	26.9
1977	497.0	1,268.4	39.2
1978	510.3	1,429.4	35.7
1979	1,110.0	2,096.2	53.0
1980	1,099.0	1,861.0	59.0
1981	780.0	1,812.0	43.0
1982	830.0	2,319.0	35.8
1983	664.0	2,110.0	31.5
1984	550.0	1,927.0	28.5
1985	650.0	n.a.	n.a.

[a]Logs plus sawn timber

Sources: Column 1, 1974–80: Gillis (1980); 1980–85: U.S. Department of State (1985: 16). Column 2: Table 3.7.

cultivation. Government policies have encouraged rapid exploitation and conversion of timber stands. Subsequent sections examine the role of individual policies, both forestry policies, as such, and non-forestry policies.

Forestry policies

Responsibility for forest-related policies in Sabah is divided between several institutions: the Sabah Foundation, the Forest Department, the Sabah Forest Development Authority (SAFODA), the Sabah Rural Development Authority, and the Office of the Chief Minister. The precise division of responsibilities, however, is unclear and perplexing to outsiders. The Sabah Foundation grants concessions. So too have Chief Ministers: an article in the *Far Eastern Economic Review* in 1977 indicated that a new Chief Minister in 1976 (Datuk Harris Salleh) cancelled concessions granted by his predecessor, covering 12 percent of total forest area under license (1977: 58). The Office of the Chief Minister also determines tax and royalty policy. SAFODA was established in 1976 in order to promote rural development in a 250,000-hectare tract of relatively poor land in western Sabah. SAFODA does not grant concessions; it oversees man-

Table 3.10. *Comparison of approximate timber royalty payments per cubic meter at various f.o.b. prices for meranti logs: Sabah, Sarawak, and Indonesia*

Country	F.o.b. log prices (U.S. dollars per m³)			
	75	100	125	150
Sabah	33.39	48.39	66.92	84.82
Sarawak (pre-1980)	5.53	5.53	5.53	5.53
Sarawak (post-1980)	9.25	9.25	9.25	9.25
Indonesia (basic and additional royalty)	6.00	7.50	9.00	10.50

agement of shifting cultivation and, with World Bank help, has reforested about 17,000 hectares of the tract (U.S. Department of State 1985: 18). The Rural Development Corporation grants credits both to farmers for improving shifting cultivation techniques and to the Sabah Foundation. The Forest Department controls harvest regulations, based on terms and conditions laid down in special licenses and concession agreements, and in the 1968 Forest Enactment and the 1969 Forest Rules. Clearly, no single entity oversees the design or implementation of policies affecting the forest sector.

Royalties, license fees, and timber cesses. Sabah's rain forest stands are among the world's richest in terms of density and quality. They are more readily accessible to loggers than in Indonesia, Zaire, or other countries with large remaining tropical forests. Not surprisingly, Sabah's timber royalty is the world's highest (Table 3.10). Indeed, the Sabah royalty exceeds the combined value of royalty plus export tax of any other nation with tropical timber endowments, including neighboring Indonesia, where fiscal charges on log exports are higher than those prevailing in Africa (Table 3.11).

The Sabah royalty is structured to yield the government an increasing share of log values as world prices increase (Table 3.12). Royalty rates were drastically increased in 1979, after the doubling of Indonesia's export tax in 1978. Whereas before the maximum rate was just over 36 percent, since 1979 the Sabah royalty structure has yielded a maximum of 57 percent of log value at very high log prices. At the maximum price prevailing since 1979, the royalty yielded a 54 percent share for the government.

Unlike most timber-exporting countries, Sabah does not employ annual area license fees based on timber concession size. Instead, sealed tender auctions have been used for one form of concession, the Special

Table 3.11. *Comparison of timber royalty plus export tax payments per cubic meter at various f.o.b. prices for meranti logs: Sabah, Sarawak, and Indonesia*

Country	F.o.b. log prices (U.S. dollars per m³)			
	75	100	125	150
Sabah	33.39	48.39	66.92	84.82
Indonesia[a]	21.00	27.50	34.00	40.50
Sarawak[b] (pre-1980)	9.28	10.53	11.78	13.03
Sarawak[c] (post-1980)	16.75	19.25	21.75	24.25

[a]For Indonesia, both the royalty and the export tax are assessed on "posted" prices ("check" prices).

[b]Sarawak imposes a variety of forest fees on volume other than royalty. These include inspection fees, cess fees, and development fees. Together these totaled M$19.00 (US$8.84) per cubic meter in 1980. Inclusion of these in this table would substantially raise the Sarawak figure.

[c]Sarawak sharply increased its forest taxes in 1980.

License. These are of smaller area and shorter duration (5 to 10 years) than regular concessions (see Table 3.6), but are more numerous and cover more total area. Tender prices ranged from M$0.20 per hectare to M$0.30 per hectare in the 1970s (Gray 1983: 201).

Sabah has long applied timber cesses (export taxes). A "special" cess of M$0.15 per Hoppus foot[1] (M$4.16 per cubic meter, or US$1.71 at September 1985 exchange rates) was the second most important source of government forest sector revenue from 1973 to 1983, accounting for between 6 and 11 percent of forest revenue throughout the period. In September 1982, a cess on sawn timber exports was also enacted, at a specific rate of M$15.00 per cubic meter (US$6.17 at recent exchange rates). In 1983, the cess alone was a greater revenue source than the Indonesian timber royalty.

Other forest fees and charges include timber extraction charges, a timber development charge, and miscellaneous others. Together, these account for less than 5 percent of forestry revenues and are not further discussed here. Sabah's timber royalty has accounted for about 85 percent of total government forestry revenues since 1979, and the timber cess for most of the remainder. As the cess now applies most heavily to sawn timber exports, its role in the rate of forest depletion will be evaluated separately.

The timber royalty on logs is basically an *ad valorem* tax imposed at a rate of 70 percent[2] upon f.o.b. value minus "presumptive" logging costs (M$63.22 per cubic meter).[3] Thus the royalty is progressive: at high f.o.b. prices the royalty takes a relatively higher proportion of log export value.

Table 3.12. *Sabah: structure of timber royalty (Malaysian dollars per cubic meter)*

If f.o.b. log price[a] per cubic meter is:	Then, royalty payment[b] per cubic meter is:			
	Before June 1979[c]		After June 1979[d,e]	
	Amount	Percent of value	Amount	Percent of value
83.19	20.80	25.0	24.96	30.0
110.92	31.89	28.8	41.60	37.5
138.65	42.98	31.0	58.23	42.0
166.38	54.07	32.5	74.87	45.0
194.11	65.17	33.6	91.51	47.1
221.84	76.26	34.4	110.92	50.0
249.57	87.35	35.0	130.33	52.2
277.30	98.44	35.5	149.74	54.0
395.03	109.54	35.9	169.15	55.5
332.76	120.63	36.3	188.56	56.7

[a]Prices and royalties originally given in values per Hoppus foot. These were converted to values per cubic meter by multiplying by 27.73 (i.e., 1 m^3 = 27.73 Hoppus feet).

[b]For all logs except Class A logs, which are subject to higher rates at low prices and lower rates at high prices.

[c]Formula for royalty prior to June 1, 1979:

$$RL = (0.4)(\text{f.o.b.} - 1.13) \quad \text{per Hoppus foot}$$

[d]Formula for royalty after June 1, 1979: (1) If f.o.b. is less than M\$8 per Hoppus foot:

$$RE = (0.6)(\text{f.o.b.} - 1.5)$$

(2) If f.o.b. is more than M\$8 per Hoppus foot:

$$RE = (0.7)(\text{f.o.b.} - 2.28)$$

[e]The above royalty schedule applies to logs only. If logs are processed locally, the royalty is $RP = (0.07)(\text{f.o.b.})$.

In Chapter 2 we saw that *ad valorem* royalties can exacerbate "high-grading" of the forest and thus worsen logging damage. Sabah's progressive royalty structure provides more incentive to "high-grade" forest stands at high log prices than at low prices. The low ratio of actual to potential commercial timber yields in Sabah indicates substantial high-grading exists. Tax-induced high-grading accentuates forest depletion from logging for two reasons. First, it requires that larger areas be cut over to obtain a given amount of log output. Second, under selective cutting methods, damage to the residual stand (trees not cut) results primarily from entry into the stand. When firms bypass stems that would have been harvested in the absence of a royalty, damage to the residual stock is scarcely less than if a larger number of trees had been harvested. Available evidence indicates that between 45 and 74 percent of trees left in the residual stands in Sabah suffer "consequential" damage after logging (FAO 1981: 301). Implementing an income tax instead of the royalty would clearly reduce such damage (see Chapter 2, Appendix A).

In contrast, reported damage to Sarawak's residual stands is often half of that of Sabah, largely because effective royalty rates are much lower than in Sabah.

The extraordinary level of Sabah's royalty (relative to stumpage charges elsewhere) doubtless leads to a somewhat greater degree of high-grading, but the precise effects of this damage associated with selective cutting are indeterminable without detailed stand-by-stand post-logging surveys, which appear unavailable after 1970. We do know that the interaction of an *ad valorem* royalty with selective cutting systems of the type prescribed in Sabah (see below) compounds damages to logged-over stands.

Apart from the timber royalty, the 1982 cess on sawn timber clearly encourages high-grading of the Sabah forest estate. The tax applies to sawn timber, at a specific rate of US$6.17 per cubic meter of *product*. Since 1.8 cubic meters of logs are typically required to yield 1 cubic meter of sawn timber, for each cubic meter of log extracted and processed into sawn timber the cess is US$3.43. As noted in Chapter 2, a specific tax based on volume is particularly conducive to high-grading. Since the amount of the cess is invariant with respect to value, it discourages loggers from selecting both merchantable trees of secondary species and defective but merchantable trees of primary species (e.g., most dipterocarp stems). More tracts are entered to secure a given volume of wood, with consequent further damage to the forest.

Reforestation and reforestation fees. The "special" cess (M$0.15 per Hoppus foot) is earmarked to finance SAFODA replanting programs. At 1981 extraction levels (10.9 million cubic meters), the cess would have yielded M$45.3 million, or at recent exchange rates (Table 3.3) about US$18.7 million. Given that the minimum cost of any effective replanting program is about US$1,000 per hectare (1985 prices) a year's revenue from the special cess would finance replanting on about 18,000 hectares, or 32 percent of the total area deforested annually in Sabah.

However, replanting has not proceeded at anywhere near that rate. Prior to 1974, felled land was not reforested at all. Between 1974 and 1981, SAFODA replanted 21,000 hectares in the Tawau district, and between 1977 and 1981 15,000 hectares were replanted in northeast Sabah (Segal 1983: 55). Although revenues from the special cess totalled about US$116 million from 1973 to 1980, sufficient to finance replanting on 116,000 hectares, only 36,000 hectares were replanted.

Length of concessions. Duration of timber licenses is typically even shorter in Sabah than in Indonesia. Indonesia generally allows all concessionaires a 20-year term, while Sabah has three types of licenses. The first is the ordinary concession. Granted by the Sabah Foundation, which itself gained title to the forest rights for 100 years in 1966,[4] these run for a term of 21 to 25 years. By 1980, 10 such licenses were held, covering an area of 869,000 hectares (Table 3.6). Regulations nominally restrict exploitation of the Foundation's total land to 1.25 percent annually, but there are conflicting indications regarding compliance. One report indicates that between 1970 and 1981, 18 percent of the area under the Foundation's control had been exploited, exceeding the 11 percent limit for this period. Yet another report indicated only 77 percent of areas awarded by the Foundation had been logged (Hepburn 1982: 401–402).

A second form of concession is the special license. These originally had a duration of 5 to 20 years. In 1976, however, the duration of all these licenses was reduced to 5 years. A total of 109 special licenses were in force in 1980, covering about 782,000 hectares altogether. The final form of timber license carries the label "Form I" license. These are restricted to one year, may not cover an area larger than 2,000 hectares, and often are granted for less than 120 hectares.

Licenses of such short duration clearly create no incentives for logging firms to undertake yield-sustaining measures or to actively preserve the forest's productive and protective functions (Whitmore 1984: 101; Ross 1984: 53). This argument, developed at length in Chapter 2, is particularly true regarding the one- and five-year licenses available in Sabah. These extremely short licenses covered exactly 50 percent of the total area under concession in 1980. In recent years, issuing shorter term licenses has been the norm; the area under them may have reached 55 percent by 1985.

Harvesting methods and allowable cut. From 1955 until 1977, the silvicultural method prescribed by Sabah was a variant of the Malayan Uniform System (See Chapter 2, Appendix A). This is a monocyclic system under which all *saleable* trees are extracted in one entry, and the time lapse before the next allowable entry (the cutting cycle) is approximately equal to the rotation age of the trees. It has been thought that the monocyclic system does more severe damage to Sabah's rich forest stands than would polycyclic methods (Whitmore 1984: 96). With polycyclic methods *selected* trees (those of 58 cm. d.b.h. or larger) are removed in a continuous series of felling cycles of a duration *less* than the rotation age of trees. In monocyclic systems, advanced growth of saplings and poles (as well as mature trees) are cut down, leaving a large, uniform canopy opening to encourage regeneration of the desired light-demanding tree species. These trees are fast-growing, with pale light timber, highly valued in world markets. Monocyclic systems rely almost entirely on seedlings to produce the next crop. If the seedling stock is poor or nil, the commercial value of the timber stand is lost.

Alternatively, polycyclic or selective systems rely on residual poles and saplings (plants to 2.7 meters tall) for future growth. Small gaps left in the canopy support the adolescent saplings and poles but are insufficient to allow seedlings to prosper (Whitmore 1984: 96–97).

Another significant feature of the monocyclic Malayan Uniform System involves silvicultural treatment of stands selected for logging. This first involves poison girdling (Whitmore 1984: 99) of big overmature, unmerchantable trees after felling to ensure a good light supply and less root competition.[5] Then, within 10 years, a diagnostic sampling is made, followed by eradication or curbing of climbing vines and creepers and, on occasion, further canopy-opening.

The Sabah variant of the Malayan Uniform System, used until 1977, retained some adolescent growth (poles and saplings), to allow intermediate yields prior to the full rotation age. Pressures for expanded harvests led to modifications in 1977; all silvicultural treatment was stopped (FAO 1981: 302). The reported reason was that silvicultural treatment "was no longer considered useful because of the very high rates of damage being caused in current logging" (FAO 1979: 31). Since 1977, Sabah (like Indonesia) has employed a de facto "zero treatment" option, although de jure current silvicultural rules call for three stand treatments: before, during, and 10 years after felling (Nor 1983: 166).

With this change, the last pretense of conservation in logging management was discarded, with the forest openly treated as a wasting asset, mined in essentially the same fashion as a zinc deposit. As in Indonesia (Chapter 2) a private sector monopolist holding secure long-term forest rights would have utilized the forest in far more prudent fashion, if the firm sought to maximize profit.

Moreover, enforcement of cutting regulations has been lax. Within the Sabah felling system, actual harvests are governed by a "coupe" system: control on felling is by area, and concessionaires are allowed to work one coupe. A coupe is specified in terms of the number of acres to be logged and minimum tree girth limits. In principle, coupes must be systematically logged according to plans approved by the Sabah Forest Department, but evidence suggests that concessionaires frequently exceed their assigned coupes, often substantially.

The combination of short duration of concessions, high *ad valorem* and specific royalties, abandonment of silvicultural treatment in 1977, and cutting beyond assigned coupes would be expected to have dire long-term consequences for the forest's economic value. But disastrous consequences struck ahead of schedule. In Sabah, as in Indonesia, forest fires consumed millions of hectares in 1983, during that year's severe *El Niño* (Chapter 2). Fires have historically been associated with the more serious droughts brought on by *El Niño* occurrences, but never in recorded history was the damage as extensive as on Borneo (Kalimantan) in 1983. Though the drought associated with the 1983 *El Niño* was the most severe in decades,[6] the area burned was well out of proportion to the drought's severity. On the Indonesian side of Borneo an area the size of Belgium was burned. Much less is known, even four years later, regarding the extent of Sabah's loss; estimates run into the millions of hectares (Leighton 1984: 4). A series of satellite photos in the author's possession indicate a vast area of Sabah under smoke cover on May 19, 1983, toward the end of the drought.

In Indonesia, destruction was particularly severe in logged-over areas; indeed the unlogged forest escaped major damage. There is good evidence that logging effects exacerbated the fire's extent and severity. Forest specialists and ecologists (Leighton 1984: 3–6; *Asiaweek* 1984) have generally attributed the more intense fires in logged-over areas in East Kalimantan to the great amount of dead wood and other combustible litter on the ground.[7] Damage was slight in undisturbed forest areas as relatively low volumes of dead wood and other ground litter failed to generate enough heat to ignite large stems (Mackie 1984: 63–75).

The causes and the extent of the Sabah fires have been much less discussed; the government has not encouraged publication of damage estimates. But given the similarities between forest endowments in Sabah and East Kalimantan, their proximity (Sabah abuts East Kalimantan) and the tendency of the appreciably higher Sabah royalty rates to generate greater forest damage than Indonesia's royalties and timber taxes, logging methods were likely a significant factor in the severity and extent of the fires in Sabah. Government policies almost certainly made the forest estate in Sabah more vulnerable to fire devastation, to a degree unforeseen before 1983.

Entry terms for foreign investment. In Sabah, as elsewhere in Malaysia, foreign firms initially entered the timber sector under traditional concession arrangements, wherein firms with 100 percent foreign equity received harvesting rights for what were called long-term (21-year) concessions, with minimal commitment for investment in processing activity. The creation of the Sabah Foundation in 1966, and its subsequent 1970 land award, materially affected forest sector involvement by foreign firms, particularly as the Foundation is ultimately to become the sole concessionaire in Sabah.

By 1987, the Foundation was expected to absorb all remaining long-term concessions granted earlier to foreign enterprises. Any future foreign involvement will be limited to joint venture with the Sabah Foundation, with contracts of one to 10 years on relatively small areas under the Foundation's control. Most of the large timber projects were cancelled in 1983. Few large foreign firms have expressed interest in the short-term special licenses and the even shorter-term Form I licenses; smaller logging contractors from Peninsular Malaysia and the Philippines are the most likely candidates. As in Indonesia, yields per hectare in Sabah have been much lower for small logging firms than for large (primarily foreign) firms. As the smaller firms gain larger shares of available tracts, faster rates of forest depletion will be required to maintain any given harvest volume.

Policies toward non-wood forest products. Forestry policy in Sabah has focused more strongly upon utilization of the forest as a source of wood and agricultural land than in Indonesia. Not surprisingly, opportunities for utilizing the non-wood products of the forest have been largely overlooked in Sabah. There are no statistics on the annual amounts and values of non-wood forest products extracted. Moreover, we find no trace of a policy toward non-wood products. The above applies also to Peninsular Malaysia and particularly to Sarawak. Where economic values of non-wood forest products are ignored, they provide no incentive to slow the cutting and conversion of the natural forest.

Non-forestry policies

Tax policy and industrialization policy. Unlike Indonesia, Malaysia has not offered income tax holidays to purely extractive timber operations. Logging firms are liable for a 40 percent corporate income tax and a special timber profits tax of 10 percent. However, income taxes are national taxes and do not accrue to the Sabah state government. Not only are there no available estimates of federal income taxes collected on Sabah timber companies; it is even unclear that such taxes are actually collected. In any case, most income from harvesting activity (after payment of royalties and other forest fees) now accrues to the Sabah Foundation, a nontaxable entity.

Like Indonesia and Peninsular Malaysia, Sabah has sought to use royalty and trade policies to encourage forest-based industrialization. However, these policies were applied later in Sabah (and Sarawak) than elsewhere in the region, and with less tenacity.

Not until 1979 did the state government establish a policy of replacing log exports with processed wood product exports. Indeed, until 1979, virtually all forest exports were in log form. Beginning in 1979, export quotas for logs were gradually applied, with the explicit objective of phasing out all log exports by 1985 (Takeuchi 1982: 54–55). As is evident from Table 3.3, there has been substantial slippage in this program.

Manipulation of royalties and cesses has been much more significant than log quotas in promoting forest-based industrialization. Whereas Indonesia stressed plywood in its forest-based industrialization policies, Sabah has emphasized sawn timber, for two reasons. First, Indonesian log quality, particularly meranti from Kalimantan, is more uniform. Large quantities of single species are available and more suitable for plywood. Sabah logs, particularly *seraya* (a dipterocarp) are more suited for sawmilling, because of their greater density and less uniform quality (Ross 1984: 35–36).

After 1979, the Sabah timber royalty provided very strong incentives for investments in sawmilling capacity. Application of the cess to sawn timber, in September of 1983, weakened the incentive. We first examine the royalty and then the cess.

The Sabah timber royalty applies on logs according to Equation 3.1 (expressed in Malaysian dollars per cubic meter):

$$RE = (0.7)(\text{f.o.b.} - \$63.22) \tag{3.1}$$

However, if the log is processed locally, the royalty is defined as follows:

$$R_p = (0.07)(\text{f.o.b.}) \tag{3.2}$$

Average f.o.b. log values in 1980 and 1981 were US\$99.61 and US\$85.45 per cubic meter respectively. We will use US\$100 per cubic meter as illustrative of the f.o.b. value of Sabah log exports in the early 1980s.

In U.S. dollars, at an exchange rate of M\$2.40 for each U.S. dollar, the royalty formulas are as in Equations 3.1 and 3.2:

$RE = (0.7)(\$100.00 - \$26.34)$ for logs

$R_p = (0.07)(\$100)$ for logs used in sawn timber

The incentive for local processing provided by the Sabah royalty is therefore US$44.56 per cubic meter (US$51.56 royalty on logs not processed minus US$7.00 on logs processed into sawn timber).

Takeuchi estimates that in the period 1977–79, the additional value-added at domestic prices in sawn timber over log export in Malaysia was between 59 and 65 percent of the value of logs (Takeuchi 1982: 29). The figure is far lower in recent years; f.o.b. values for sawnwood, in round-wood equivalent, have not exceeded 122 percent of f.o.b. log prices in Sabah. We will assume that a value-added figure of 25 percent is representative for the early 1980s.

On this basis, given log export values of US$100 per cubic meter, sawn-wood export values would be $125 in Sabah. The royalty structure then implies that Sabah foregoes US$44.56 in forest revenue for each US$25 in additional value-added gained from processing logs into sawn timber. By processing the logs, firms save a royalty equal to 178 percent of the incremental value-added in sawmilling; the royalty provides effective protection of 178 percent. (See Chapter 2, Appendix B.) This is a relatively high protection rate, though lower than the 221 percent rate provided domestic plywood producers in Indonesia concurrently. It does indicate that Sabah sawmills may remain profitable even though 1.78 times more inefficient than sawmills in log-importing countries.

In September 1982, Sabah reduced effective protection[8] rates to value-added in its sawmills when a cess was applied to sawn timber exports at a rate of M$18 per cubic meter of sawn timber, or US$8 per cubic meter. Thus, Sabah sawmills then paid a combined royalty and cess of US$15 on log inputs with an f.o.b. value of US$100, reducing the effective protection afforded to domestic sawmills to 146 percent (total taxes saved by processing as a percent of value-added).

Investors clearly responded to higher protection rates available to sawmills before 1982: by 1981, more than 200 sawmills had been established, compared to about a dozen before 1978. Nevertheless, even at effective protection rates of 146 percent, a third of the mills established between 1978 and 1981 had ceased operations by early 1983 (Segal 1983: 55). The reduction in protection arising from the imposition of the cess on sawn timber in September 1982 could only have hastened the demise of inefficient mills.

Sabah has not enforced restrictions on log exports as tenaciously as Indonesia. Although log exports were to be phased out entirely in 1985, planned log export volumes for 1982 were actually 27 percent higher than in 1980, and sawn timber production was only 6 percent of total log production.

Forest-based industrialization policies in Sabah will likely accelerate forest depletion rates, for the same reasons as in Indonesia. Once milling is at full capacity for Sabah's entire log production, sawmilling alone will involve 30,000 jobs.[9] At upwards of 6 percent of the 1986 labor force, it is a share too large by itself for any government to ignore. Together, logging and milling jobs will constitute 13 percent of the labor force. Policy-makers may continue logging activities at high levels even during periods of slack world demand for sawn timber, to provide domestic processing feedstock to keep the mills going. As in Indonesia, domestic log processors will have claims on log harvests that no foreign mill could ever exercise.

Another aspect of Sabah industrialization policy has created pressure on the natural forest, albeit to a lesser degree than the effort to promote sawmilling. In the late 1970s, policy-makers decided to construct a pulp-and-paper mill complex at Sipitang to go on stream in 1987 to compete in world markets. A major project in any setting, this US$560 million undertaking is quite large relative to Sabah's small economy.

Pulp-and-paper plants typically use long-fibered softwoods, whereas Sabah's timber endowment consists primarily of short-fibered tropical hardwoods. The Sipitang plant, however, was designed to utilize a mix of 70 percent hardwoods and only 30 percent softwoods. Natural stands of long-fiber softwoods grow only in relatively inaccessible areas of the eastern coast; thus, thousands of hectares of natural forest were cleared to enable plantings of scotch pine to feed the plant (Berthelsen 1985: 7). The pines will not mature for six to 12 years, and the plant was built far away from natural softwood stands; the project is unlikely to survive against international competition. Once again, the natural forest cleared to support the paper mill has been treated as a low-value asset, to permit an industrialization project of dubious value to the economy.

Resettlement programs. Sabah has nothing comparable in scope to the Indonesian transmigration program. However, two major agencies operate resettlement programs: SAFODA (the Sabah Forest Development Program) and SLDB (the Sabah Land Development Board).

SAFODA's largest resettlement program is located in the remote Bengkoka peninsula in northern Sabah. This scheme involves M$250 million (about US$103 million), partially funded by the World Bank. Each family is given a two-room house and a six-hectare lot of fully developed forest plantation (Segal 1983: 55). This project will ultimately involve 2,000 families; the cost per family will be about US$52,000. Since most forest exploitation in the 1970s was in southern Sabah, it is not clear whether the SAFODA plantation was developed on logged-over forest or the undisturbed natural forest was cleared for the plantation.

SLDB is the largest settlement agency and had accumulated debts of M$421 million by November 1982. We know little of the details of SLDB's resettlement programs, partly because internal management problems have overshadowed the agency's operation for years. Indeed, by September 1981 problems had worsened to the point that SLDB management was replaced by a team from the private sector conglomerate Sime Darby. The federal government owns 51 percent of Sime Darby. This management change thus indirectly increased federal influence over a Sabah government agency that influences forest use.

Other regulations on foreign investment. The largest foreign holders of Sabah timber concessions are North Borneo Timber, Sabah Timber, and Weyerhauser. The first two are British-owned, the last U.S.-owned. Weyerhauser has operated in Sabah since 1967, while the two British firms have roots in Sabah predating the Second World War. The concessions of these enterprises, and those of most smaller foreign firms, are slated to expire by 1988 (Rowley 1977: 59).[10]

The implications of this concession expiration for deforestation are unclear. The large foreign firms have professed stronger interests in reforestation programs and have conducted some research to this end. Indeed, both Weyerhauser and North Borneo Timber have entered joint ventures involving replanting programs with the Sabah Foundation (Segal 1983: 55). Whether either firm will maintain a long-term presence in Sabah following expiration of their concessions will be known only in 1988.

Rents and rent capture. Cost estimates for logging and sawmilling in Sabah differ by as much as 50 percent depending on the source; in some cases the estimates include forest taxes, in others they are exclusive of taxes, and in others it is unclear whether taxes are included or excluded (Takeuchi 1982: 64). Nevertheless, such estimates are required in any computation of forest rents. We will arbitrarily assume that Sabah's logging and milling costs are essentially identical to those in Indonesia. At least, Indonesian figures have been cross-checked and vetted; Sabah's proximity to Indonesia and similarities between the forest estates in Sabah and East Kalimantan in Indonesia additionally justify this procedure.

Table 3.13. *Sabah: average f.o.b. values, logs and sawn timber, 1979–84*

	Average f.o.b. value (Malaysian dollars per m³)	Average f.o.b. value[a] (U.S. dollars per m³)	Average f.o.b. value, roundwood equivalent[b] (U.S. dollars per m³)
Logs			
1979	216.08	93.96	93.96
1980	222.25	96.63	96.63
1981	188.77	84.27	84.27
1982	212.88	91.75	91.75
1983	178.57	76.31	76.31
Sawn timber			
1979	475.87	206.90	103.45
1980	542.94	236.06	118.03
1981	349.74	156.13	78.07
1982	353.58	152.41	76.21
1983	372.86	159.34	79.67
1984	373.53	158.95	79.48

[a]Converted at average exchange rates prevailing for the year (see Table 3.3).

[b]Assumes a conversion rate of 50 percent (two logs required for each cubic meter of sawnwood exports). Note, however, that the average conversion rate for Sabah sawmills was only 45 percent in the mid-eighties and only 40 percent in the late seventies.

Sources: Tables 3.3 and 3.4; Sabah Forest Service, personal communication (June 1980).

Using Indonesian logging costs will doubtless result in underestimating Sabah timber rents given the greater density and accessibility of Sabah's stands.

Average f.o.b. values and estimated rents per cubic meter are shown in Tables 3.13 and 3.14. Aggregate rents are derived in Table 3.15, using the same procedures as followed in Chapter 2.

Table 3.16 presents an estimate of timber rents captured by the government of Sabah over the period 1979 to 1983. Taxes and forest fees collected by the government amounted to almost 81 percent of estimated rent in 1979 and rose to a striking level of 92 percent in 1980. The proportion of rent captured by the government declined over the next two years as tropical hardwood prices weakened on world markets. However, with the enactment of the new cess on sawn timber in late 1982, the government's share in rent in 1983 was restored to 1980's very high level.

Without question Sabah's policies toward timber rent capture in the late 1970s and early 1980s were the world's most aggressive. Unlike Indonesia, where the government rent share has fluctuated between one-third

Table 3.14. *Sabah: per unit rents in tropical timber, 1979–83 (U.S. dollars per cubic meter)*

	(1) Average f.o.b. value, roundwood equivalent[a]	(2) Logging costs, roundwood equivalent[b]	(3) Sawmilling costs, roundwood equivalent[b]	(4) Potential rent [Col. 1 − (Cols. 2 + 3)]
1979				
Logs	93.96	29.84	—	64.12
Sawn timber	103.45	29.84	10.36	63.25
1980				
Logs	96.63	34.24	—	62.39
Sawn timber	118.03	34.24	11.92	71.87
1981				
Logs	84.27	37.93	—	46.34
Sawn timber	78.07	37.93	13.21	26.93
1982				
Logs	91.75	41.00	—	50.75
Sawn timber	76.21	41.00	14.28	20.93
1983				
Logs	76.31	45.92	—	30.39
Sawn timber	79.67	45.92	16.00	17.75

[a]From Table 3.13.
[b]From Chapter 2, Table 2.16.

and one-half, it cannot be argued that Sabah sold its forest resources too cheaply. The resources were, however, sold quickly. One set of government figures on forest depletion suggest that between 1973 and 1983, the remaining undisturbed forest area fell from 55 percent to just under 25 percent of the total land area, the most pessimistic figures on forest use yet released. The acting Director-General of SAFODA remarked in 1983 that "by 1986, there will be a sudden drop in timber production because of the areas being depleted" (Segal 1983: 55). Indeed, Sabah's southern forests were reported to have been largely logged out by 1983 (Segal 1983: 55).

Sabah, like Indonesia, has also destroyed some rent while encouraging forest-based industrialization. After 1980, when Sabah began seriously to promote domestic sawmilling, the average f.o.b. value of sawn timber, in log equivalent, was less than the average f.o.b. export value of logs in every year except 1983. As in Indonesia, "value-added" in domestic processing has tended to be "added national costs" instead, for many of the

Table 3.15. *Sabah: aggregate rents in tropical timber, 1979–83*

	Logging			Sawmilling			
	(1)	(2)	(3)	(4)	(5)	(6)	(7)
Year	Rent (U.S. dollars per m³)	Volume of log exports (million m³)	Total log rent (million U.S. dollars) (Col. 1 × Col. 2)	Rent (for sawn timber, roundwood equivalent) (U.S. dollars per m³)	Volume of sawn timber exports (million m³)	Total rent in sawn timber (million U.S. dollars) (Col. 4 × Col. 5)	Total rent, all forest-based activity (million U.S. dollars) (Col. 3 + Col. 6)
1979	64.12	9.72	623.2	63.25	0.19	12.0	635.2
1980	62.39	8.21	512.2	71.87	0.29	20.8	533.0
1981	46.34	8.69	402.70	26.93	0.39	10.50	413.2
1982	50.75	9.83	498.87	20.93	0.64	13.39	512.3
1983	30.39	9.48	288.10	17.75	0.94	16.69	304.8

Sources: Tables 3.3, 3.4, 3.7, and 3.14; Sabah Forest Service, personal communication (June 1980).

Table 3.16. *Sabah: rent capture by government, 1979–83*

Year	Total aggregate rents in forest-based activity (logging plus milling) (million U.S. dollars)[a]	Total royalties and taxes on forest-based activity (million U.S. dollars)[b]	Proportion of rent captured by government, percent (Col. 2 ÷ Col. 1)
1979	635.2	511.6	80.5
1980	533.0	488.4	91.6
1981	413.2	342.1	82.8
1982	512.3	360.9	70.4
1983	304.8	283.8	93.1

[a]From Table 3.15.
[b]From Table 3.9.

same reasons offered in Chapter 2. The major difference between Sabah and Indonesia is that Sabah has destroyed less rent, as it has not forced forest-based industrialization to the degree of Indonesia.

Sarawak

Sarawak, located on the northwest portion of Borneo adjacent to Brunei, covers an area 68 percent larger than nearby Sabah and nearly as large as all of Peninsular Malaysia (Table 3.1).

Forest endowments, forest utilization

As in the rest of Malaysia, all of the 9.4 million hectares of forest land belongs to the state. The permanent forest estate covers about one-third of total forest area; the ill-defined "statelands" cover the remaining two-thirds (Table 3.17).

The permanent forests are under the full control of the Sarawak Forest Department. They are subdivided into three categories: reserve forest, protected forest, and communal forests. Forest reserves are set aside for productive forestry; entry requires a license. Protected forests cannot be harvested, but traditional hunting, fishing, and gathering activities are permitted. Commercial activities are prohibited in communal forests.

On statelands, Forest Department authority is not well defined. While logging on these lands is licensed by the Department, their systematic management is legally infeasible, owing to customary rights held by local Dayak groups.

Two major types of forest exploitation take place in Sabah: that in the

Table 3.17. *Sarawak: forest resources, 1980 (thousand hectares)*

	Forest type			
	Hill forest[a]	Swamp forest	Mangrove	Total
A. Permanent forest[b]	2,453	684	41	3,174
1. Forest reserve	386	338	28	748
2. Protected forest	2,040	342	13	2,395
3. Communal forest	27	4	0	31
B. Statelands[c]	5,337	790	133	6,260
C. Total forest	7,790	1,474	174	9,434

[a]Hill forests include mixed dipterocarp and *keranga* forest types.

[b]Permanent forests are under the full control of the Sarawak Forest Department.

[c]The statelands have unsettled status. These are partly forested and partly non-forested. Logging in the forested areas of statelands is nominally under the control of the Forest Department; systematic management is legally infeasible. In the secondary growth areas, customary rights of local Dayak groups exist and are exercised.

Source: FAO (1981: 316–318).

swamp forest, which covers 1.5 million hectares (16 percent of the forest area), and in the hill forest (83 percent of the forest area). Table 3.17 indicates areas of both, plus mangrove forests. Swamp forests have been logged for decades, and by 1981 were largely depleted of valuable ramin (*Gonystylus bancanaus*), but still contain sizeable quantities of dense, pure stands of desirable *Shorea albida* (Nor 1983: 164). Virtually all of the swamp forest is or has been under logging license (Table 3.18).

From 1950 through 1980, logs from the mixed swamp forest constituted the mainstay of the state's economy (Seng 1982: 485). After 1970, however, logging in the hill forest accelerated, and by 1978 the log harvest of mixed dipterocarp species from the hill forest, at 3.1 million cubic meters, was marginally larger than the swamp logs harvest. By 1980, only 41 percent of the hill forest estate was under concession agreements (FAO 1981: 317– 319).

Available evidence on deforestation in Sarawak is conflicting and difficult to interpret. The minority view among foresters maintains that shifting cultivation has not been a major factor in deforestation in Sarawak (Callahan and Buckman 1981: 33).[11] Other evidence suggests that the total land area under shifting cultivation, already 2.3 million hectares in 1965, possibly reached 3.7 million by 1985 (FAO 1981: 326). This information suggests that areas under shifting cultivation may be as high as

Table 3.18. *Sarawak: forest utilization and deforestation*
(thousand hectares)

A. Total forest area (1979)	9,434
B. Total area under logging license (1978)	3,459
1. Permanent forest	1,974
2. Stateland	1,485
C. Total area under license by type of forest (1977)	
1. Hill forest	3,166
2. Swamp forest	1,313
D. Deforestation—average annual deforestation	
1. 1965–75	50
2. 1976–80	80
3. 1981–85	89

Source: FAO (1981: 316–318).

one-fourth of Sarawak's total land area, or one-third of the total forest area. One published estimate placed deforestation due to shifting cultivation as high as 60,000 hectares annually between 1966 and 1976, and indicated that for every log exported, fires started by shifting cultivators claimed another (Tiing 1979: 419). Official statistics, however, show only 2.25 million hectares under shifting cultivation.[12] Forest clearing for shifting cultivation is believed to account for not less than one-third of annual deforestation (80,000 hectares per year) in the period 1976–1980. Remaining deforestation has been attributed to shifting cultivation on logged-over areas.

It would be surprising if shifting cultivation were not a major source of deforestation, since rural poverty is widespread. One source placed the number of people living entirely or basically on shifting cultivation in 1979 at 230,000, or over 20 percent of the total inhabitants (Tiing 1979: 420). Sarawak per capita incomes are about half that of the rest of the country. The rural interior areas are primarily populated by indigenous Dayak peoples (Iban, Bidayuh, and others) who make up 44 percent of the population. Government figures show that two-thirds of the Bidayuh and almost half (49 percent) of the Iban households fall below the poverty line, as that threshold is defined for all Malaysia (Clad 1985: 36). In general, indigenous groups in Sarawak are reported to experience chronic malnutrition comparable to the poorest groups of indigenous people in Latin America (Tiing 1979: 419).

In sharp contrast to Peninsular Malaysia, conversion of natural forests to permanent agriculture has not been a significant source of deforestation in Sarawak. However, major hydroelectric projects planned for the

Table 3.19. *Sarawak: export values: logs and processed products, 1971–84 (million U.S. dollars)*

Year	Log export earnings	Processed timber export earnings (sawn timber, plywood)	Total timber exports
1971	50.3	19.6	69.9
1972	34.8	21.2	56.0
1973	56.3	35.3	91.6
1974	49.2	26.3	75.5
1975	29.0	28.3	57.3
1976	112.6	54.4	167.0
1977	118.3	41.3	159.6
1978	147.3	36.0	183.3
1979	161.9	41.4	202.3
1981	362.5	72.8	435.4
1982	543.5	72.9	616.4
1983	467.1	64.1	531.2
1984	522.1	69.0	591.1

Sources: 1971–79: Malaysian Timber Board; 1981–84: Table 3.3.

coming decade would submerge vast areas south of the capital of Kuching, and particularly in the Rajang river valley, where a US$3.3 billion project involving two dams is now ready for tenders (Clad 1985: 36).

The role of logging in deforestation has not been quantified in Sabah. However, as in Indonesia, logging shares the responsibility with shifting cultivation for two-thirds of the deforestation occurring in logged-over areas. Although logging has been and continues to be the principal activity in the forest-based sector, recent years have witnessed an increase in investments in wood processing. Nevertheless, wood processing in Sarawak remains relatively undeveloped relative to Peninsular Malaysia and Indonesia; Sarawak has not placed restrictions on log exports. By 1984 the Sarawak wood-processing sector consisted of only 76 active sawmills, three veneer/plywood mills, 15 moulding factories, and about 20 other processing firms (*Asian Timber* 1984: 37).

Benefits

Forest-based activity has long formed a sizeable proportion of the state's total gross regional product. In the period 1976–84, the forest sector grew strongly, and export earnings accelerated sharply. By 1984, foreign exchange earnings from the sector, principally logs, had reached nearly US$600 million, a seven-fold increase over 1971 (Table 3.19).

Logging and related activities have been the driving force behind the

Table 3.20. *Sarawak: government revenue from forests and total tax revenue, 1970–82 (thousand Malaysian dollars)*

Year	Revenue from forests	Total revenue	Percentage of revenue from forests
1970	19.3	n.a.	n.a.
1971	23.8	n.a.	n.a.
1972	21.8	88.2	24.7
1973	21.9	93.8	23.3
1974	27.7	90.1	30.7
1975	15.9	206.3	7.7
1976	42.8	243.3	17.6
1977	45.6	201.1	22.7
1978	54.3	238.8	22.7
1979	57.6	n.a.	n.a.
1980	82.6	n.a.	n.a.
1982	256.9	n.a.	n.a.

Sources: 1970–1980: Sarawak Forest Service, *Angarram Hasil dan Perbelanjuan Bagi Tahun* (various issues); 1982: *Asian Timber* (1984).

spread of the money economy in Sarawak and account for a significant share of total employment. In 1978, between 10,000 and 15,000 people were employed in the timber sector; by 1984, this figure had risen to 22,000, or 9 percent of the total labor force (Sarawak Study Group 1986: 3).

Another significant benefit accruing from the timber sector has been revenue from taxes and forest charges. In the mid-1980s, timber revenues typically accounted for between 20 and 25 percent of total state government revenues (Table 3.20). A sharp increase in the Sarawak royalty in 1980 raised timber's share in total taxes to nearly 50 percent. Before that, Sarawak had clearly underutilized the forest sector as a revenue source relative to both Indonesia and Sabah. Forest sector fiscal receipts averaged only 13 percent of total export value (Table 3.21), well below Sabah and nearby Indonesia. Even after the 1980 royalty rate hikes, forest taxes were still only 18 percent of export values, compared to 37 percent for Sabah and 28 percent for Indonesia (Table 3.22).

Given Sarawak's lower forest quality, depletion of valuable ramin stands, and difficult hill forest accessibility (partially due to steep slopes) we would expect forest taxes to be a lower percentage of export values than in, say, Sabah. The ratio in Sarawak was one-fourth that of Sabah in 1979, and significantly below Indonesia's despite comparable dipterocarp endowments.

Table 3.21. *Sarawak: fiscal receipts from forest sector as percent of export values,*
1970–82

Year	Total forest taxes[a] (million U.S. dollars)	Total export values[b] (million U.S. dollars)	Forest taxes as percent of export values[c]
1970	8.05	n.v.	n.v.
1971	9.90	69.9	14.1
1972	9.09	56.0	16.2
1973	9.10	91.6	9.9
1974	12.59	75.5	16.7
1975	7.23	57.3	12.6
1976	19.90	167.0	11.9
1977	21.21	159.6	13.3
1978	24.90	183.3	13.6
1979	26.18	202.3	12.9
1980	36.72	200.7	18.3
1982	111.70	616.4	18.2

[a]From Table 3.20.
[b]From Table 3.19.
[c]Estimated collections based on application of 1980 tax changes to representative meranti logs with f.o.b. value of US$100 per m^3.

Table 3.22. *Taxes and other forest charges as a percentage of exports:*
Sarawak, Sabah, and Indonesia

Year	Sabah[a]	Indonesia[b]	Sarawak
1974	27.1	15.8	16.7
1975	25.3	15.2	12.6
1976	26.9	14.8	11.9
1977	39.2	19.1	13.3
1978	35.7	25.7	13.6
1979	53.0	22.7	13.5
1980	59.0	32.6	18.3
1982	36.9	27.6	18.2

[a]Chapter 3.
[b]Chapter 2.

This suggests that, at least until 1980, government rent capture in Sarawak was extraordinarily low.

Forestry policies
Responsibility for forestry policies in Sarawak rests primarily with the Conservator of Forests in the Forest Department; however, all nego-

Table 3.23. *Sarawak: fiscal receipts from the forest sector (million Malaysian dollars)*

Year	(1) Export duty	(2) Forest product fee	(3) License fees and royalty	(4) Timber cess	(5) Other forest	(6) Total forest tax revenue
1970	1.93	n.a.	16.59	0.87	0.0	19.32
1971	0.12	n.a.	23.02	0.63	0.0	23.77
1972	0.84	n.a.	21.33	0.42	0.0	21.83
1973	2.43	n.a.	18.97	0.46	0.0	21.84
1974	7.24	0.09	19.99	0.45	0.0	27.69
1975	4.30	0.42	11.30	0.29	0.0	15.90
1976	11.44	0.04	29.62	0.85	0.0	42.78
1977	12.97	0.35	28.45	1.00	0.0	45.61
1978	15.66	0.40	34.15	1.52	2.55	54.28
1979	20.00	0.50	33.00	1.10	1.00	57.60
1980	34.00	0.12	46.00	2.50	n.v.	82.62
1982	n.v.	n.v.	n.v.	n.v.	n.v.	256.9

Sources: Sarawak Forest Service, *Angarram Hasil dan Perbelanjuan Bagi Tahun* (various years), and Table 3.8.

tiations with foreign investors since 1973 have been handled by the Sarawak Timber Industry Development Corporation. Further, policies toward taxes and royalties on timber are formulated in the Chief Minister's office, along with non-forestry policies of the state government.

Royalties, license fees, and taxes. Nearly all of Sarawak's timber levies, except the export taxes, are volume-based (specific) taxes, a fact which explains why government revenues from forestry were so low until 1980. In 1980, virtually all specific charges were increased by 50 percent.

Royalties are far the most important forest charges in Sarawak, followed by the export tax (Table 3.23). The timber royalty is a specific tax and highly differentiated by species, as shown in Table 3.24. In addition, the timber cess and the development premium are both levied on a volume basis: the cess is fixed at M$1.80 (US$0.75) per cubic meter, and the timber development premium at M$3.60 per cubic meter (US$1.50).

The only significant *ad valorem* charge on timber in Sarawak is the export tax. It was imposed at a rate of 5 percent until 1980, when it was doubled to 10 percent. The tax only applies to logs, creating an incentive for domestic log processing.

The Sarawak system of forest fees results in tax revenues per cubic

Table 3.24. *Sarawak: specific timber tax on volume: timber royalty, timber cess, and timber development premium (after 1980)*

	Rate of levy	
	Malaysian dollars per Hoppus ton	U.S. dollars per m^3
A. Royalty		
1. *Shorea* spp. (meranti)	30	7.50
2. *Ramin* spp.	24.00–36.00	6.00–9.00
3. Bindang (*Agathis*)	30	7.50
4. 27 other species	9.00–22.50	2.20–5.60
B. Cess (all hill species)	1.00	0.75
C. Timber development fund (all swamp species)	3.60	1.00

Sources: Royalty figures from Gray (1983: 91). Other data from Malaysian Timber Board (personal communication).

meter of approximately US$19.25 for dipterocarp logs, at f.o.b. values of US$100 per cubic meter. At US$150 per cubic meter, the government receives US$24.25, only 29 percent of the government take from a similar log at the same price in Sabah.

Though the Sarawak forest fee system is a poor means of rent capture, it should result in less damage to logged-over stands than systems in Indonesia and Sabah, for two reasons. First, the overall effective royalty rate is lower than in Sabah or Indonesia, implying less high-grading. Second, the royalty (though not the export tax) is highly differentiated by species, with rates 20 to 60 percent lower on low-valued species. This too tends to reduce high-grading (see Chapter 2, Appendix A).

Available evidence does indicate lighter damage to logged-over stands in Sarawak. Whereas in Sabah broken or missing trees account for as much as 61.7 percent of the stand after logging, in Sarawak the proportion has been estimated at only 14.6 percent (FAO 1981: 301, 323). Also, in Sabah, one estimate places the number of trees with "inconsequential or no damage" at only 26.1 percent. For Sarawak, 58.1 percent of trees suffered "no damage" (FAO 1981: 301, 323).[13]

Clearly the system of forest taxes can indeed play a significant role in discouraging needless tropical forest depletion. These findings also indicate that an income tax would be Sarawak's most efficient incremental revenue source, rather than further *ad valorem* royalties such as the export tax, though problems of administering income taxes in a regional government will likely preclude this option. The true cost of administering income taxes (as applied to forestry) consists both of lost revenues and higher depletion rates.

Reforestation and reforestation fees. There is no evidence, published or otherwise, of any systematic reforestation program in Sarawak. The lack of such programs may partially be a carryover of views from the days when most logging was conducted in the swamp forest, and loggers relied upon natural regeneration assisted by removal of damaged, defective, and undesirable stems (Seng 1982: 485).

Concessions policy. Logging concessions are granted by the Conservator of Forests, the head of the Forestry Department. For the 111 licenses granted for logging in hill forests, areas conceded ranged from 2,000 to 4,000 hectares (FAO 1981: 321). Two kinds of licenses are awarded in Sarawak: timber logging licenses (for dipterocarp forest) and mangrove licenses. Timber-logging licenses are subdivided into the annual license and the "long-term license" (divided into five- and 10-year licenses) (Johnson 1984: 2). The longer-term license nominally requires management plans to be established, a requirement which firms have followed to a degree. Where annual licenses are granted, a felling plan is required. The clear-cutting system is generally permitted and practiced: all the wood is used, including trees small in diameter, and clear-cut land is later used as farmland.

Maximum concession length is 10 years. Many concessions, however, are restricted to one and five years in length. Such concession lengths, again, are insufficient to promote optimum forest depletion rates.

Harvesting method. Except for one-year licenses in dipterocarp forests, selective cutting is generally allowed and practiced in both the depleted mixed swamp forests and the mixed dipterocarp forests.

For the longer-term licenses in the mixed dipterocarp forest, harvesting is limited to trees with a minimum d.b.h. of 46 cm. (Seng 1982: 2). The cutting cycle is 25 years, although the longest license is 10 years. A modified form of the Malayan Uniform System is prescribed, but the system is neither well-defined nor much enforced (Nor 1983: 166; Seng 1982: 486–489). Virtually no silvicultural treatment is practiced, except that loggers are nominally required to prevent excessive tree deaths caused by root damage when streams are dammed by logging road construction (Seng 1982: 487).

Sustained logging in the permanent mixed dipterocarp forest in Sarawak began only in 1969. Except for information gained from small-scale experimental plots established in the 1970s, there has been virtually no published research on the effects of logging under the present system (Seng 1982: 489–490). There is, therefore, little basis for projections of the impact of large-scale logging on the future value of the forest estate.

Entry terms for foreign investment. Foreign investment in forest-related activity, whether logging, wood processing, or agro-conversion, has been much more limited in Sarawak than in the rest of Malaysia. Even the wood moulding industry, the largest of its kind in Malaysia, is run by Sarawakians.

Negotiations with foreign investors are handled by the Sarawak Timber Industry Development Corporation (STIDC), established in 1973. Since timber taxes and royalties in Sarawak are much lower than the rest of the Asia-Pacific region, particularly Sabah, we might reasonably expect foreign investor interest in the state's timber to intensify in the future.

Policies toward non-wood forest products. Non-wood forest products have traditionally been an economic mainstay of Sarawak's indigenous people, no less than in other countries with tropical forests.

While many of the same fruits, nuts, barks, and pharmaceutical compounds as in Indonesian natural forests are harvested by local people, wild game has been particularly important to the subsistence of Sarawak's indigenous groups. Detailed studies have shown that hunting produces 35,500 tons of meat, primarily wild boar and deer, valued at US$82 million per year (Whitten 1986: 13). In no year has the value of processed wood exports from Sarawak exceeded the value of meat harvested in the forest.

Little is known of the aggregate economic value of non-wood products available in Sarawak's forests. It is known, however, that 102 categories of such products exist in typical rain forests (Jacobs 1982: 3768–3782). Two animal categories generated $82 million in one year in Sarawak. If each of the other 100 categories, including rattan, fruit, and nuts, had an annual value of but $2 million, the total annual value of non-wood products would be $182 million. This plausible figure is too large to overlook in forest policy design and implementation, which has been geared solely to maximizing the value of wood products.

Non-forestry policies

Forest-based industrialization policies. Through 1986, Sarawak policies toward forest-based industrialization remained much less ambitious than those of the rest of Malaysia; neither log export quotas nor export prohibitions (à la Indonesia) have been imposed.

Sarawak's export tax policy does closely resemble those of Indonesia and Sabah. Sarawak increased the *ad valorem* export tax on logs from 5 percent to 10 percent in 1980, but as in Indonesia the tax does not apply to sawn timber, plywood, or moulding. We have seen that this type of export tax structure strongly protects value-added in wood processing. The Sarawak export tax on logs, at 10 percent, is only half that of Indonesia; thus, effective protection to processing in Sarawak is likely considerably lower.

Consider the effects of the tax on protection for a plywood mill using dipterocarp veneer logs from the hill forests of Sarawak. At an f.o.b. price of US$100 per cubic meter the government gives up US$10 in tax for every log exported in plywood form. If, as in Indonesia, additional value-added *per log* used in plywood is around 10 percent, then additional value-added allowed by the tax is equivalent to the taxes foregone by the government. Therefore, the effective protection to value-added in plywood is 100 percent.

While this incentive for forest-based industrialization is high, it is well below that provided by the tax and royalty structures in Sabah and Indonesia. Therefore, incentives for more rapid forest depletion would be correspondingly lower in Sarawak.[14]

Infrastructure policy. To date, there is no evidence that government policies toward infrastructure have directly led to substantial deforestation.

The very large (US$3.3 billion) hydrocomplexes planned for the Rajang river valley could have important future effects on the forest, for two reasons. First, vast areas would be submerged, including an indeterminate amount of forest land. Second, flooding large land areas will reportedly require resettlement of 83 percent of the population above the upper Rajang catchment area (Clad 1985: 42). To the extent that poor families are resettled into forest areas, additional shifting cultivation may cause further deforestation.

Peninsular Malaysia

Forest endowments, forest utilization [15]

Peninsular (West) Malaysia, with a land area nearly twice as large as Sabah and slightly larger than Sarawak, contains 81 percent of Malaysia's entire population. Whereas more than 90 percent of East Malay-

Table 3.25. *Land use, Peninsular Malaysia, 1910–80 (thousand hectares)*

	1910	1930	1950	1980
A. Forest land, wood land	10,169	8,794	8,252	7,167
B. Arable land	490	1,836	2,126	3,663
1. Permanent crops	246	1,528	1,703	2,731
2. Other arable	244	308	423	932
C. Wetlands	1,585	1,353	1,303	1,060
D. Other land	924	1,185	1,487	1,278

Total area: 13,168,109 ha

Source: Reber and Richards (1985: Appendix Table 1).

sian timber production is exported in the form of logs, virtually all of West Malaysia's timber exports have been in processed form since 1978.

Peninsular Malaysia deforestation has proceeded at a rapid rate since the turn of the century. Most of the factors in deforestation in East Malaysia and in Indonesia are also present in Peninsular Malaysia (hereafter, West Malaysia), but the order of importance is different. Historically, the leading cause of deforestation in West Malaysia has been conversion of natural forest areas to permanent agriculture. Logging was more significant in West Malaysia's deforestation from 1955 to 1980 than in Sabah, Sarawak, or Indonesia, largely because the latter two were late entrants (circa 1969) into extensive logging. Deforestation due to shifting cultivation has been much less significant in West Malaysia than in Sabah, Sarawak, or Indonesia (FAO 1981: 290).[16] As late as 1970, wood harvests for purposes other than timber (fuelwood, etc.) were 15 percent of total production, but forest depletion for fuelwood has declined steadily in the past 15 years.

Table 3.25 summarizes land use patterns in West Malaysia since the turn of the century. Fully 30 percent of the 1910 forest area had disappeared by 1980. Even from 1950 to 1980, the natural forest estate shrank by 13 percent. Still, by 1980 roughly 7.5 million hectares remained under forest cover or in wetlands, or about 57 percent of the land area. Views differ as to the size and composition of the forest estate (Table 3.26). Reserve forest areas, those set aside for commercial exploitation over the long term, are among disputed areas. Agricultural conversion has been even more striking than shrinkage of the overall forest estate (Line B.1 of Table 3.25). By 1980, the land area under permanent crops was 11 times higher than in 1910. From 1950 to 1980, areas under permanent crops increased by 60 percent.

Table 3.26. *Peninsular Malaysia: three estimates of size and composition of the forest estate, 1953–84 (million hectares)*

Year	Total forest area			Reserve forest total			Productive forest			Forest area outside reserve		
	I	II	III	I	II	III	I	II	III	I	II	III
1953	9.7			3.2			2.1			6.5		
1960	9.5			3.5			2.3			6.0		
1965	8.6			3.5			2.3			5.1		
1968	8.1			3.4			2.2			4.7		
1970	8.0			3.3			2.3			4.7		
1976	7.2			2.9			2.3			4.3		
1980		7.5			5.3						2.2	
1984			7.2			3.0			2.1			4.2

Sources: Estimate I, 1953–76, Reber and Richards (1985: Table 5); Estimate II, 1980, FAO (1981); Estimate III, 1984, Malaysian Forestry Department (1986: 54–56, Appendix Tables 1 and 3).

As early as 1977, government officials openly predicted exhaustion of commercial natural forests before 1990, should mid-1970s logging and agro-conversion rates continue (Ministry of Finance 1977).

Deforestation in West Malaysia has claimed about 90,000 hectares per year since 1976 (FAO 1981: 290), about equal to that for Sarawak in recent years. Agricultural conversion has been estimated to account for about 90 percent of this, and mining, hydroelectricity, and highway construction for the remainder.[17] Agro-conversion has primarily been in rubber, oil palm, and sugar cane. Logging was not mentioned as a cause of deforestation in the late 1960s in an authoritative 1981 FAO report (FAO 1981: 289). However, annual log harvests over the period 1967 to 1977 were estimated to involve 207,000 hectares (Rowley 1977: 47).

As in Sabah and Sarawak, several state governments in Peninsular Malaysia hold property rights in the national forest. The most important timber-bearing states in West Malaysia have been Pahang (35 percent of forest area), Johore, Trengganu, Kelantan, Kedah, and Perak. Even Perlis state (with 21,000 hectares of forest reserves) has experienced rapid forest conversion since 1970.

Benefits

Where the forest has been utilized primarily as a wood products source, it may be acceptable to focus on the logging and timber products industry in assessing forest exploitation benefits. In West Malaysia, however, agro-conversion has been responsible for at least as much forest area

Table 3.27. *Peninsular Malaysia: wood production, 1953–84 (million cubic meters)*

Year	(1) Timber output (logs)	(2) Other wood harvest (fuel, poles)	(3) Total wood production from forest (Col. 1 + Col. 2)
1953	1.289	0.551	1.840
1960	2.248	0.567	2.815
1965	3.223	0.573	3.796
1968	5.075	0.707	5.782
1970	6.541	1.199	7.740
1972	8.925	n.a.	n.a.
1974	8.629	n.a.	n.a.
1976	9.954	n.a.	n.a.
1978	9.767	n.a.	n.a.
1984	9.182	1.737	10.919

Sources: For 1953–70, Reber and Richards (1985), Table 5; data given in cubic feet, converted to cubic meters (1 ft^3 = 0.02832 m^3). For 1972–78 and 1984, Malaysian Forestry Department (1986: Table 6).

activity as has logging. Thus, discussion of forest exploitation benefits must consider both the wood products industry and the agricultural sector (especially rubber, oil palm, and sugar).

Wood products. Tropical hardwood harvests expanded rapidly in West Malaysia from 1953 through 1976; in 1978 output declined for the first time in nearly 25 years (Table 3.27). The region's production peaked in 1976, when log harvests alone reached nearly 10 million cubic meters, similar to Sabah's harvest. By the mid-1970s, the forest-based sector accounted for 2 to 3 percent of GDP (Alexandratos, et al. 1982: 50).

By 1971 timber exports had reached US$122 million. Over the next seven years, West Malaysian export earnings from wood and wood products tripled, to US$369 million (Table 3.28), about 7 percent of total regional exports. By that time, employment in the forest-based sector had reached about 43,500 jobs, rising to 67,183 persons by 1984, including 17,000 in logging, 35,000 in sawmills and about 16,000 in plywood and veneer mills (Malaysian Forestry Department 1986: Table 13). In sharp contrast to East Malaysia, however, timber sector employment was but 0.1 percent of the labor force in 1984.

Agro-conversions. Seventy-five years of rapid forest conversion to permanent tree crops made Peninsular Malaysia the world's leading

Table 3.28. *Peninsular Malaysia: value of log and wood product exports, 1971–84 (million U.S. dollars)[a]*

Year	Logs	Wood products: sawn timber, plywood, chips, etc.	Total export value
1971	42.1	79.6	121.7
1972	41.4	114.6	155.7
1973	23.4	233.8	257.2
1974	24.5	202.7	227.2
1975	17.2	178.2	195.4
1976	16.4	397.2	413.6
1977	10.9	343.3	354.2
1978	5.6	363.7	369.3
1981	9.4	554.0	563.4
1982	0.9	512.9	513.8
1983	4.7	543.2	547.9
1984	2.2	446.3	448.5

[a]Figures originally in Malaysian dollars, converted to U.S. dollars at following exchange rates: 1971–73, M$2.4 = US$1; 1974–76, M$2.2 = US$1; 1976–77, M$2.15 = US$1; 1978, M$2.18 = US$1; 1979–82, M$2.28 = US$1; 1981–84, see Table 3.3.

Sources: 1971–78: Malaysian Timber Board (communications); 1981–84: Table 3.3.

exporter of both natural rubber and oil palm by 1970. From 1970 to 1980, the areas devoted to permanent tree crops increased by 28 percent. The gross benefits from agro-conversion have undoubtedly been substantial but have not been systematically quantified. Moreover, the forest conversion over the last eight decades does not represent an environmental disaster. Perennial agricultural tree crops do provide effective soil protection and catchment area cover.

Extensive forest conversion has undoubtedly been a major factor in bringing Malaysia's per capita income to US$2,000 in 1985 (World Bank 1987: Appendix Table 1). The incidence of rural poverty is appreciably lower in Peninsular Malaysia than in East Malaysia (Segal 1983: 54), perhaps because of the long history of agro-conversions in the former. Notably, shifting cultivation spawned by rural poverty is the principal source of deforestation in East Malaysia and in Indonesia. Therefore, serious evaluation of the effects of agro-conversion on the Peninsular Malaysia tropical forest estate requires determination of how much forest loss *would* have occurred from shifting cultivation without agro-conversions, an issue not yet addressed.

With appropriate fertilizer application, converting forest land to tree crops may indeed be a more productive use than logging it perpetually, especially given current logging practices. The clearest net loss from rapid forest conversion in Peninsular Malaysia has been not in the forest's productive value but in its protective value and the irreversible loss of the value of the forest's non-wood products. Tree crop cover may indeed furnish effective soil protection and water catchment, but does not duplicate the natural forest's full range of protective services. They too have economic value, but cannot easily be commercially appropriated.

Policies

Forestry policies: the domain of the states. We stressed earlier that one may not meaningfully speak of *Malaysia* tropical forest policy. The autonomy of the 14 states in matters relating to forest utilization is as complete as possible within a federal state. Political forces outside of a state's boundaries, no matter how powerful they may be in Kuala Lumpur, have virtually no influence on state land use policies. Indeed, power over land disposition represents "the only real power of the states in the federal system" (Das 1977: 55; Callahan and Buckman 1981: 31–36), and because for many states timber is the major source of budget revenue, this authority is jealously guarded by state governments.

We have seen that one may loosely speak of "forestry policy" in each of the two separate Eastern Malaysian states of Sabah and Sarawak. While authority over forest use may be fragmented in both states, there are nonetheless identifiable policies associated with each state. "Forestry policy" in Peninsular Malaysia, however, defies description, much less systematic analysis. Each of the 12 Peninsular states imposes and collects its own forest fees and taxes, and at different rates. Only fragmentary data exist on actual timber tax collections in each state.[18] Each state awards concession licenses and has authority to determine concession size and duration as well as harvesting methods. States have separate reforestation regulations and taxes, although two states (Pahang and Trengganu) have imposed forest regeneration standards compatible with those of the National Forestry Council,[19] a strictly advisory body. Finally, state governments have ultimate authority over agro-conversion and the creation and size of both state and national parks.

In the past decade efforts have been made to harmonize forestry policies on Peninsular Malaysia. The National Forestry Council (NFC) proposed a Uniform National Forestry Policy in 1977 to apply to West Malaysia. The National Land Council approved this in April of 1978.[20] Important policy elements include:

- Designation of five million hectares of land in Peninsular Malaysia as the permanent forest estate (3.3 million hectares were considered productive forest, 62 percent of which was already logged and 2 million hectares of which was virgin forest).
- Imposition of maximum annual forest removals of 76,000 hectares of new forest (removals totalled 65,000 hectares in 1984).
- Proposed limitation of agro-conversion to 68,000 hectares for 1976–1980.
- Uniformity across states in granting of licenses.
- Uniformity in allowable harvesting methods.
- Uniformity in forest regeneration policies.

Foresters, environmentalists, and forest ecologists widely supported the uniform forestry. It nevertheless remains true that individual states in West Malaysia (not to mention East Malaysia) still exercise legal authority over forest use. Individual states decide which land is to be designated as permanent forests. Enforcement of all other provisions of the Uniform Law also remain the prerogative of individual states.

Separate forest tax and fee structures in the Peninsular states, together with the lack of published data on government revenues and timber exports by state, render it impossible to estimate government rent capture. Accordingly no estimate was attempted.

General agricultural policies. By 1979, about 20 percent of the Peninsular Malaysia land area was under agriculture, versus less than 5 percent in both Sabah and Sarawak. From 1970 to 1980, an average of 200,000 acres of forest land were cleared annually for crop cultivation and settlements mostly in Peninsular Malaysia (Ayub 1979: 348–353). No fewer than three government land development agencies continue to seek land for agricultural development; the pace of agro-conversion is unlikely to slow in the 1980s.

Whether benefits from the policy of conversion, in the form of poverty eradication and employment, will exceed the economic and environmental costs of forest clearing cannot be addressed here. It is known that not all agro-conversion actually results in new cultivation. According to the Secretary General of the Agriculture Ministry, 810,000 hectares (27 percent of the area of the logged-over forest) had been cleared but were not under cultivation by 1979 (Ayub 1979: 350). Some conversion clearly involves minimal return, as was true historically.

Industrialization policies. Peninsular Malaysia first established policies restricting log exports in 1972. An export ban was applied to 11 of the hundreds of species exported and later extended to 16 of the most important species (Takeuchi 1982: 54). By 1978, the ban was virtually complete: only very small volumes of logs of secondary species were exported (FAO 1981: 277).

As in Indonesia, and to a lesser extent Sabah, log export restrictions were intended to promote forest-based industrialization. In terms of new sawmills, plymills, and veneer mills, the policy was a distinct success. By 1984, nearly 600 sawmills and 40 plywood/veneer mills were in operation. However, the net effect of policy-induced forest-based industrialization is clearly to increase stress on the forest estate in Peninsular Malaysia, as it did in Indonesia and Sabah, with some special twists for West Malaysia. First, processing in Peninsular mills is characterized by lower recovery rates than in, say, Japan or in Europe (Takeuchi 1982: 68).[21] More forest acreage must be cut to obtain a given timber volume when timber is processed in Malaysia. Second, the log export ban created substantial milling capacity, and thus a strong constituency for continuing high domestic log harvests. As in Indonesia, policy-makers will not likely stand by and allow bankruptcy of mills in periods of slack demand or low world prices for wood products. However, Malaysia's federal government has fewer levers available than Indonesia to force a continuing supply of East Malaysian logs to West Malaysian mills given its minimal influence in forest utilization. Therefore, future depletion on this account may not be as significant as is likely for Indonesia.

Federal policies affecting forest utilization in all Malaysia. The only significant influence (aside from moral suasion and political favors) exerted by the federal government over forest utilization arises from its control of exchange rate policies, and even here, any effects on forest utilization were largely unintended and were not helpful. Malaysia's currency, the Malaysian dollar (or ringget), has been consistently *undervalued* since the mid-1960s (Gillis and Dapice, in press). The Malaysian dollar was clearly undervalued during the late 1970s and early 1980s when national timber production peaked.

Whereas the Indonesian rupiah's overvaluation effectively reduced log export incentives, and therefore log production, undervaluation of the Malaysian ringget had the opposite effect. Not only in Sabah but in Sarawak and Peninsular Malaysia, log exporters earned more local currency per dollar of export value due to currency undervaluation.

Although short- and medium-term supply elasticities for logs are not likely to have been higher in Malaysia than in Indonesia,[22] Malaysia's undervaluation persisted for an extended period. If investors in timber exports thereby viewed undervaluation as a persistent theme of Malaysian policy, exchange rate policy may have significantly increased the deforestation rate. In addition, the ringget's undervaluation made imported wood products artificially expensive, thus encouraging excess consumption of wood products made locally from Malaysian logs.

Endnotes

1. 1 cubic meter = 27.73 Hoppus feet.

2. The *ad valorem* rate is 60 percent when the f.o.b. value is less than M$221.84 per m^3 (US$91.00 at the mid-1985 exchange rate).

3. At 1985 exchange rates, presumptive logging costs are US$26.10 per m^3.

4. The original grant to the Sabah Foundation was 100 years. Exploitation was restricted, in theory, to 1 percent of the grant per year. In 1981, however, loggers were allowed to exploit up to 1.25 percent of the total grant area per year.

5. Poison girdling is undertaken only in lowland rain forests, not in hill forests.

6. A drought in 1972 was, however, nearly as severe as that of 1983, but in 1972 the logged-over area was relatively small.

7. This ground litter consisted of a large volume of dead wood from stumps, crowns, and limbs of harvested trees, as well as trunks of unharvested trees toppled during felling, bucking, skidding, and transportation.

8. For a discussion of the meaning of "effective protection" see Chapter 2, Appendix B.

9. Based on forecast production of 12 million logs per year, a conversion rate of 50 percent in milling, and five employees per 1,000 m^3 of annual sawnwood production. Figures on employment implications of sawmills are based on Takeuchi (1982: 79).

10. In the mid-1970s, 1988 was the date mentioned for expiration of most large foreign concessions (Rowley 1977: 59).

11. These two observers concluded in 1981 that "Sarawak does not have a shifting cultivation problem."

12. Only 1.3 million hectares were actually under various cycles of shifting cultivation. The remainder (0.95 million hectares) was in exhausted soils.

13. The Sarawak estimates in both cases refer to trees of less than six feet in girth. Damage to trees larger than six-foot girth was even smaller.

14. By 1983, two and a half years after the doubling of the tax, exports of processed wood products had risen to 13 percent of export value of logs, a substantial increase over pre-1980 levels, and in fact above the share of processed products in Sabah's wood exports in 1983.

15. Work on this section of this paper was greatly facilitated by a recent paper by two Duke colleagues on Malaysian land use since 1900. See Reber and Richards (1985).

16. "Shifting cultivation, fuelwood collection and burning have had much less effect on the permanent forest estate in Peninsular Malaysia than in East Malaysia" (FAO 1981: 290).

17. Projected official estimates of deforestation do not include the 350-square-mile swath of land scheduled for clearing along the Penang–Kota Baru road, to be cleared for security reasons.

18. According to one source (Rowley 1977: 47), Pahang and Trengganu states, regarded as prolific timber states by Peninsular Malaysia standards, collected US$17 million and US$6 million in timber charges in 1977. Together, this was less than 10 percent of the total for Sabah the same year.

19. The National Forestry Council consists of the chief ministers of West Malaysian states plus the Deputy Prime Minister of the federal government.

20. Provisions of the Uniform National Forest Policy are summarized in detail in *The Malaysian Forester* 1980: 1–6.

21. One reason for lower recovery rates in Malaysian plywood than in Japanese plywood is that Malaysian plants produce mainly 4 × 8 sheets, while Japanese plants produce plywood primarily in 3 × 6 sheets. Wood recovery in Malaysia is 45–50 percent, and in Japan it is 67 percent. Still, recovery rates for 4 × 8 plywood in Japan would be 8–10 percent higher than in Malaysia.

22. See Chapter 2 of this volume.

References

Alexandratos, N., A. Condos, and P. Wardle. 1982. A Model for the Evaluation of Forest Sector Development in Peninsular Malaysia. In *Agricultural Sector Analysis,* ed. E. Thornbecks. Rome: FAO.

Ayub, Arshad. 1979. National Agricultural Policy and its Implications on Forest Development. *The Malaysian Forester,* Vol. 42, No. 4.

Asian Timber. 1984. Logging Still Dominates Timber Industry in Sarawak. July/August.

Asiaweek. 1984. Fire in the Earth. July 13.

Berthelsen, John. 1985. Sabah Chief Inherits Troubled Economy. *Wall Street Journal,* June 5.

Callahan, R. Z., and R. E. Buckman. 1981. *Some Perspectives of Forestry in the Philippines, Indonesia, Malaysia, and Thailand.* Washington, D.C.: U.S. Department of Agriculture, Forest Service, December.

Chong, P. W. 1980. Perspectives of LDCs on the Tropical Hardwood Sector. Paper presented to FAO/CTC Pacific Regional Workshop, Pattaya, Thailand.

Clad, James. 1985. Dayaks Want Their Share of Riches Now. *Far Eastern Economic Review,* May 30.

Das, K. 1977. Pressure on Park Loggers. *Far Eastern Economic Review,* December 2.

Davies, Derek, and George Lauriat. 1980. Spicing Up Sabah's Recipe. *Far Eastern Economic Review,* July 18.

Dornbusch, Rudiger. In press. Overvaluation and Trade Deficits. Chapter 2 in *A Policy Manual for the OPEN Economy,* ed. Rudiger Dornbusch.

Draft Sabah Regional Planning Study. 1979. Kota Kinabalu, Malaysia.

FAO. 1979. The Effects of Logging and Treatment on the Mixed Dipterocarp Forests of South-East Asia. FAO/MIS/79/8. Rome: FAO.

FAO. 1981. Tropical Forest Assessment Project. *Forest Resource of Tropical Asia.* Rome: FAO.

Fox, J.E.D. 1968. Logging Damage and the Influence of Climber Cutting Prior to Logging in the Lowland Dipterocarp Forests of Sabah. *Malaysian Forester,* Vol. 31, No. 4.

Gillis, Malcolm. 1980. Fiscal and Financial Issues in Tropical Hardwood Concessions. Development Discussion Paper No. 110. Cambridge, Mass.: Harvard Institute for International Development, December.

Gillis, Malcolm. 1981. Foreign Investment in the Forest-Based Sector of the Asia-Pacific Region. Report prepared for FAO. Rome: FAO, July.

Gillis, Malcolm, and David Dapice. In press. External Adjustments and Growth: Indonesia Since 1965. Chapter 6 in *A Policy Manual for the OPEN Economy,* ed. Rudiger Dornbusch.

Gray, John W. 1983. *Forest Revenue Systems in Developing Countries.* Rome: FAO.

Hepburn, A. 1982. The Possibility for Sustained Yield Management of Natural Forest in Sabah With Reference to the Sabah Foundation. *Malaysian Forester,* Vol. 42, No. 1.

Jacobs, M. 1982. The Study of Minor Forest Products. *Flora Malesiana Bulletin,* No. 35.

Johnson, Brian. 1984. Lumbering in Forest Ecosystems: Easing Pressures of the Tropical Timber Trade on Forest Lands. IIED Policy Paper No. 4. London: International Institute for Environment and Development, February 1.

Lee, H. S. 1982. The Development of Silvicultural Systems in the Hill Forests of Malaysia. *Malaysian Forester,* Vol. 45, No. 1.

Leighton, Mark. 1984. The El Niño-Southern Oscillation Event in Southeast Asia: Effects of Drought and Fire in Tropical Forest in Eastern Borneo. Unpublished. Harvard University, Department of Anthropology.

Mackie, Cynthia. 1984. The Lessons Behind East Kalimantan's Forest Fires. *Borneo Research Bulletin,* Vol. 16, No. 2.

The Malaysian Forester. 1980. Vol. 43, No. 1.

Malaysian Forestry Department. 1986. *Annual Report on Forestry in Peninsular Malaysia 1984.* Kuala Lumpur: Kementarian Perusahaan Utama.

Ministry of Finance. 1977. *Annual Report.* Kuala Lumpur: Government of Malaysia.

Nor, Salleh Mohd. 1983. Forestry in Malaysia. *Journal of Forestry,* March.

Reber, A. Lindsey, and John F. Richards. 1985. Land Use Changes in West Malaysia 1900–1980. Paper presented at the Southeast Asian Studies Institute, Ann Arbor, Michigan, August 1–3.

Ross, M. S. 1984. *Forestry in Land Use Policy for Indonesia.* Ph.D. thesis, University of Oxford.

Rowley, Anthony. 1977. Expanding a Market in Malaysia. *Far Eastern Economic Review,* Vol. 96, No. 15: 46–47.

Sabah Forest Service. 1979. *Angarram Hasil dan Perbelanjuan Bagi Tahun.* Kota Kinabalu: Division of Statistics, Department of Forestry, Government of Sabah.

Sarawak Forest Service. Various years. *Angarram Hasil dan Perbelanjuan Bagi Tahun.* Bintulu: Division of Statistics, Department of Forestry, Government of Sarawak.

Sarawak Study Group. 1986. *Logging in Sarawak: The Belaga Experience.* Petaling Jaya: Institute for Social Analysis.

Segal, Jeffrey. 1983. A Fragile Prosperity. *Far Eastern Economic Review*, April 14.

Seng, Lee Hua. 1982. Silvicultural Management in Sarawak. *Malaysian Forester*, Vol. 45, No. 4.

Simon, Julian, and Herman Kahn, eds. 1984. *The Resourceful Earth*. New York: Basil Blackwell.

Takeuchi, Kenji. 1982. *Mechanical Processing of Tropical Hardwoods in Developing Countries: The Asia Pacific Region*. World Bank Working Paper No. 1922–1. Washington D.C.: World Bank, January.

Thang, H. C. 1984. Timber Supply and Domestic Demand in Peninsular Malaysia. Kuala Lumpur: Forestry Department, July.

Thornbecks, E., ed. 1982. *Agricultural Sector Analysis*. Rome: FAO.

Tiing, Lau Buong. 1979. The Effects of Shifting Cultivation on Sustained Yield Management in Sarawak's National Forests. *The Malaysian Forester*, Vol. 42, No. 4.

U.S. Department of State. 1985. Malaysian, Sabah Forest Products Report. Kuala Lumpur: U.S. Embassy, September 12.

Whitmore, T. C. 1984. *Tropical Rain Forests of the Far East*. Oxford: Clarendon Press.

Whitten, Anthony J. 1986. Indonesia's Transmigration Program With Special Reference to Sumatra. Paper presented at Environmental Studies Association Workshop on Role of Applied Ecology in Economic Development, Dalhousie University, Halifax, Nova Scotia, August.

World Bank. 1985. *World Development Report, 1985*. Washington, D.C.: World Bank.

World Bank. 1987. *World Development Report, 1987*. Washington, D.C.: World Bank.

4   Incentive policies and forest use in the Philippines

EUFRESINA L. BOADO

History of forest use

History of deforestation

At the start of the century, the Philippine archipelago was covered with rich dipterocarp forests. Today, forests have disappeared in many places, and those that remain are concentrated on a few islands: Mindanao, Palawan, Samar, and pockets of eastern Luzon. In 1982, the government reported that 16.6 million hectares, or 55 percent of the country's land area, were forest lands,[1] of which 11.2 million hectares were forested. About 9 percent of the country, or 2.7 million hectares, was said to be virgin forests. Yet the rate at which virgin forests declined is astounding. From 1971 to 1980, they decreased by 1.7 million hectares, with 1.1 million converted permanently to nonforest uses (Reyes 1983). Table 4.1 summarizes the status of the forest area of the Philippines.

However, these government statistics have been disputed in recent years. One study (Revilla 1984), using Landsat photos, estimated forested lands at only 8.9 million hectares, 0.6 million hectares of them in alienable and disposable lands. In another study (Revilla 1984) Landsat photos for 1976 suggested that only about 8.5 to 9 million hectares of forest lands were forested in 1976, and that the total had been reduced to 7.8 to 8.3 million hectares by 1983. This study estimated there were 2 to 2.5 million fewer hectares of virgin forest than the government's figure of 2.7 million hectares. These discrepancies may be corrected with the completion of the second national forest inventory, begun in 1981. Meanwhile, the lower figures seem more credible.

Deforestation in the Philippines began to accelerate in the post-independence era following the Second World War. From the late 1950s through 1973, the annual rate of deforestation ran as high as 172,000 hectares. In 1960 the government claimed that more than 5 million hectares needed rehabilitation or reforestation. Over 1 million hectares of

Table 4.1. *Status of forests in the Philippines, 1982 (thousand hectares)*

Forest type	Total	Luzon	Visayas	Mindanao
Dipterocarp	8,610	2,805	1,799	4,006
Virgin	2,625	682	701	1,233
Second growth	5,984	2,123	1,098	2,773
Mangrove	211	17	87	107
Virgin	10	—	5	5
Second growth	201	17	82	102
Pine forests	189	189	—	—
Virgin	151	151	—	—
Second growth	38	38	—	—
Mossy forests	1,726	910	471	345
Man-made plantations	468	257	119	92
Total	11,204	4,178	2,476	4,550

Source: Ministry of Forestry (1982).

these areas were in critical watersheds.[2] More recent estimates of degraded forest land area are 8 to 10 million hectares.

The loss of forests was due partly to the deliberate conversion of forest lands through the land classification system instituted by the Forestry Bureau in 1919. Under this system, 42 percent of the country's area, mainly land with slopes above 18 percent, was to be classified and reserved as permanent forests, and the rest classified as alienable and disposable lands for agriculture and other uses. In 1982, 5.6 million hectares of the 16.6 million hectares of forest lands remained to be classified.[3] The slow classification process allowed forest lands, even those above 18 percent in slope, to be treated as nonforest lands. People often enter forests ahead of land classification teams and cultivate both logged-over areas and virgin lands. Devoid of vegetation, these areas become easier to release as alienable and disposable lands.

The major persistent causes of forest loss are destructive logging and shifting cultivation. In the absence of effective government supervision of logging operations, many forest areas have been destroyed. Destructive logging and the wasteful practices of "kaingineros"[4] are mainly responsible for the millions of hectares of forests now out of production. An estimated 80,000 to 120,000 kaenginero families have cleared or destroyed about 2.3 million hectares of forest land. The recent influx of kaingineros into forest lands is due largely to the government's inability to provide economic alternatives for rural populations. In the earlier periods, extensive squatting in forest lands was due more to land speculation.

The role of social and cultural factors in deforestation cannot be ignored. Most Filipinos lack conservation ethics. The people generally avoid responsibility for conserving anything public. Moreover, in early years, forests were considered a hindrance to agricultural development. In fact, government efforts to develop agriculture concentrated on land expansion through forest conversion. While some of these converted areas became productive croplands, many were left with degraded soils, unsuitable for sustained agriculture. The result has been increased erosion and siltation of watersheds and a growing shortage of forest products. Yet the most significant social factor is the rapid rise in population, which exerts heavy pressure on a finite area.

Equally significant are political and administrative factors. There is still no sustained political support for conserving forests. For example, two or three years after forest conservation programs are launched, the logistical support usually dries up until the programs are supplanted by other politically attractive programs. Forestry administration is also afflicted with the usual woes of government agencies in developing countries, including the lack of adequately trained staff, poor logistics, corruption, and low employee morale.

After World War II, logging proceeded at phenomenal rates. In a few years, the country resumed exporting logs and some processed products. Timber harvests reached 5 million cubic meters per year in the late 1950s. The fledgling government, in need of resources to spur economic development, saw enormous potential in further forest exploitation as a way to raise revenues.

In the 1960s, authorized timber harvesting increased further as the industry mechanized, the number of licenses increased, and the market for Philippine logs proved lucrative. The timber boom continued until the first half of the 1970s. In 1971, over 10 million hectares of forest lands were licensed for harvest, double the 1960 figure. Concession after concession was awarded to multinational corporations and to local rich minority and political powers who became entrepreneurs. Subsidiaries of large U.S forest products companies, such as Weyerhauser, Georgia Pacific, Boise Cascade, and International Paper were major concessionaires. Both the traditional landed elites and those who came recently to power in association with ex-President Marcos or the military acquired substantial logging interests (Porter 1987). Timber production almost tripled from 1955 to 1968, rising from 3.8 million to 11 million cubic meters, and was sustained through 1974. From 1975 on, production began to decline, and in 1984 it had dropped to the 1955 level. Table 4.2 shows the trend of authorized forest exploitation.

Forest product exports followed the production trend. Logs dominated

Table 4.2. *Actual production of timber, 1958–84 (thousand cubic meters)*

Year	Production	Year	Production
FY 1958–59	5,452	FY 1971–72	8,416
FY 1959–60	6,315	FY 1972–73	10,446
FY 1960–61	6,596	FY 1973–74	10,190
FY 1961–62	6,772	FY 1974–75	7,332
FY 1962–63	7,668	FY 1975–76	8,546
FY 1963–64	6,536	CY 1977	7,874
FY 1964–65	6,175	CY 1978	7,169
FY 1965–66	8,047	CY 1979	6,578
FY 1966–67	7,843	CY 1980	6,352
FY 1967–68	11,114	CY 1981	4,514
FY 1968–69	11,584	CY 1982	5,400
FY 1969–70	11,005	CY 1983	4,468
FY 1970–71	10,680	CY 1984	3,849

exports. Between 1969 and 1971, the volume of log exports averaged 8.6 million cubic meters a year. This volume, about 77 percent of the total production, was about a sevenfold increase from the 1955 level. Since 1974, however, the average has dropped to only 57 percent of that year's 10.2 million cubic meter production and continued to decline. In 1984, only 22 percent of the year's 3.8 million cubic meter production was exported. Japan was the most important market for Philippine logs.

The Philippines is a net exporter of forest products, and forestry has long been an important part of its export economy. In 1968, forestry contributed 32.5 percent of total foreign exchange earnings. This rate dropped steadily, however, and in 1974 forestry accounted for only 10 percent of the total. A top dollar earner for many years, forestry products now rank behind nontraditional manufactures and mineral, coconut, and sugar products. The value of forest products exports has been generally stagnant at an average of $300 million per year at current prices, except for a few years when values topped $400 million. The growth of the forestry sector's exports lags behind that of other sectors in both volume and value. This poor performance is due to shrinking log exports and the fairly stagnant wood products industry (lumber, plywood, and veneer).

In addition to their contribution to foreign exchange, forests are a source of revenue to government through taxation and rents. There is consensus, however, that the government has not collected an adequate share of the resource value of timber harvested (Puyat 1972). Charges assessed on volumes harvested, the major form of forest taxation until the 1970s, were set at very low rates. Revenues increased in 1970, when export and domestic sales taxes were imposed on forest products. Suspension of the export tax on logs between 1974 and 1978 reduced government revenues. Its reimposition in 1979, together with the increased rates of consolidated forest charges in 1981, created a significant increase in collection, although it was still not sufficient even to cover the government operating budget for supervising selective logging, scaling, timber stand improvement, forest protection, and reforestation.

The authorized cut by timber licensees only partially accounted for the total timber drain. According to the first forest inventory, using aerial photographs from 1957 to 1963, the annual timber drain in Mindanao alone amounts to 20.1 million cubic meters, of which authorized logging accounted for only about one-fourth. The industry has argued that even if there had been no legitimate logging, three-fourths of the timber drain would have occurred. This is not entirely true, as destructive logging, overcutting, and high-grading characterize their harvesting. High-grading occurs if forest charges are based on the volume of timber removed, not on the volume of merchantable timber in the tract. This encourages licensees to take only the most valuable stems, so that a larger area must be harvested to meet timber demand, and the remaining trees are usually damaged during logging. Besides, as logging roads open forests, other destruction, especially shifting cultivation, is encouraged.

Selective logging – to safeguard future harvests and provide forest cover for conservation of soil and water – has been prescribed since 1954. The system involves three phases: tree marking, inventory of remaining trees, and timber stand improvement. More than 30 years of selective logging show disappointing results, however, and many areas have been overcut, high-graded, and left with heavily damaged residuals. Poor implementation of rules and regulations, especially the tree marking and residual inventory phases, has contributed to the overcutting of dipterocarp forests. Tree markers, who determine the trees to be cut and those to be left, are often inexperienced and easily fall prey to bribes. In many cases, poor supervision of company tree markers has abetted high-grading. The residual inventory phase is often not conducted because personnel and funds are lacking, and, when it is correctly conducted, the resulting fines imposed for damaged residuals are too low to deter wasteful logging. Studies show that for every cubic meter of tree cut, about one cubic meter is wasted.

Logging has been wasteful and destructive. During the timber boom, from the late 1950s to the early 1970s, the harvest was far above sustainable limits. Furthermore, because logging was inefficient, more of the forest was exploited than was warranted. Improper logging practices such as high stumps, improper bucking, and unusually high skidding and loading losses caused excessive waste, averaging 30 percent and ranging as high as 70 percent of the total timber stock (Sy and Lasmarias 1983). Logs that could not be hauled were often abandoned to rot or to be carried downstream during floods.

After sustaining great damage during logging, logged-over forests were usually left unmanaged. Timber stand improvements, intended to ensure the quality and volume of future growing stock, were often not executed at all, or only in meager show-window areas. The result is second-growth forest with unreliable yields containing more noncommercial and weed species than crop trees. Stand deterioration was exacerbated by inadequate protection from many destructive elements. Most kaingins were carried out in logged-over forests, destroying the already damaged trees. In some cases, uncontrolled kaingin fires encroach into virgin forests. The government response – increasing the number of forest guards from 800 in 1976 to 3,000 in 1980 – hardly improved the situation. The number of kaingineros remained high.

In addition to kaingin-making, timber smuggling is a potent threat to the future of second-growth forests. Despite the creation of a law enforcement group in the forestry agency, deputization of barangay captains (a barangay is the smallest political unit in the Philippines) as forest officers, and help from the military in apprehending timber smugglers, smuggling was unabated. In fact, local politicians, military officers, and forestry officials have frequently been accused of complicity in illegal timber operations (Porter 1987). All these problems make second harvests highly uncertain.

Serious environmental problems have long been observed, especially in the Ilocos, Central, and Bicol regions of Luzon, and on Cebu, Bohol, and some parts of Mindanao. On the island of Cebu, where 3 million people live, critical shortages of potable water have become commonplace. Devastating floods, frequent in many parts of the country, result in loss of lives, property, and livestock and destruction of important infrastructure. In recent years, some parts of Luzon and Mindanao have suffered more severe droughts that have hampered the cultivation of prime crops, worsening the already volatile situation of many poverty-stricken families. Minor dust storms and the advancing sand dunes in the Ilocos region have also been attributed to loss of forest cover.

The loss of forest cover has also caused heavy soil erosion, which not only decreases the inherent productivity of the uplands, but also increases the sediment loads of rivers, creeks, and streams, with serious consequences to hydroelectric reservoirs, irrigation facilities, and other infrastructure. In Luzon, the three most important reservoirs have lost an estimated half of their lifespans to excessive sediment.

Many rivers and streams have either dried up completely or become irregular in their flows. During the rainy season, the Agno River in Luzon overflows its banks onto croplands, turning them into river beds. During the dry season, the river flow abates so much that only a small portion of the expanded river bed is used. This irregular flow affects not only the adjacent croplands but also downstream lowlands that depend on the river for irrigation. Fish and other aquatic life have become scarce or have disappeared in some areas. Rural people who depend on the fisheries from these affected bodies of water are left without sources of food or livelihood.

The effects of deforestation are more telling on the people from the upland, whose cultural minorities (numbering about a million in 1978), depend on forest fruits, tubers, and wildlife, supplemented by their harvests of yams, vegetables, and other annual crops. Now their survival is threatened by the loss of the forest habitat.

The disappearance of some wildlife has also been blamed on the loss of habitat. Already, more than 20 species are considered threatened or endangered, including the monkey-eating eagle and the tamarau. Many plant species, some of which may be unknown to science, have also vanished as forests have been lost.

History of forest policies

The history of forest policies in the Philippines can be divided into four major periods: a period of low exploitation during colonial, wartime, and postwar eras; a period of increased exploitation for development during the post-independence era; a peak period of logging and concession exploitation during the 1960s and 1970s; and one of building a forest products industry in the 1970s and 1980s.

Period of low exploitation. Forestry in the Philippines formally started in 1863 with the creation of the *Inspección General de Montes*. This office intervened in all matters pertaining to cutting and extracting timber and minor forest products, awarding concessions for opening up mountain lands and virgin forest lands. Three main goals of forest policy could be discerned in the Spanish royal decrees: (a) providing Spanish civil and naval needs for timber, (b) contributing to government revenue, and (c) perpetuating the forest resources. These goals were not met. Both commercial forest exploitation for timber and government revenues from forest use were low. Abuses and violations of regulations designed to protect forest resources were common. In fact, as early as 1874, kaingin-making was banned and cutting for commercial purposes in Cebu and Bohol was a crime.

Under the Spanish regime, no forest land was allowed to be sold unless it was properly surveyed, had its boundaries marked out, and was certified as alienable and disposable by the *Inspección General de Montes*. While timber could be used freely under a permit, illegal cutting of trees and cultivation in forest lands increased among the natives. However, with a small population, pressure on forest lands was negligible, and forest loss not extensive. In fact, when the United States took over the Philippines in 1898, the country was still almost blanketed with forests, except for Cebu and Bohol. The first annual report by the American-appointed director of the Forestry Bureau described the lush forest vegetation in Mindanao, Palawan, Samar, and Luzon as intact resources "waiting to be explored by American capitalists." Paramount among the Forestry Bureau's goals were developing and perpetuating the forest industry. Among the important work of the Bureau were inventory, construction of volume tables, and botany. Mechanized logging was also introduced, although it was not widespread until after World War II. Extensive government ownership was enforced. All forestry activities were intensified, necessitating the importation of American foresters and the creation of a local forestry school to train Filipinos. The Forest Act, enacted by the U.S. Congress in 1904, became the bible of Philippine forestry and was the basis of all forestry regulations until 1975.

During the American regime, the forest industry flourished. The country began to export logs and lumber to the United States, and before World War II broke out the lumber industry ranked fourth in value of production, second in employment, and third in monthly payments among Philippine industries. Annual government revenue from forest charges averaged 2.5 million pesos.

While the industry flourished, the forests started to suffer, from both destructive logging and kaingin-making. Laws to prohibit kaingin-making and illegal entry into public forests were enacted. These laws proved worthless and difficult to enforce because of the size of population, the lack of enough forest rangers, and the enormous area of forest lands.

During the Japanese occupation, starting in 1942, all districts and forest stations in occupied territories were allowed to operate. The Japanese took every opportunity to exploit the resource for war purposes. Severe destruction of forests and a devastated industry resulted. Of 163 sawmills operating before the war, 141, or 87 percent, were completely destroyed.

Just after the war, Commonwealth Act (C.A.) No. 720 extended the time for accomplishing or complying with the terms, conditions, or stipulations involving forests and other public lands that existed in 1942 and could not be performed by reason of the war.

Period of increased exploitation. The Constitution of the new Philippine Republic, established July 4, 1946, provided that all timberlands belonged to the state. This provision reinforced the system of ownership established during the Spanish and American regimes. Forest policy did not change much during the period. The U.S.-passed Forest Act of 1904 continued to guide forest management. If there was any change in policy, it was only in exploiting more forest areas. This meant more revenues, which the country needed to help accelerate its rehabilitation and development.

After independence, the forest industry was rehabilitated. Early logging was primitive and only partly mechanized. Felling, bucking, and squaring of logs were done by axes or cross-cut saws. Hauling was by carabao and manpower. Americans helped in the complete transformation of the logging industry to mechanized systems. Large-scale logging grew after the war, in response to the U.S. market demands. In a few years, the country resumed exporting forest products. In 1949, forest products exports had a value of 3.3 million pesos and accounted for only 1.5 percent of total exports, but wood processing was also starting to grow. Later, wood exports – mainly logs – would become important to the national economy. In the period 1955–1957, wood products constituted 11 percent of the value of Philippine exports.

One notable policy that evolved in 1954 was the application of selective logging to the dipterocarp forests. Previously, forests were logged without much concern for future harvests; the Forestry Bureau's role was oriented toward exploitation. Its tasks were mostly regulatory, managing concessions and collecting forest charges.

Peak period of logging exploitation. The period from 1961 to 1976 marked the peak of logging exploitation in the Philippines. During that time the government reported that authorized licenses cut about 132 million cubic meters of timber, an average of 8.8 million cubic meters per year.

During this period more and more forests were licensed for exploitation. In 1960 there were only 5.5 million hectares under license. That area nearly doubled, to 10.59 million hectares in 1971 (Table 4.3), mainly in response to Japanese market demands.

American and Japanese capital is deeply entrenched in the wood indus-

Table 4.3. *Forest area under license, 1959–84*
(thousand hectares)

Year	Licensed area
FY 1959–60	4,485
FY 1961–62	6,554
FY 1963–64	7,928
FY 1965–66	6,745
FY 1967–68	8,302
FY 1969–70	8,979
FY 1971–72	10,598
FY 1973–74	8,452
FY 1975–76	10,137
CY 1978	8,768
CY 1980	7,938
CY 1982	7,539
CY 1984	6,346

try in the Philippines. Some companies are 100 percent American-owned, but most have more than 40 percent Philippine equity, including Lianga Bay, Zamboanga Wood Products, Agusan Wood, Nasipit Lumber Company, and Philippine Wallboard. The role of multinational corporations in exploiting Philippine forests cannot be overemphasized. Without these corporations, the rich forest would not have been cut so rapidly. Huge profits excited these multinationals as much as they did wealthy local investors whose own capital was invested in the logging business. The norm of most forest operators was to "cut and get out." The forests were mined. Forest products became the largest category of exported goods, reaching a peak of 33 percent of all exports in 1969. The forestry agency was mainly concerned with more forest exploitation. In fact, the national government took pride in the ever-increasing timber harvest, which meant more forest revenues and foreign exchange earnings. The intensity of exploitation was matched by widespread and larger scale alienation of forest lands. Demographic pressure and social justice policies gave rise to larger scale conversion of forest lands to provide agricultural lands for growing rural populations.

During this period, professional Filipino foresters and experts from other countries warned that exploitation and alienation were proceeding too rapidly, and that soon the once-lush forest would become a desert. Yet government policy and inaction did not encourage control.

Building a forest products industry. During the 1960s and 1970s, as much as 80 percent of the recorded log production was marketed in raw form. Logging was the most profitable wood operation, and lumber and plywood were considered almost residual industries. Prime logs were directed to the export market, and only the un-exportable were fed to the mills.

Attempts to build a forest products industry started in the 1960s, when the Bureau of Forestry preferentially awarded concessions to applicants who would set up processing plants. This initiative was given more substance in January 1967, with a directive from President Ferdinand Marcos requiring licensees to establish processing plants and reduce their log exports by 10 percent every year until 1971, when only 40 percent of a licensee's allowable cut could be exported. This policy gave rise to many small and poorly located mills. Many licensees put up processing plants to comply with regulations, while continuing to concentrate on log exports. The result was an unusually low utilization of mill capacities, which companies blamed on the lack of markets for processed products. As of 1977, the government reported that sawmills were operating at only 29 percent of capacity, veneer mills at 64 percent, and plywood mills at 35 percent (Presidential Committee on Wood Industries Development).

Beginning in 1975, the government designed and began to implement a rationalization program for the forest products industry. Its most important component was a ban on log exports. That ban had two major goals: to encourage wood processing and to reduce forest destruction. Although it was to begin in 1976, the ban did not materialize because of strong opposition from the timber licensees. Instead, Presidential Directive (P.D.) 865 authorized a partial and selective export ban. In 1979, P.D. 1159 established a ceiling on log exports of 25 percent of the total annual allowable cut. A complete ban, which was to have taken effect in 1982, was postponed indefinitely. While reported volumes of log exports were as low as 11 percent of actual production in 1980, rampant underreporting, as evidenced by trade data from other countries, indicates that actual exports were much higher.[5]

The rationalization program also called for phasing out poorly located and undersized mills, mostly circular and band mills with rated capacities of less than 10,000 board feet per day, owned by operators who were not concession holders and situated in log-deficit areas. Between 1975 and 1979, 149 of 174 existing small mills were phased out.

The wood industry's most imminent problem is the dwindling resource base. To remedy it, the government has had to implement measures that, combined with unfavorable external factors, slowed the growth of the industry. Production and exports of major forest products have not significantly increased.

Attempts at conservation and regeneration. In the 1970s the results of several decades of exploiting the forests were inescapable even to the government, especially during the formation of President Marcos' New Society[6] in 1973. Yet conflicting goals made policy choices difficult. Conservation, revenue generation, promotion of foreign exchange, and development of the industry were equally important national goals. Consequently, the means adopted to regenerate and conserve the dwindling forest resource were tempered and sometimes of little use.

The focus of forest policy during this period was to reaffirm the sustained-yield management concept that lost meaning during the timber boom, and to introduce the multiple-use forestry concept to satisfy societal wants. Tighter control of logging and emphasis on forest protection were paramount objectives. Later, government foresters gave equal emphasis to community forestry, participatory schemes in reforestation, industrial plantations, and other regenerative techniques. In its early years, the New Society also attempted to promote professional and ethical values in the public service, including forest service employees, but these values were lost in the large-scale graft and corruption of later years.

Several regulations were enacted in the 1970s to conserve the forests. The most controversial was the ban on log exports, introduced in the belief that lucrative log exporting encouraged overcutting and destructive logging. In 1985, logging was banned in three regions of the country and many other provinces in other regions. Affected were 70 timber licensees with (1983) a total annual allowable cut of 2.6 million cubic meters and production of 771,000 cubic meters. Theoretically, banning these licensees should have saved about 500,000 cubic meters of timber for the next two years' harvests, excluding another 500,000 cubic meters in logging damage. Yet these savings did not materialize because of rampant timber smuggling in banned areas.

Another measure, the least effective, was the reduction of allowable cuts granted to licensees by revising the allowable annual cut (AAC) formula, which had overestimated future yields from second-growth forests because it was based on limited data representing only better quality stands. While the new formula effectively reduced the computed allowable cut, it had no effect on the actual cut. Reported timber harvests had been far below the allowable cuts granted except in 1965. In fact, even with subsequent reductions issued by the President, the gap between AAC and actual production was wide (Table 4.4).

Under the martial law rule of President Marcos, cancellation and suspension of erring licensees became government policy. Between 1976 and

Table 4.4. *Variance between annual allowable cut and actual production (thousand cubic meters)*

Year	Production	Annual allowable cut	Variance
FY 1959–60	6,315	7,486	1,171
FY 1961–62	6,772	8,721	1,949
FY 1963–64	6,536	11,018	4,482
FY 1965–66	8,047	9,358	1,311
FY 1967–68	11,114	11,603	489
FY 1969–70	11,005	15,491	4,486
FY 1971–72	8,416	16,440	8,024
FY 1973–74	10,190	20,912	10,722
FY 1975–76	8,646	21,885	13,239
FY 1977–78	7,169	18,672	11,503
FY 1979–80	6,352	14,001	7,469
FY 1981–82	4,514	14,001	9,487
FY 1983–84	3,849	9,764	5,915

1984, the decline in the number of licenses, from 371 to 157, was ascribed primarily to cancellation or suspension. Many suspended licenses were taken over by Marcos cronies, leading to a concentration of timber holdings (Porter 1987). For example, although the average concession size was 30,000 hectares in 1977, Juan Ponce Enrile, President Aquino's political opponent and erstwhile defense minister, obtained at least four concessions averaging 80,000 to 90,000 hectares each. The total area covered by licenses has been reduced from 10.2 million to 6.3 million hectares, but the threat of possible cancellation or suspension did not significantly improve forest management. Except for some window-dressing, most licensees generally evaded compliance with many decrees and directives that would have added to production costs.

Yet another measure was to scrap the 1- to 4-year timber licenses. Now, all timber licenses are for 25 years, renewable for another 25 years after a satisfactory performance review. The longer duration was thought to make licensees more conservation-conscious. It is doubtful that the desired effect has been achieved.

The environmental crusade finally caught up to Philippine forestry in the late 1970s. Letter of Instruction (LOI) 917, enacted in 1979, requires each timber concessionaire to set aside 5 percent of his concession area as wilderness area not subject to logging. Potentially, this decree withdraws about 300,000 hectares of virgin forests from commercial production. However, inadequate on-the-ground supervision and administrative weakness of the forestry agency have limited compliance to a paper demarcation only.

Three major government programs were implemented to regenerate the resource: reforestation, industrial tree plantation, and social forestry.

Reforestation, once solely a government undertaking, was intensified in 1976 to include private industry and the whole citizenry. Among the significant programs were (a) P.D. 1153, which required each Filipino ten years old or older to plant 12 trees a year for five years; (b) the Forest Ecosystem Management Program, establishing one municipal nursery for each of the 1,000 municipalities and increasing the role of the Bureau of Forest Development in reforestation; and (c) the Energy Farm program, which required each barangay to plant at least two hectares of trees as a community fuel reserve. The government reports that reforested areas increased substantially as a result. By 1983, the government reports, more than 78,000 hectares had been reforested, more than 42,000 by the government and 36,000 by private industry. The average tree survival rate is reported to be about 50 percent.

These figures, though, are based on over-optimistic, bloated reports. President Aquino's former Minister of Natural Resources, Ernesto Maceda, revealed that 90 percent of concessionaires had violated the terms of their agreements by failing to carry out reforestation programs (Porter 1987). Performance in reforestation is still low, for a variety of reasons. Reforestation sites are far from ideal; forest planting is generally relegated to marginal lands – usually grasslands inhabited by cogon (*Imperata cylindrica*), where site preparation is costly and difficult (cogon impoverishes the soil). Climatic extremes, such as droughts and floods, hinder the establishment and growth of seedlings. Seedlings must be balled to survive the first hot season and require fertilizers to provide nutrients the poor site cannot supply. And, after seedlings were planted, the problems were far from over. Intentional fires or those due to grazing practices kill many planted seedlings.

These technical difficulties are coupled with financial and managerial problems. Lack of funds ranks high on the list. Despite significant increases in reforestation budgets starting in 1976, funding falls far short of the magnitude and number of programs that are to be implemented. The cost of reforesting a hectare has tripled since the 1970s. Corruption and bureaucratic red tape at many levels have persisted, exacerbating the shortage of funds. Appropriated funds arrive too late, or sometimes not at all. Delayed funds means late plantings, resulting in low seedling survival rates.

Most of the other programs were intended to remedy the lack of government funds for reforestation. P.D. 1153, for example, was meant to provide 1.5 billion seedlings for planting 720,000 hectares in five years. After seven years (1977–1983), only 32,593 hectares were planted. Many people are able to secure the required certificate of planting without planting a single tree. The efforts of the few who complied were wasted. Inexperienced and often unsupervised, they planted the free seedlings from the Bureau of Forest Development poorly – along roads, in plazas, and in back yards instead of on the denuded hillsides where the need is greatest – or failed to plant them at all.

The efforts of timber licensees have also been relatively poor. Exacting compliance from them is just as difficult as it is from the ordinary citizen (Payer 1982). Even President Marcos admitted that only about 50 percent of loggers complied with reforestation and protection directives. In fact, this estimate greatly exceeds the actual compliance rate. Despite the barrage of decrees and programs to promote reforestation, it is still conducted poorly.

The most recent program, presented as the ultimate solution to the problem of shifting cultivation, is the Integrated Social Forestry Program. Under this program, project areas are selected after examination of their sociopolitical, economic, and ecological characteristics. How these can be ascertained is questionable, since the surveys are often made by inexperienced foresters. The beneficiaries are issued stewardship certificates giving them tenure over the land they occupy or develop for 25 years. The Bureau of Forest Development reports that by 1984 it had issued 28,574 certificates, representing a total of 77,628 hectares. The participants supposedly develop their lands into productive operations. In support of the program, the government established more than 500 nurseries that produce over 12 million seedlings for participants in 1983 alone.

The program has many flaws. Eleven ministries are involved. Each has a limited budget and chronic personnel shortages, leaving no resources to spare for the social forestry program. Moreover, the lands offered by government as social forestry sites are not those now occupied by kaingineros.

In addition to the regular government and private sector reforestation programs, another program implemented to regenerate the dwindling forest resource was the establishment of industrial tree plantations, tree farms, and agroforest farms pursuant to Executive Order No. 725.

Guidelines were released in 1981. The minimum area that can be leased is 100 hectares, and of that not more than 30 percent of a given license can be converted to industrial plantation. The 25-year lease is renewable for another 25 years.

Industrial plantation ventures are given ample government incentives, including minimal rental, reduced taxes, help in obtaining low-interest loans, and no export restrictions. In addition, these ventures can be registered with the Board of Investments to obtain the incentives described below. In 1984, there were 81 industrial plantation leases, covering 284,705 hectares. Only two registered with the Board of Investments between 1981 and 1984.

Tree farm and agroforest farm leases are also available. As of 1984, 108 tree farms covering 15,000 hectares and 80 agroforest farms with an area of 78,000 hectares have been reported.

Cumbersome procedures, bureaucratic red tape, and administrative problems have prevented the processing and approval of many applications for these leases. It is not known how much of the area leased is really used for industrial tree plantations, tree farms, or agroforests.

Key issues in forest policy

Revenue system and rent collection

About 98 percent of the forests of the Philippines are publicly owned. The government delegates exploitation of these resources to the private sector through a licensing system. Timber license agreements are awarded to qualified applicants for periods of 25 years. Licenses to harvest minor forest products are issued for specified volumes in specified times. As the resource owner, the government imposes taxes and compulsory conditions of use.

The main purpose of forest taxation is to generate government revenue, although forest taxes enacted recently are designed primarily to influence forest management. Other tax objectives include promoting domestic processing, developing forest industries, increasing or decreasing people's participation in certain activities, and promoting conservation.

For a long time forest charges based on the volume of timber cut represented the main rent available to the government, the resource owner. Established by a congressional act in early years, the rates were set at 3.5, 2.5, 1.25, and 0.6 pesos per cubic meter of timber harvested for four classes of timber. These rates have not been adjusted for inflation and became absurdly low when the windfall profits of concessionaires were extremely high. These rates have remained unchanged because passing a new law in the old Congress, where most members represented landed elites involved in the logging business, was difficult. Accordingly, in the 1970s, the Bureau of Forest Development, responsible for imposing and collecting these fees, reclassified almost all commercial species to the highest class to increase total revenue collection.

Other charges, also based on volume cut, are levied to finance various forestry activities. The reforestation fund was instituted to provide funds for the Reforestation Administration in 1960. The forest information fund was set up to finance the forestry extension program of the University of the Philippines College of Forestry. The FORPRIDE fund was instituted for forest products research and development undertaken by the Forest Products Research and Industries Development Commission. The latest additions were special deposits for the working units for management and protection and the research and development trust fund to finance forest production and economic research in the early 1970s.

In early 1980, these volume-based charges (including the regular charges) ranged from 6.35 to 9.35 pesos per cubic meter for logs used domestically, and from 10.85 to 13.25 pesos for logs exported.[7] Later that year, these charges were replaced by a uniform charge of 20 pesos per cubic meter of timber cut. In 1984, this consolidated forest charge was raised to 30 pesos per cubic meter.

The government also collects an annual license fee on concessions, computed at 5 percent of the value of the allowable cut (determined by multiplying the allowable cut by the average forest charge: 30 pesos for timber, 5 pesos for pulpwood, and 2 pesos for residues). This fee is in addition to the license application fee, a once-only charge at the time of application that is computed at 1 peso per hectare of the area applied for.

Other relatively important forest taxes are the manufacturer's sales tax for domestic and export sales and the municipal graduated sales tax.

The domestic sales tax is computed at 10 percent of the gross value of sales of wood and wood products. The export tax is levied on logs, lumber, veneer, and plywood at 20 percent of f.o.b. value for logs, 4 percent for lumber, and 4 percent for veneer. The former 4 percent export tax for plywood was eliminated in 1984. The export tax rate for logs was originally set at 10 percent and was later reduced to 4 percent. In 1974 it was eliminated, but was reimposed in 1979.

The municipal graduated sales tax is imposed on both domestic and export sales of wood and wood products. For domestic sales, the tax is 40,000 pesos for the first 20 million pesos in sales; for sales in excess of 50

Table 4.5. *Actual collection of forest charges, service fees, and export tax, 1970–82 (thousand pesos)*

Year	Forest charges	Service fees	Total	Export tax[a]	Total
1970	26,026	41,221	67,247	139,382	206,629
1971	32,803	39,683	72,486	162,795	235,281
1972	35,117	30,663	65,780	58,216	123,996
1973	97,856	27,755	125,611	102,258	227,869
1974	77,088	29,087	106,175	28,897	135,072
1975	62,960	20,928	83,888	27,397	111,285
1976	50,655	37,410	88,065	39,201	127,266
1977	75,256	11,228	86,484	38,452	124,936
1978	73,119	9,850	82,969	50,886	133,855
1979	61,508	21,023	82,531	336,911	419,442
1980	85,525	10,426	95,951	198,437	294,388
1981	148,618	21,134	169,752	168,742	338,494
1982	143,011	22,496	165,507	201,051	366,558
Total	969,542	322,904	1,292,446	1,552,625	2,845,071

[a]Estimated by the author.

million pesos, a tax of 250 pesos per 500,000 pesos excess is imposed. For export sales, the tax is 20,000 pesos for the first 1 million and 200 pesos per 1 million excess.

A realty tax is also levied on timberland. Established in 1975, it is computed at 1 percent of the assessed value of the land. The assessed value is set at 40 percent of the market value of the annual allowable cut. A market value set at 200 pesos per cubic meter in 1975 is still in force.

The stumpage value of rich dipterocarp forests like those in the Philippines can be considerable. Yet the government, which owns the stumpage, captures only a small part of its value through all these taxes and fees. Timber operators pocket the larger portion. The government has persistently refused to implement stumpage valuation, despite its adoption as official policy in 1975. Instead, the government has continued to rely on forest charges, sales taxes, license fees, and the realty tax.

Before export and domestic sales taxes were imposed in 1970, the only major sources of rent for the government were forest charges and service fees. Actual collections of forest charges (including service fees) for the period 1970–1982 totalled 1,292 million pesos, ranging from 65 million to 169 million pesos annually. With the imposition of export taxes, government revenues for the period more than doubled and could have been higher if the export tax on logs had not been eliminated between 1974 and 1978 (Tables 4.5 and 4.6).

Table 4.6. *Taxes as percentage of reported export values of logs, lumber, plywood, and veneer, 1970–82*

Year	Taxes from forest sector (million pesos)	Total reported exports (million pesos)	Taxes as percent of exports
1970	207	1,666	12.4
1971	235	1,623	14.5
1972	124	1,455	8.5
1973	228	2,788	8.2
1974	135	1,859	7.3
1975	111	1,682	6.6
1976	127	2,006	6.3
1977	125	1,975	6.3
1978	134	2,404	5.6
1979	419	3,757	11.2
1980	294	3,405	8.6
1981	338	2,715	12.4
1982	367	2,479	14.8

Table 4.7. *Timber rents in the absence of domestic processing, 1979–82*

Year	Potential rent per m³ (U.S. dollars)	Total log production (thousand m³)	Total available rent without domestic processing (thousand U.S. dollars)
1979	69	6,578	453,882
1980	78	6,352	495,456
1981	55	4,514	248,270
1982	57	5,400	307,800
Total		22,844	1,505,408

Potential rents from the forest sector were estimated at more than $1.5 billion for the period 1979–1982 (Table 4.7). Yet the rents actually realized were slightly more than one billion dollars (Table 4.8). The difference represents a loss due to conversion of an increasing volume of exportable logs to plywood in inefficient mills. The low conversion rates of Philippine plywood mills meant that each log exported as plywood brought a lower net return over cost than the same log exported as saw timber or in raw form. As shown in Table 4.9, between 1979 and 1982 the potential rent per cubic meter in roundwood equivalent of either logs or lumber was as high as seven times that of plywood. In fact, in 1981 through 1983, the potential rent of plywood dipped to negative figures.

From 1979 to 1982, forest charges and export taxes represented 11.5 percent of the value of forest products exports. In the preceding nine

Table 4.8. *Aggregate rents in timber: logs, lumber, and plywood, 1979–82*

Year	Logs Net production (thousand m³)	Rent per m³ (U.S. dollars)	Aggregate rent (thousand U.S. dollars)
1979	1,496	69	103,224
1980	1,265	78	98,670
1981	360	55	19,800
1982	1,680	57	95,760

Year	Lumber Production in RWE[a] (thousand m³)	Rent per m³ (U.S. dollars)	Aggregate rent (thousand U.S. dollars)
1979	2,699	75	202,425
1980	2,538	84	213,192
1981	2,023	75	151,725
1982	1,992	58	115,536

Year	Plywood Production in RWE[a] (thousand m³)	Rent per m³ (U.S. dollars)	Aggregate rent (thousand U.S. dollars)
1979	1,166	10	11,660
1980	1,282	27	34,614
1981	1,060	−3	−3,180
1982	979	−11	−10,769

Year	Total rent (thousand U.S. dollars)
1979	317,309
1980	346,476
1981	168,345
1982	200,527

[a]RWE = roundwood equivalent.

years (1970–1978), they came to only 8.2 percent (Table 4.6). In the 1960s, when timber harvests were also high, the government's share of the rents generated must have been much lower, considering the low rates of forest charges imposed during that period, and the absence of sales and realty taxes on timberland.

From 1979 to 1982, the government's total revenue from forest charges and export taxes on logs, lumber, and plywood was $170 million, only 11.4 percent of the total potential available for capture by exploiters of the forest resource. Undoubtedly, timber operators profited greatly from the

Table 4.9. *Potential rent of logs, lumber, and plywood in roundwood equivalent (RWE) in U.S. dollars*

	Average value per m³		Production cost per m³, RWE	Potential rent per m³, RWE
	Product	RWE		
1979				
Logs	116	116	47	69
Lumber	217	131	56	75
Plywood	263	113	103	10
1980				
Logs	127	127	49	78
Lumber	242	145	61	84
Plywood	326	141	114	27
1981				
Logs	106	106	51	55
Lumber	230	138	63	75
Plywood	280	120	123	−3
1982				
Logs	111	111	54	57
Lumber	209	125	67	58
Plywood	280	120	131	−11
1983				
Logs	93	93	59	34
Lumber	204	123	74	49
Plywood	260	112	146	−34

forest resource. Granted that the government's capture might have been higher by another 5 percent owing to unaccounted domestic sales and realty taxes, the profit margin of timber operators would still remain high.

The government's inability to capture a large share of the available rent from forest resources has contributed to the rapid exploitation of the forests. The very low forest charges and fees (the main revenue source before 1970) implied that the government was almost giving away the stumpage to the operators. In the 1950s and the 1960s this fact attracted many fly-by-night loggers who cut, made high profits, and got out.

There were timber booms everywhere in Luzon, Visayas, and then Mindanao. The area under license more than doubled between 1958 and 1970, from 4.6 million to 9.4 million hectares. The allowable cut also more than doubled during that period, from 7.2 million to 15.5 million cubic meters. Luzon was mostly covered with timber in the 1950s and 1960s, but timber started to run out in the 1970s. By the mid-1970s, most of the logging areas in central and western Luzon were either abandoned or included in logging bans.

In 1959, low forest charges and the buoyant log export market saw timber production rise to 5.4 million cubic meters, 55 percent exported as logs. The country became a major exporter of tropical logs, supplying almost one-third of the world market. Log production more than doubled from 1959 to 1970, from 5.4 million to 11 million cubic meters. High production continued until the early 1970s, but now most of the cut was in Mindanao.

The interaction of forest taxation with other government forest policies promoted the rapid exploitation of Philippine forests. The policy of granting short-term timber licenses of one, two, four, and 10 years from the 1950s into the early 1970s did little to encourage licensees to practice sustained-yield management, since the second cut would not come until at least 30 or 35 years later. Even the current 25-year license is not long enough to ensure that young trees that mature in 70 years or more will be conserved. Nor is the renewal of licenses for another 25 years (making a 50-year maximum limit) automatic. Renewal required the licensee to go through the same process as when applying for a new license. Depending on the operator's past performance, the whims of approving officials, and political considerations, the renewal might not have been granted. Such laws created insecurity of tenure that encouraged forest operators to exploit every available marketable product and all the timber to recoup their investments in the shortest possible time. The result was an acceler-ated harvest schedule with less concern for future productivity.

Weak enforcement of the license terms, especially with respect to forest protection, selective logging, and timber stand improvement, exacer-bated the effects of inappropriate forest taxation policies. The need for personnel and logistic support for effective supervision was not met as operations expanded during the timber boom.

Political factors added to the negative effects of forest taxation and other policies. Because a significant number of timber operators were politicians, it was almost impossible to enforce policies that would reduce the operators' profit margins. Concessions that belonged to these influen-tial people became almost untouchable by mere foresters, so many young foresters fell victim to corruption. Moreover, the lack of political support has been the major impediment to the drive to conserve the remaining forests. Several programs and policies enacted and implemented in the 1950s and 1960s lacked sustained support because of simple ignorance and apathy among policy-makers.

Deforestation has also accelerated because of the manner in which charges are assessed. Reclassifying almost all commercial species to first class, in order to charge the highest rate, theoretically encourages high-grading of the forests, as loggers seek to avoid paying high rates for low-value timber. In practice, however, in view of the low charges for timber of even the highest class, overcutting was more of a problem than high-grading.

In the 1960s, however, the consolidation of the six individual forest charges and the regular charges into a single charge may have encouraged high-grading. Since the consolidated charge is fixed, there is no added premium for quality or efficiency in harvesting and no corresponding reduction for a decrease in value resulting from inaccessibility or other high production costs. Therefore, it is likely that the consolidation encouraged high-grading. High-grading was also promoted by the fact that the forest charges are based on the volume cut and not on the volume of merchantable timber, enabling licensees to select the trees of highest value. As a result, a larger forest area would have to be harvested to meet production goals. Since forest charges are based on volume and not on value, they actually decreased in real terms as inflation continued during the last decade.

Low fixed and consolidated forest charges, combined with poor implementation of selective logging in all its three phases, helped promote deforestation. The first phase, tree marking, is meant to ensure adequate residual growing stock. The second phase, residual inventory-taking after logging, ascertains the size and condition of the residual stand to predict the second cyclic cut and also determines any punitive action that may be imposed. The third phase, eradication of undesirable competing species, ensures the timely renewal of the dipterocarp forests. Once the forest has been opened by logging, rapid growth of intolerant species and weeds can suppress dipterocarps and intermediate-size trees. However, these three phases have rarely been implemented correctly, if at all. Besides, supervision during logging, which could prevent excesses or mistakes by loggers, has typically been absent. Because of unsupervised logging, the logged area could hardly recover after sustaining severe damage. All these factors contributed to the overcutting and destruction of dipterocarp forests in the Philippines.

Protection of the forest products industry

In the 1970s, the government, hard-pressed by economic difficulties and the dwindling forest resource, began once again to promote domestic wood processing. The main goal was a smooth rationalization program for the wood industry to shift from raw wood to finished product exports.

Early attempts to build the wood processing industry did not prosper.

In 1967, the then Bureau of Forestry issued Forestry Administrative Order No. 11–17, which stipulated, among other things, the gradual reduction of log exports. However, the policy was suspended in 1969 to alleviate the country's balance of payments deficit. Four years later President Marcos again announced that log exports would be phased out. An interagency group recommended that 40 percent of log production be retained for domestic processing in 1973–1974, and an additional 20 percent every year until all production was retained by 1976–1977. Instead, P.D. 428, issued in April 1974, prohibited log exports effective January 1976. The directive allowed concessionaires to export up to 80 percent of their log production in calendar years 1974 and 1975 to generate sufficient funds to expand or establish processing plants. The Revised Forestry Code (P.D. 705), issued on May 19, 1975, reiterated a complete ban of log exports beginning January 1976. However, the policy-makers' ambivalence was evident, since the code allows the National Economic Development Authority to recommend, and the President to grant, exemptions from the ban.

As predicted, the wood industry succeeded in securing another suspension in 1975, when the government issued P.D. 865, allowing temporary log exports on a limited and selective basis. The reason given for suspending the ban was to avoid what the President called "unnecessary economic dislocation" or adverse effects on the country's economic conditions and industry stability. By memorandum, the Ministry of Natural Resources decided on January 4, 1979, that the selective log export ban policy was to continue only until 1981, but selective log exports were still allowed as late as 1985.

The decision to permit continued log exports was due partly to pressure from Japanese importers. Japan, the number one market for Philippine logs, refused to increase its finished products imports. Tariffs and unnecessary product specifications made it almost impossible to compete in the wood products markets there. With the entry of Indonesia, Sabah, and Sarawak into the log export market, Japan would not be affected if the Philippines curtailed log exports. As a result, the Philippines was forced to continue exporting logs.

The grant of log export quotas to qualified licensees was tied to processing. Processing, however, was undertaken mainly to gain the benefit of log export quota allotments rather than to earn profits from efficient processing. This required tie-in has been an inept and costly means of encouraging processing. As shown earlier, rents would have been greater if logs converted in inefficient plywood mills were instead marketed as either logs or lumber.

Table 4.10. *Wood-processing plants as of 1982*

Type	Number
Sawmill	190
Plywood	35
Veneer	11
Blockboard	18
Fiberboard	2
Particle Board	2
Pulp and Paper	6
Pulp	4
Paper	16
Wood treating plants	44
Total	328

Even with the mill rationalization program, lumber production increased by only 40 percent between 1971 and 1982; plywood production decreased by 15 percent; and plywood exports fell by 45 percent. However, from a very small base, veneer production increased by 173 percent, and veneer exports increased by 245 percent.

Between 1976 and 1980 the number of active sawmills shrank from 325 to 209 and corresponding log requirements declined from 7.868 million to 4.71 million cubic meters. In 1982, there were 190 sawmills, requiring 4.41 million cubic meters. This decline was partly caused by the phase-out of uneconomic and poorly located mills. In 1980 the 209 plywood mills had an annual log requirement of 3.392 million cubic meters, and the 23 veneer mills required 0.94 million cubic meters. In 1982, the remaining 35 plywood mills required 3.12 million cubic meters, and the 11 veneer mills required 0.61 million cubic meters. As of 1982, the wood industry had 328 processing plants (Table 4.10).

The government tried to encourage more domestic processing for four major reasons: (1) to earn and save foreign exchange, (2) to add more value, (3) to gain more employment, and (4) to use more completely the dwindling resource base.

As a foreign exchange earner, the wood industry's contribution to the Philippine economy is significant. Up to 1975, lumber, plywood, and veneer averaged 25 percent of the forestry sector exports, but their share increased to 78 percent in 1981, owing mostly to increased lumber and veneer exports. The average export price of all products continued to increase until 1980. Total export value started to decline in 1980 because of reduced exports of logs, lumber, and plywood (Table 4.11).

Table 4.11. *Value of forest products exports (in million U.S. dollars)*

Year	(1) Total	(2) Logs	Processed products (3) Lumber	(4) Plywood	(5) Veneer	(6) Total (Cols. 3 + 4 + 5)	(7) Processed products as percent of total (Col. 6 ÷ Col. 1)
1970	277	237	13	20	7	40	14.4
1971	256	215	11	24	6	41	16.0
1972	213	164	10	34	5	49	23.0
1973	403	304	35	58	6	99	24.6
1974	275	216	30	26	3	59	21.4
1975	227	167	27	21	12	60	26.4
1976	264	135	68	43	17	128	48.4
1977	262	134	67	41	20	128	48.8
1978	324	145	85	72	22	179	55.2
1979	461	144	198	85	34	317	68.7
1980	391	92	181	103	15	299	76.5
1981	342	76	125	110	31	266	77.7
1982	290	80	123	67	20	210	72.4
1983	330	78	149	76	27	252	76.4

Although the government intended licensees to use the 1974 and 1975 proceeds of log exports to set up processing plants for the anticipated expansion of the industry beginning in 1976, estimated capital investment in forest-based industries did not change much in 1976. While sawmill investments increased more than fourfold, capital investments in pulp and paper declined by two-thirds. Employment in the industry increased from almost 61,000 in 1975 to over 98,000 in 1982 (Table 4.12).

Three types of incentives were available for the wood industry: investment, trade, and public spending.

In 1981, P.D. 1789, known as the Omnibus Investment Code and later amended by "Batas Pambansa Bilang 391," defined the new incentive system available to enterprises. To become eligible, enterprises had to comply with the activities listed in the Investment Priorities Plan approved annually by the President, and had to register with the Board of Investments. The incentive package differed for agricultural producers, domestic producers, new or expanding export producers, existing export producers, export traders, service exporters, and producers falling under the scope of industry rationalization or energy-saving programs.

The annual Investment Priorities Plans for 1981 through 1985 provided three categories in which forestry firms could be listed: tree farms and tree plantations, wood products manufacturing, and paper products. Table 4.13 lists the specific activities under each category.

Table 4.12. *Change in employment, 1976 and 1982*

	Employment	Production (thousand m³ RWE)	Employment per unit of output
1976			
Logging	35,311	8,646	4.08
Sawmilling	10,665	2,670	3.99
Plywood	11,250	965	11.65
Veneer	3,761	773	4.86
1982			
Logging	43,560	5,400	8.06
Sawmilling	28,780	1,992	14.44
Plywood	20,890	979	21.33
Veneer	5,280	821	6.43

From 1981 to 1984, only 108 forestry firms registered with the Board of Investments (Table 4.14). Of these, only 26 are registered export traders; the rest produce mostly for export. Most of these firms are considered non-pioneer; they are expanding existing operations. Of the 84 wood products firms, 37 are manufacturing furniture from rattan, wood, and other material.

Table 4.15 presents the specific incentives available to registered forestry firms. Generally, investment incentives include tax credits, exemptions, and tax deductions. Tax credits of 5–10 percent are given on net export value earned, and credit of up to 10 percent on net local content of export sales, on domestic capital equipment, on taxes and duties on raw materials used in export production, and on withholding tax on interest. Tax exemptions of 50–100 percent are given for export fees and imposts, including wharfage fees, and specific sales taxes. Tax deductions are given to export traders on 20 percent of total export sales, and an additional 1 percent is given on incremental export sales of new brand-name export products, and for expenses of establishing and maintaining offices abroad. Other forms of incentives include net operating loss carry-over, permission to hire foreign nationals, protection from anti-dumping measures, freedom from government competition, and protection of patent and other proprietary rights.

Investment incentives have significant effects on the rates of return, user costs of capital, and labor costs of eligible enterprises. Studies of past incentive programs under Republic Act (R.A.) 6135 (Export Incentives Act) and R.A. 5186 (Investment Incentives Act) can be used to approximate the effects of the present incentive package. Under the old program the rate of return of registered firms could increase substantially, by 6 percentage points for a nonexporting firm and 16 percentage points for a

Table 4.13. *Specific forestry activities listed in the investment priorities plan for 1981–85*

1. Forestry: tree farms and plantations
   a. Fast-growing species for timber/pulpwood (pioneer and non-pioneer)
   b. Rattan (pioneer)
   c. Bamboo (pioneer and non-pioneer)
   d. Others (pioneer and non-pioneer)
2. Wood products manufacturing
   a. Builders' woodwork, including prefabricated and sectional buildings and other components of wood (non-pioneer)
   b. Sawmilling and kiln-drying operations (only if integrated with other wood-manufacturing operations)
   c. Plywood (non-pioneer) but limited to expansion/modernization of existing plants except when part of a processing center or for processing timber from plantations
   d. Blockboard (non-pioneer) (must be integrated with primary wood-processing plants)[a]
   e. Particle board (non-pioneer) processing wood waste or timber from plantations[a]
   f. Furniture (non-pioneer)[a]
   g. Other wood products for export such as tongs, chopsticks, lacquered wood products, etc., including combination of wood and other materials (non-pioneer)
3. Paper and paper products
   a. Pulp from indigenous raw materials (pioneer and non-pioneer)[b]
   b. Integrated production of pulp and paper (pioneer and non-pioneer)[b]
   c. Fiberboard from indigenous raw materials (non-pioneer)
   d. Packing containers (non-pioneer), those types that will cater primarily to the export market requirements

[a]Eligible only if exported as of 1984.
[b]Dropped in 1984. Replacing them were modernization or rationalization of pulp and paper mills (pioneer and non-pioneer) and paper products such as notebooks, diaries, albums, and stationery for exports.

Table 4.14. *Breakdown of registered forestry firms according to major types of products, 1981–84*

Types of products	1981	1982	1983	1984	Total
Plantations	4	1	0	0	5
Wood products manufacturing	22	30	17	15	84
Paper and paper products	3	7	6	3	19
Total	29	38	23	18	108

Source: Compiled by the author from raw data obtained from the Statistics Division, Board of Investments.

Table 4.15. *Incentives granted to registered firms under P.D. 1789 as amended by Batas Pambansa Bilang 391*

1. New and expanding export producers
   a. Tax credits
      (1) on net value earned to the extent of 10 percent for pioneer projects;
      (2) on net local content of export sales to the extent of 10 percent;
      (3) on domestic capital equipment equal to the value of exemption from taxes and duties that would apply for imported equipment;
      (4) on taxes and duties paid on raw materials used in export production;
      (5) on witholding tax on interest for pioneer projects to cover foreign loans.
   b. Tax exemptions
      (1) on imported capital equipment;
      (2) on export fees and impost, including wharfage fee.
   c. Other incentives
      (1) net operating loss carry-over;
      (2) employment of foreign nationals;
      (3) anti-dumping protection;
      (4) protection from government competition;
      (5) protection of patents and other proprietary rights.
2. Existing export producers
   a. Tax credits
      (1) on net local content of export sales to the extent of 10 percent;
      (2) on taxes and duties paid for raw materials used in export production.
   b. Tax exemption
      (1) from export taxes and fees.
3. New and expanding domestic producers
   a. Tax credits
      (1) on net value earned to the extent of 10 percent for pioneer projects and 5 percent for non-pioneer projects;
      (2) on net local content to the extent of 10 percent;
      (3) on domestic capital equipment equal to the value of exemption from taxes and duties that would apply for imported equipment;
      (4) on witholding tax on interest for pioneer projects to cover foreign loans;
      (5) on taxes and duties on raw materials used for export products.
   b. Tax exemption
      (1) on imported capital equipment to the extent of 100 percent for pioneer projects and 50 percent for non-pioneer projects.
   c. Other incentives included in number 1 together with post-operative tariff protection for pioneer projects.
4. Export traders
   a. Tax credits
      (1) equivalent to the specific and sales taxes on registered export products bought from export producers and subsequently exported.
   b. Tax exemptions
      (1) from export taxes, duties and fees for registered products bought from registered export producers;
      (2) from specific and sales taxes on export products.
   c. Tax deductions
      (1) equivalent to 20 percent of total export sales;
      (2) additional tax deduction of 1 percent of total export sales when the export trader extends financial assistance to export producers;
      (3) additional tax deduction of 1 percent of incremental export sales when a new brand name for an export product is used which distinguishes it from products manufactured abroad.

typical exporting firm (Bautista and Power 1979). The user cost of capital could be reduced by 39 percent to 42 percent for a typical firm registered under R.A. 5186. For firms registered under R.A. 6135, the user cost of capital can be reduced by 15 percent (for new projects) to 42 percent (for expansion of pioneering firms). These estimates assume a 20-year asset life and use of imported capital equipment only. Labor costs are also reduced, by 4 percent for registered firms and 22 percent for exporting firms. Added up, the incentives available to a registered firm represent a substantial subsidy.

In practice, however, relatively few firms received such assistance. As with the old program of incentives, few firms have registered with the Board of Investments. The totals for forestry firms for 1981, 1983, and 1984 were 29, 23, and 18, respectively. These few registered forestry firms were mostly small-scale enterprises making furniture, woodworks, and other small products.

Before P.D. 1789, LOI 1040, issued by ex-President Marcos, granted registered export traders access to rediscounting facilities with more favorable interest charges and maturities. It also gave them access to more liberal export payment terms and preferential export credit for nontraditional export products. Credit to favored sectors at artificially low rates is a modest implicit subsidy in the range of 2 to 5 percent of output value. Some forestry firms benefited from this subsidy. At present, only industrial plantations and tree farms are eligible for artificially low rates of interest (Serna 1985).

To determine the effects of price intervention policies such as tariffs, discriminating sales taxes, and quotas, protection rates can be estimated using Nominal Protection Rates (NPR) and the Effective Rates of Protection (ERP). NPR is the proportional difference between the domestic and border prices of a product. EPR is a measure of protection that takes into account protection accorded to inputs as well as outputs of a particular activity. Since the calculation of effective rates of protection assumes that inputs are tradable internationally or mobile among domestic activities, it is more appropriately applied to processing activities than to primary extraction stages of production, in which rents to fixed resource endowments are significant.

One study (Power and Tumaneng 1984) found that the nominal rate of protection to logging decreased from minus 6 percent for 1970–1975, to minus 26 percent in 1976–1978, to minus 46 percent in 1979–1980, largely because of export taxes and bans on log exports. The increasing discrepancy between domestic and world prices that resulted provided enormous incentives for log exporters to evade the bans and quotas. That this took place on a large scale is indicated by the difference between the volume of recorded log exports to Japan, the main importer, and the 40 percent higher recorded volume of log imports from the Philippines in Japanese trade statistics.

For lumber, plywood, and veneer, the only wedge between domestic and world prices has been the export tax. Therefore the reduction in domestic log prices has provided a large measure of effective protection to those processing industries. An earlier study (Bautista and Power 1979) found that pulp, paper, and paper products enjoyed one of the highest effective rates of protection of all Philippine industries in the late 1970s. This author believes that the pulp and paper industry has been a factor in the rapid exploitation of forest resources. For example, the Paper Industries Corporation of the Philippines was granted an exception to the government's selective logging policy, to allow it to clear-cut thousands of hectares of logged-over dipterocarp forests to meet the wood requirements of its pulp mill. After clear-cutting, the area is to be developed into pulpwood plantations. In Luzon, the Cellophil Resources Corporation is responsible for the rapid harvest of pine forests in the Abra region at high costs, because of a faulty feasibility study regarding minor transport. But these costs to the company are small compared to the loss incurred by local people displaced as a result of the operations of this pulp mill.

Inefficient processing has also characterized the emerging forest products industry. Philippine plywood mills need 2.32 cubic meters of logs to produce one cubic meter of plywood, a conversion factor of only 43 percent – far below Japan's 55 percent conversion rate, or the 50 percent rate in other Asian countries. Inefficient processing reduces rents from the Philippine forests by absorbing potential profits in higher costs. In 1982 and 1983, the potential rent of plywood per cubic meter of round-wood equivalent of plywood was negative. In 1983, one cubic meter of plywood in roundwood equivalent, which cost $146 to produce, had a value of only $112, leaving a potential rent of minus $34. This situation could only arise because of the industry's highly protected status. It implies not only economic losses but waste of natural resources; every cubic meter of log fed to plywood mills meant foregone revenue equivalent to the export tax of logs. In 1983, this was equivalent to $18 lost per cubic meter of log fed to a plywood mill.

Sawmilling was more efficient than either plywood or veneer making. Its reported conversion factor is 60 percent, i.e., one cubic meter of lumber required only 1.66 cubic meters of log. The potential rent for lumber is higher than that for logs exported in raw form. In 1979–1983, the potential rent for lumber per cubic meter of roundwood equivalent ranged from $49 to $84, compared to $34–$78 for log exports.

The country seems to have comparative advantages in both logging and sawmilling, provided that sustained-yield management can be carried out, but destructive logging and other forest practices inflict heavy ecological and economic losses that should be included as domestic resource costs. As for paper products, even with sustained-yield forestry, the industry seems to have no comparative advantage. The domestic resource costs are much higher than the shadow exchange rate of foreign exchange. The industry thrives only because of government protection.

Conversion of forest land to other uses

The loss of forests in the Philippines has been accelerated by deliberate government policies to open virgin forest lands and distribute public lands to landless and impoverished households. From the time of the first public land act in 1903 to the late 1960s, the government opened publicly owned virgin forest lands and encouraged landless people to settle there, to broaden the agricultural base of the economy. During the early part of the American regime, when the population was only about 18 million, there was much land to give. But toward the end of World War II, as the population began to grow rapidly, the clamor for more lands mounted. After the war, the land disposal program was given more impetus by the U.S. foreign aid program.

But the policy that had the most pervasive effect on forest lands was the "land for the landless" policy implemented in the 1950s and the 1960s. This was politically more acceptable than altering the high degree of concentration in the ownership of existing agricultural lands. The policy was mainly responsible for converting a huge area of forest lands for agricultural purposes. From 1959 to 1963, the government razed an average of 100,000 hectares yearly under this policy (Puyat 1972). The Manahan law, enacted in 1964, had less effect; it allowed people of cultural minorities who had occupied alienable or forest lands since July 1955 to apply for free patent titles. The Homestead Act also allowed the conversion of occupied and cultivated forest lands for agricultural use. Conversion of forest lands in the 1950s and the 1960s reached 200,000 hectares per annum.

The widespread practice of shifting cultivation affected natural forests. There are two types of shifting cultivators or kaingineros. Uplanders, the traditional kaingineros, make their living by clearing and cropping forest lands and observing fallow periods. Their practice seemed sustainable until they became so numerous that long fallow periods could no longer be maintained. For lowlanders, clearing and cropping is a part-time occupation. Because of increasing landlessness and poverty in rural lowland areas, these poor people are forced to clear patches of forest lands to eke out a living. Additional clearing results from land speculators who pay kaingineros and then apply for reclassification of cleared forest land as alienable and disposable. The slow pace of land classification has led to the preemption of vast forest lands for agriculture by this pernicious practice. Shifting cultivation intensified with the opening of virgin forest lands during the logging boom. Logging provided easy access to once impenetrable areas. The more destructive logging was, the easier it was for the kainginero. In this sense, kaingin-making and destructive logging are intertwined as causes of deforestation in the Philippines.

Until 1975, shifting cultivators or kaingineros were prosecuted, but jail sentences or fines did not deter them. Prosecuted kaingineros often returned and established other kaingin plots that they thought were beyond the eyes of law enforcers. In 1975, the government finally realized that kaingin-making was not a matter of law enforcement, but rather a social, economic, and political problem. Kaingineros occupying forest lands prior to May 19, 1975, were given an amnesty from prosecution. Under a prescribed management system requiring them to plant trees and agricultural crops, kaingineros were at last given permission to occupy up to five hectares of the land they till. This permit, called the forest occupancy permit, is good for 25 years and is renewable for another 25 years. In view of the unfavorable terrain, continuous cultivation, and poor farming and soil conservation practices, most kaingin plots quickly lose their fertility to erosion and depletion of soil after heavy downpours and are usually abandoned after two or three cropping seasons. After abandonment, the areas are taken over by grasses and weeds.

Natural grasslands are classified as forest lands, and most are interspersed with timberlands. There were only slightly more than a million hectares of grasslands in the country as of 1976. By 1982, 535,000 hectares were covered by pasture leases or permits. Because pasture ownership in the Philippines is a status symbol, most of these areas are controlled or owned by powerful politicians and rich families. This indirect use of kaingineros to clear patches of forest lands for conversion into pasture was common in the past. After they secured pasture leases, the kaingineros were driven away. In some areas, the spread of pastures resulted from the natural conversion of abandoned kaingin areas into grasslands. Often, the practice of burning pastures during the dry season to enhance the growth of new grass resulted in wildfires that spread to nearby forests. Pastures formed in this way are not suitable for cattle raising, since cattle expend so much energy grazing on such steep terrain that they produce little meat.

Another form of loss that has wrought havoc to the Philippine forests is conversion of most of the mangrove forests into fishponds. Most of these forests were illegally denuded, and are eventually released to the Bureau of Lands for disposition into fishponds. A total area of 100,000 hectares has already been converted into fishponds.

Conclusion

Lessons from the Philippine experience

The Philippine experience provides some important lessons for other developing countries trying to exploit forest resources to bring more benefits and minimize costs to their societies. The rapid rate at which the Philippines' forests have been depleted shows government's lack of foresight and inability to manage resources. The Philippine government, especially during the timber boom, gave the impression that all it wanted was to liquidate the forests. Excessive allowable cuts beyond the capacity of authorized licensees to exploit have been granted. Had licensees used up their allowable cuts, the results would have been even more devastating.

The government's decision to open up vast forest areas for exploitation is incomprehensible from the conservation viewpoint. Delegating forest management through the licensing system can only succeed if there is effective government supervision. Yet, until the 1970s, and to a certain extent in the 1980s, government documents revealed chronic shortages of personnel to supervise licensees' operations. With too few people to supervise them, licensees could do whatever they wanted. They overcut and inflicted heavy damage on soil and vegetation through careless operations. The shortage of government personnel was exacerbated by inadequate logistic support for effective execution of their duties as police officers and managers of the forest resource. Most important, such licensing regulations can be successfully enforced only where the people have general respect for the law. Unfortunately, in the Philippines the saying, "lawmakers are lawbreakers," apparently held true. A complete lack of sustained political support for conserving forest resources made the licensing system almost completely unworkable.

The unfavorable political setting was reinforced by inappropriate government incentive and taxation policies that discouraged rather than promoted forest conservation. The mode of assessment and rate of forest charges that constituted the main form of forest taxation policies until the 1970s encouraged overcutting and high-grading. Fly-by-night operators whose operations were very destructive dominated the logging industry.

The cost of these operations to the government were high. Not only had it sacrificed its fair share of conversion return, resulting in low government revenues, and promoting quick exploitation, but it had to pay for the resulting environmental damage too.

The Philippine case also shows how a government's objective of social justice through forest land disposal could be deleterious. Like many other governments, that of the Philippines chose to expand agricultural acreage at the expense of forests rather than redistribute already cultivated land more equally. Indiscriminate conversion of about 100,000 hectares of forest lands per annum for almost two decades has been the result of government's deliberate policy of land distribution. Sadly, most of these areas turned out to be unsuitable for sustained agricultural production. Because of steep terrain, loss of forest cover, and characteristically heavy downpours during the rainy season, most of these areas have been heavily eroded and have become virtually wastelands where weeds like the impoverishing *Imperata cylindrica* now thrive.

The rapid loss of forests in the Philippines was also a classic example of the government's inability to balance or set priorities for its national development goals. Opening up vast area of forests in such a short period was intended to satisfy several goals: agricultural expansion, foreign exchange earnings, employment, and government revenues. Yet most of the gains have gone not to the whole society, as planned, but only to a few operators. The converted lands were mostly unsuitable for farming and ended up as pasture lands for the affluent. Government revenues have been extremely low and in fact were not even sufficient to fund the forestry agency for forest management and reforestation. As for employment, the numbers did not increase much in view of declining log exports and harvests and the lack of integration in the forestry industry. The only significant contribution was to foreign exchange earnings, but this was a short-lived gain, for log harvests began to decline in the mid-1970s. Besides, with the domination of the industry by multinational corporations and greedy local entrepreneurs, the net foreign exchange earnings remaining in the country were far lower than the reported gross values of exports.

The Philippine case can also provide a good example of how minimal the contribution of forestry can be when industrialization is forced through government protection. Attempts to build an industry carried big social costs. While incentives were ample, there were gross economic and physical inefficiencies in the mills, especially the plywood mills. Actual rents on processed logs were far below the potential rents, because an increasing volume of exportable logs was converted in inefficient plywood mills. Aside from reduced rents, heavy protection granted to processing industries, especially to the pulp and paper industry, pushed the margin of the forest harvest further than would have been feasible under normal market conditions.

Overall, what made the big difference in the Philippines was the significant role that political forces played in shaping and implementing forest policies.

New directions in forest policy

In reshaping forest policy in the Philippines, experience of the past, along with future needs, provides the best guide to satisfying society's needs from the remaining forest resources.

The main problem in the future will be meeting the rising demand for forest products. Virgin forests will have been logged off in the mid-1990s, and second-growth forests will be expected to fill the demand. But, their lower yields mean that harvests will have to be supplemented from other sources. Log importation is out of the question, since Indonesia and other neighboring countries will have fully curtailed log exports. Two things can then happen. One is that the remaining forests will be intensively logged so that even lower diameter classes not presently harvested will be cut. The government will not likely tolerate this approach, in view of its conservation concern. Therefore, the only realistic alternative is for the industry to shrink. Only firms that can still operate economically with logs from plantations will remain. As local plantations expand, the industry may again flourish, owing to rising domestic demand.

In view of this log supply scenario, the main activity of the forestry sector in the next decades will be to establish and maintain plantations. More and more forest lands will be devoted to industrial tree plantations; in fact, poorly stocked second-growth forests may give way to plantations. Natural forest management will constitute a minor activity of the forestry agency, limited to improving growth and yield of the better-stocked second-growth forests through silviculture and protection of an increasing area of forest reserves. The amount of natural forests requiring timber management will be lower than was planned because of the predicted conversion of second-growth forests into fast-growing plantations. Little logging will take place in the second-growth forests. It is doubtful that much can be relogged economically, in view of the lower quantity and quality and the greater distance between harvestable trees. During the first quarter of the next century, it is possible that no significant harvests will come from the natural forests. Beyond the first quarter, there may still be no logging in the second-growth forests. It is likely that all will be declared as reserves, wilderness areas, protection areas, and national parks serving different purposes than today.

During the transition period the forestry industry will contribute less to the national economy. As plantations become fully developed, though, and as the industry fully adapts to the change from big logs (from natural forests) to smaller plantation logs, the industry can become a major sector in the Philippine economy.

With this scenario and in light of past experience, forest policy change should concentrate on seven matters.

First, forest charges and fees should be substantially increased so that, in combination with other forms of taxes, their total will approximate the stumpage value per unit of product. It follows that the export tax for logs must be further raised to discourage further exports, but quota restrictions should be lifted. Because evasion of these restrictions was rampant, the measures were useless yet costly.

Second, the government should tighten implementation of direct conservation measures. The policy granting no new concessions and banning logging in specific areas should be fully implemented. Under President Marcos these policies were only haphazardly executed. Several new concessions were awarded as exceptions to the policy. In the same manner, some licensees operating in banned areas were able to obtain exemptions from the ban, allowing them to continue their operations in the guise of preventing economic dislocation.

Third, other forestry policies of the Marcos regime that sought to boost agricultural production at the expense of forest lands should be halted. These policies include P.D. 472, which authorized logging companies to devote parts of their concessions to food production; P.D. 861, which authorized pasture lessees and permittees to devote portions of their pasture areas to food production; and P.D. 262, which authorized the use of virgin lands for crop production. Likewise, no further conversion of mangrove swamps to fishponds should be tolerated.

Fourth, the government should enact policies that improve better-stocked second-growth forests and at the same time encourage the establishment of industrial plantations where they are economical.

Fifth, research and development of products using small-diameter logs from plantations should now be given impetus.

Sixth, the social forestry program should be strengthened, as the shifting cultivation problem is a very real one. Defects of the present program should be eliminated and the program should be fully implemented.

Finally, the forestry service should rid itself of corruption and other ills of public service that usually afflict poor developing countries.

Endnotes

1. Forest lands are lands in the public domain that have not been declared alienable or disposable. The term includes public forests, permanent forests or forest reserves, forest reservations, timberlands, grazing lands, game refuges, and bird sanctuaries.

2. Critical watershed is here defined as a land area drained by a stream or fixed body of water and its tributaries having a common outlet for surface runoff where such bodies of water are harnessed for hydroelectric purposes.

3. Classification of lands into alienable, disposable, or forest land was stopped in 1982.

4. *Kaingin* is a portion of the forest land, occupied or not, subjected to shifting or permanent slash-and-burn cultivation. A kaingin has little or no provision to prevent soil erosion. A *kainginero* is a person who makes a kaingin.

5. Japan's imports of Philippine logs for the period 1977–1980, according to Philippine data, totalled 4.6 million cubic meters. Japanese data say the total was more than 5.8 million cubic meters.

6. The "New Society" was decreed by President Ferdinand Marcos when he declared martial law in 1972.

7. With an exchange rate of 7.79 pesos to one dollar in 1980, the forest charges were equivalent to between $0.81 and $1.20 per cubic meter for logs used domestically, and between $1.39 and $1.70 per cubic meter for exported logs.

References

Bautista, Romeo, and John Power. 1979. *Industrial Promotion Policies in the Philippines*. Manila: Philippine Institute for Development Studies.

Ministry of Forestry, Government of the Philippines. 1982. *Philippines Forestry Statistics*. Manila.

Payer, Cheryl. 1982. *The World Bank: A Critical Analysis*, p. 414. New York: Monthly Review Press.

Porter, D. Gareth. 1987. Resources, Population and the Philippines' Future. Unpublished report to the World Resources Institute, Washington, D.C.

Power, John, and Tessie Tumaneng. 1984. Comparative Advantages and Government Intervention Policies in Forestry. Working Paper No. 83–05. Manila: Philippine Institute for Development Studies.

Puyat, Jose, Jr. 1972. Prospects and Problems of the Wood Industries. In *Philippine Economy in the Seventies*, pp. 94–111. Quezon City, Philippines: Institute of Economic Development and Research, University of the Philippines.

Revilla, Martin. 1984. Forest Land Management in the Context of National Land Use. Submitted at the seminar-workshop on Economic Policies for Forest Resources Management sponsored by the Institute of Philippine Studies, February 17–18, at Club Sorviento, Calamba, Laguna. Unpublished.

Reyes, Martin. 1983. Intensifying Conservation of the Philippine Forests. Forest Research Institute, Philippines. Unpublished.

Serna, Cirilo B. 1985. Chief of Planning and Evaluation Division, Bureau of Forest Development, personal communication, September 5.

Sy, Manolito, and Victoria Lasmarias. 1983. The Philippine Dipterocarp Forest: A Post-Harvest Scenario. *Philippine Lumberman,* Vol. 29, No. 2: 34–38.

5 Price and policy: the keys to revamping China's forestry resources

LI JINCHANG, KONG FANWEN,
HE NAIHUI, AND LESTER ROSS

For many years China has suffered from an acute shortage of forest resources. According to the most recent data, based on a nation-wide inventory completed in 1981, China has 115,277,000 hectares of forested land. This ranks China sixth in the world in total forest area, but in relation to China's large surface area and enormous population the resource is small. Only 12 percent of the country's surface area is forested, barely half the global average of 22 percent. China has only 0.12 hectares of forested land per capita, just 18.5 percent of the worldwide average of 0.65 hectares. Table 5.1 shows that timber inventories and reserves are equally scarce. China has 10.26 billion cubic meters of reserves (only 9.03 billion if marginal stands are excluded), an average of only 10 cubic meters per capita, or 15 percent of the global average (Table 5.1).

China's paucity of forest resources has hampered the country's economic progress, forcing industry to use substitutes or simply to do without. The country's paper and paperboard production, for example, falls well short of demand despite rising output and suffers from poor quality because three-quarters of domestic output is not based on wood pulp. Industry is subject to stringent guidelines, codified and strengthened in 1983, to minimize wood used in construction, public utilities, transportation, mining, furniture manufacturing, packaging, and other sectors, in favor of brick, concrete, plastics, and other substitutes (Regulations for Economic and Rational Application 1983).

One response to the shortage of forest products has been an increase in imports to over 9 million cubic meters a year in 1985 (Table 5.2). Were it not for constraints on foreign exchange, especially in 1986 when imports declined as part of a nationwide effort to balance foreign trade accounts, import volumes would certainly be substantially higher.

The effects of this resource scarcity on ordinary households are more poignant. In 1980, nearly half of all rural households suffered from acute

Table 5.1. *Provincial forestry resources, 1981*

	Land available for forestry use		
Province	Area (thousand hectares)	Percentage of total forest land	Rank among provinces
Beijing	627.9	0.23	26
Tianjin	53.1	0.02	29
Hebei	6,213.2	2.33	16
Shanxi	5,769.3	2.16	19
Inner Mongolia	44,094.0	16.51	1
Liaoning	6,710.3	2.51	15
Jilin	8,769.4	3.28	13
Heilongjiang	21,537.1	8.06	3
Shanghai	9.6	0.01	30
Jiangsu	503.8	0.19	28
Zhejiang	5,898.3	2.21	18
Anhui	3,549.5	1.33	21
Fujian	8,874.7	3.32	12
Jiangxi	10,578.3	3.96	10
Shandong	1,936.7	0.72	25
Henan	3,839.0	1.44	20
Hubei	7,402.7	2.77	14
Hunan	11,730.2	4.39	8
Guangdong	12,244.4	4.58	7
Guangxi	13,963.2	5.23	5
Sichuan	19,030.9	7.13	4
Guizhou	9,010.0	3.37	11
Yunnan	26,123.8	9.78	2
Tibet	11,721.3	4.39	9
Shaanxi	12,486.6	4.68	6
Gansu	6,131.5	2.30	17
Qinghai	3,037.3	1.14	22
Ningxia	620.9	0.21	27
Xinjiang	2,693.7	1.01	23
Taiwan	1,969.5	0.74	24
Total	267,130.2	100.00	—

	Volume of standing timber		
Province	Volume (thousand m³)	Percentage of timber volume	Rank among provinces
Beijing	3,920.7	0.04	28
Tianjin	2,079.0	0.02	29
Hebei	48,095.0	0.47	23
Shanxi	53,382.0	0.52	22
Inner Mongolia	946,173.1	9.22	5
Liaoning	108,559.4	1.06	18
Jilin	711,017.6	6.93	6
Heilongjiang	1,551,926.1	15.12	1

Table 5.1. (*continued*)

| Province | Volume of standing timber | | |
	Volume (thousand m³)	Percentage of timber volume	Rank among provinces
Shanghai	616.3	0.01	30
Jiangsu	15,120.3	0.15	26
Zhejiang	98,740.0	0.96	19
Anhui	69,754.9	0.68	20
Fujian	430,559.1	4.20	7
Jiangxi	302,610.8	2.95	8
Shandong	24,258.6	0.24	24
Henan	68,218.6	0.66	21
Hubei	117,826.7	1.15	17
Hunan	198,879.3	1.93	14
Guangdong	231,833.9	2.26	12
Guangxi	265,875.2	2.59	10
Sichuan	1,152,928.3	11.24	4
Guizhou	159,409.0	1.55	16
Yunnan	1,321,313.8	12.88	3
Tibet	1,436,261.6	14.00	2
Shaanxi	279,346.3	2.72	9
Gansu	173,057.3	1.69	15
Qinghai	23,031.8	0.22	25
Ningxia	4,221.6	0.04	27
Xinjiang	234,736.5	2.29	11
Taiwan	226,846.0	2.21	13
Total	10,260,598.8	100.00	—

| Province | Forested land | | |
	Area (thousand hectares)	Percentage of national total	Rank among provinces
Beijing	143.8	0.12	27
Tianjin	29.9	0.03	29
Hebei	1,676.8	1.45	20
Shanxi	810.0	0.70	24
Inner Mongolia	13,740.1	11.92	2
Liaoning	3,652.7	3.17	14
Jilin	6,078.9	5.27	7
Heilongjiang	15,294.4	13.27	1
Shanghai	7.9	0.01	30
Jiangsu	324.7	0.28	25
Zhejiang	3,428.9	2.98	15
Anhui	1,791.6	1.55	18
Fujian	4,496.4	3.90	11
Jiangxi	5,462.3	4.74	9
Shandong	904.7	0.79	23
Henan	1,419.9	1.23	21

(*continued*)

Table 5.1. (*continued*)

Province	Forested land Area (thousand hectares)	Percentage of national total	Rank among provinces
Hubei	3,779.0	3.28	13
Hunan	6,872.3	5.96	4
Guangdong	5,878.6	5.10	8
Guangxi	5,227.2	4.54	10
Sichuan	6,810.8	5.91	5
Guizhou	2,309.3	2.00	16
Yunnan	9,196.5	7.98	3
Tibet	6,320.3	5.48	6
Shaanxi	4,471.4	3.88	12
Gansu	1,769.0	1.53	19
Qinghai	194.5	0.17	26
Ningxia	95.1	0.08	28
Xinjiang	1,120.9	0.97	22
Taiwan	1,969.5	1.71	17
Total	115,277.4	100.00	—

Province	Volume of standing timber on forested land Volume (thousand m³)	Percentage of national total	Rank among provinces
Beijing	1,465.9	0.02	28
Tianjin	187.0	0.00	29
Hebei	26,495.1	0.29	23
Shanxi	33,339.9	0.37	21
Inner Mongolia	847,776.3	9.39	5
Liaoning	100,393.5	1.11	17
Jilin	656,974.5	7.28	6
Heilongjiang	1,436,628.4	15.91	1
Shanghai	18.8	0.00	30
Jiangsu	3,226.4	0.04	26
Zhejiang	79,183.3	0.88	19
Anhui	54,583.5	0.61	20
Fujian	296,379.8	3.28	7
Jiangxi	236,328.0	2.62	9
Shandong	4,837.4	0.05	25
Henan	31,885.1	0.35	22
Hubei	98,604.1	1.09	18
Hunan	160,210.2	1.77	15
Guangdong	203,406.1	2.25	12
Guangxi	220,661.8	2.45	11
Sichuan	1,048,804.2	11.62	4
Guizhou	126,405.0	1.40	16
Yunnan	1,097,033.0	12.15	3
Tibet	1,400,524.6	15.51	2

Table 5.1. (*continued*)

Province	Volume of standing timber on forested land		
	Volume (thousand m³)	Percentage of national total	Rank among provinces
Shaanxi	251,532.7	2.79	8
Gansu	164,020.5	1.82	14
Qinghai	17,154.2	0.19	24
Ningxia	2,769.7	0.03	27
Xinjiang	200,278.3	2.22	13
Taiwan	226,846.0	2.51	10
Total	9,027,953.3	100.00	—

Province	Forest cover	
	Percent of land area	Rank among provinces
Beijing	8.1	20
Tianjin	2.6	26
Hebei	9.0	18
Shanxi	5.2	22
Inner Mongolia	11.9	17
Liaoning	25.1	9
Jilin	32.2	7
Heilongjiang	33.6	4
Shanghai	1.3	28
Jiangsu	3.2	25
Zhejiang	33.7	3
Anhui	13.0	15
Fujian	37.0	2
Jiangxi	32.8	5
Shandong	5.9	21
Henan	8.5	19
Hubei	20.3	13
Hunan	32.5	6
Guangdong	27.7	8
Guangxi	22.0	11
Sichuan	12.0	16
Guizhou	13.1	14
Yunnan	24.0	10
Tibet	5.1	23
Shaanxi	21.7	12
Gansu	3.9	24
Qinghai	0.3	30
Ningxia	1.4	27
Xinjiang	0.7	29
Taiwan	55.1	1
Total	12.0	—

Source: Dangdai Zhongguo de Linye (1985: Appendix 12).

Table 5.2. *Volume of timber imports, 1950–86*

Year	Volume (thousand m³)
1950–54 (average)	14
1955–59 (average)	51
1960–64 (average)	343
1965–69 (average)	685
1970–74 (average)	393
1975–79 (average)	574
1980	1,812
1981	1,553
1982	4,827
1983	6,498
1984	7,910
1985	9,710
1986	7,152

Sources: Zhongguo Tongji Nianjian (1983: 438; 1986: 572); data for 1986 supplied directly by authors.

fuel shortages for up to six months a year, due to lack of fuelwood and other biomass.

The environmental consequences have also been very serious. The depletion of forest resources has aggravated soil erosion, desertification, and stream sedimentation, with attendant effects on water quality and soil fertility. About 1.5 million square kilometers have eroded, of which one-quarter has eroded since the People's Republic was founded in 1949. Estimated soil losses total about 5 billion tons a year, representing a lost nutrient value of about 40 million tons of chemical fertilizer, about double the quantity annually applied by farmers. The net loss of phosphorus, potassium, and trace elements is larger, because of the strong bias toward nitrogen in China's fertilizer mix, a deficiency that has only recently been officially acknowledged (Dowdle 1987). While erosion has multiple causes, some of them natural, about one-third of the soil loss is attributed to fuel-hungry households uprooting plants and removing trees, thus depriving the soil of tilth and nutrients. Other major causes are overgrazing, improper land reclamation, and inappropriate development projects, many of which involve deforestation.

An example is Pengshui County in southeastern Sichuan, an agriculturally based county with a population of more than half a million. Because only 15 percent of farmland is level and of good quality, the farmers must till sloped land. Over 40 percent of this farmland is considered severely sloped. As in some other poor upland areas in southern China, swidden agriculture is still practiced. Forest cover declined from an estimated 28 percent of surface area in 1958 to only 7 percent in 1982, owing to poor agricultural practices and severe deforestation. Erosion is estimated at 480 tons of soil per square kilometer and carries away nitrogen equivalent to about three-quarters of all chemical fertilizer applied. Many ecologists and foresters hope that the official policy restricting cultivation where slopes exceed 25 degrees (itself a very liberal standard) will be enforced, especially in areas like Pengshui that are responsible for a disproportionate share of China's erosion problem. Yet in reality such enforcement will be virtually impossible, because so many people depend on this farmland for their livelihood (Li Ketong 1985; State Council 1982).

Many Chinese scientists also attribute China's tragic history of flooding over the centuries to the long-term decline in forest cover. Although flood damage is a product of a several factors, of which deforestation is rarely the most important, the removal of vegetation can be a major factor in small watersheds. Deforestation can also aggravate flooding in large watersheds by increasing stream sedimentation, which in turn aggrades stream beds and obstructs drainage channels. The most notorious example is the Yellow (Huanghe) River, which flows through the largely treeless loess plateau of northwestern China and transports over one billion tons of sediment annually, aggrading its stream bed in the lower reaches by one centimeter or more per year. The Yangtze (Changjiang) River in central China has a much greater flow, but scientists and officials still fear that its load of 500 million tons of sediment a year is contributing to the flood danger as well. In any event, there can be little doubt that sedimentation is a major problem for the country's inland and marine navigation systems, necessitating massive investments in dredging to keep shipping channels and harbors clear. It impairs reservoir storage capacity to such an extent that some reservoirs silt up almost as soon as they are built.

China's forest deficit antedates the modern era. The pattern of causation is complex and not always responsive to government intervention. Contributing factors include increased aridity in northern and western China, enormous population growth that has put added pressure on fragile soils, and even cultural traditions such as the widely practiced art of calligraphy which used ink manufactured from pine trees (Schafer 1962).

From the time of its founding in 1949, the government of the People's Republic of China recognized an obligation to remedy the forest deficit in the interest of conservation, national security, economic development, and human welfare. Massive afforestation campaigns were begun in the years after 1949, with total plantings averaging 4 to 5 million hectares a year. Some notable achievements were obtained. These include substantial progress in planting the "four arounds" (around houses and villages, and alongside roads and waterways) for timber, fuel, and amenity purposes, especially since this program was given higher priority in the early 1970s. There have also been significant strides in planting shelter belts, the largest of which is in the "three norths" (northeastern, northern or north-central, and northwestern China) where the pace accelerated in the late 1970s. Both of these programs have fulfilled conservation functions, and the village woodlots in particular have increased the local output of timber and other forest products, reducing the burden borne by traditional forestry areas. For example, one-third of timber production in the east-central province of Anhui, traditionally regarded as a very poor province, now comes from woodlots on the plains rather than forests in the mountains (*Renmin Ribao* 1986).

Despite such accomplishments, however, the overall picture is less favorable. Official statistics report that forest cover increased from 8.6 percent of surface area in 1950 to 12 percent in 1981. While this is substantial, it includes a decline from 12.5 percent in 1976. More importantly, the increase seems to have been overstated by extrapolating from incomplete survey data gathered in 1950 (*Dangdai Zhongguo de Linye* 1985: 27). Although very large afforestation programs have been conducted since 1950, with striking success in some areas, the overall survival rate for new plantations is only about 30 percent, a figure including some very marginal plantations. High mortality results in a considerable waste of material and effort and an attendant decline in morale among many program participants (ibid.: 179). Afforestation campaigns have also been partially negated by losses from fire, disease, pest infestation, poor management, and poaching; and by the conversion of forest land to other uses like agricultural reclamation, urbanization, and water resources development projects.

Moreover, the total of consumption and mortality nationwide appears to exceed the volume of new growth by a sizable margin. This is not the intended result of the accelerated logging of old-growth timber, but rather a consequence of rising consumption on a narrow resource base. According to a survey of selected sampling sites, estimated consumption and mortality amounted to about 290 million cubic meters annually between 1976 and 1980. During the same period, new growth totaled only 230 million cubic meters per year. Thus, there was an average net decline of 60 million cubic meters per year in the late 1970s (ibid.: 34).

Naturally, there are substantial regional variations. The state-owned forest districts of northeastern China and eastern Inner Mongolia were in the worst shape. These are China's principal forestry regions, rich in valuable softwoods. Consumption there was found to exceed growth by about 40 percent per year. Within the state forest districts of Heilongjiang, the leading timber producer within the country as a whole as well as the region, forest land declined by an estimated 20 percent during the thirty-plus years ending in 1982. This change resulted from extension of agriculture and human settlements into previously remote sites and to the slow regeneration of sites remaining in forestry use (*Guangming Ribao* 1986; Xie 1985). Similar problems on a smaller scale have occurred in the state forest districts of the southwest.

The problem has been less serious in other areas of the country. Consumption exceeded growth by a more modest 10 percent per year in the mostly collectively owned forest districts of south China, although these areas were seriously damaged during the collectivization drives of the late 1950s and early 1960s. In non-forest districts located primarily in agricultural areas, growth actually exceeded consumption by a small margin in the late 1970s owing to the expansion of small woodlots and shelter belts noted earlier.

The wood-processing industry and distribution sectors have also been subject to serious problems. Because most of China's forests are state-owned and its timber and many other forest products are subject to centralized state planning and strict controls on circulation, state planning and investment policy are critical influences on performance. However, the state planning process has been incomplete, covering only the planned sector of the economy, which included major industrial and commercial users. Within the planned sector, commodity prices were set at artificially low levels without sufficient allowance for species, quality, and other differentials. This resulted in serious distortions, because consumers lacked incentives to use wood efficiently, despite the shortage of forest resources, and producers tended to supply forest products without concern for consumer needs, producing too many small logs and semi-manufactures of low and uneven quality while producing too few longer logs and other desired items.

Although the planned sector accounted for some high priority uses of the resource, it nevertheless accounted for only about one-quarter of total timber output. In principle, the off-plan sector, which included local and noncommercial wood consumption, was permitted more flexible marketing arrangements. In reality, trade constraints impaired efficiency even within the off-plan sector. Producers and traders were subject to especially severe sanctions during periods of heightened ideological tension like the Cultural Revolution. Paradoxically, however, segmenting the plan and off-plan sectors may have misled planners attempting to set timber production goals on a non-declining sustained-yield basis for the nation as a whole, by excluding from their calculations three-quarters of all production. Moreover, the existence of the off-plan sector enabled producers to divert their output to lower priority but more remunerative uses by stratagems like sawing logs into below-grade lengths.

State investment failed to compensate for low commodity prices. Since the sum of profit remissions by state-owned enterprises to the state treasury and tax assessments actually exceeded state investments, there was net disinvestment in the forest sector. Investment priorities were also skewed. The wood panels industry was badly neglected despite its increasingly important worldwide role in improving the efficiency of wood fiber use. As late as 1979, total production of plywood, fiberboard, and particle board represented only 6 percent of sawnwood output by volume, a far smaller proportion than in more industrialized countries. Combined with management and technological shortcomings, this deficiency resulted in the waste of enormous quantities of useable wood fiber. Only an estimated 37 percent of fiber was converted into useful commodities, with large quantities of wood left behind in the forests, discarded in timber yards and sawmills, or diverted to lower priority uses.

There were also organizational problems. Although forest management and forest industry have been formally joined for administrative purposes at the ministry level, and often at lower levels as well, in fact there has been a sharp division between the two. Lacking both direct ownership and management influence over standing timber, state forest industry units had few incentives to facilitate regeneration. Despite enjoying 85 percent of state investment in forestry, the industrial sector was still unable to finance its approved construction projects. Although timber production within the plan sector rose from 12.33 million cubic meters in 1952 to over 63 million cubic meters by 1984, about half of which came from the state forests, the rate of growth lagged well behind that for industry as a whole. Logging was handicapped by inadequate road systems that restricted access to remote sites. Therefore, as timber production rose, more accessible sites were overlogged, while elsewhere timber rotted on the stump. Of the 131 state forestry bureaus or administrations, 61 have virtually exhausted their commercial potential, and another 25 are in imminent danger of doing so. Overconcentrated logging partly explains why regeneration has lagged behind by more than 10 percent despite regulations that mandate prompt reforestation of cutover areas.

Price and policy

Chinese officials and scholars in the 1960s and 1970s increasingly perceived the need for corrective action. Despite serious longstanding problems, forestry's renewable nature allowed a turnaround with proper countermeasures. Production could rise on a sustained-yield basis to benefit both man and the environment. When the political leadership changed in the late 1970s following the death of longtime Communist Party Chairman Mao Zedong, it became possible to analyze forestry problems as part of the "Four Modernizations" (of agriculture, industry, science and technology, and national defense). The Four Modernizations called for rapid economic development, through improving economic efficiency and making greater use of intellectual capital, unlike earlier modernization drives, which had been predicated on the extensive mobilization of underused resources. Naturally, all solutions had to be considered in the context of prevailing political conditions, usually defined in terms of building "socialism with Chinese characteristics." They included retention of state ownership of land and a continuing role for the government as a purchasing and distribution agent.

Analysts considered pricing to be the most fundamental problem. Put simply, the state plan was the dominant element in the economy, but planners' prices were set too low for growers to acquire sufficient incentive to reinvest in forestry. Sometimes they were actually forced to absorb losses, while consumers received perverse encouragement to engage in inefficient consumption patterns. Planners' prices in the unified procurement system introduced in 1953 deprived collective and state growers of forest rent. As shown in Table 5.3, retail prices actually remained constant from 1955 to 1972, and did not rise appreciably until 1979.

Not only were prices set too low, timber price differentials for variations in timber species, quality, and size were also too narrow. With valuable hardwoods costing only twice as much as poplar, and with price premiums of only about 10 percent for larger sizes, consumers had no reason not to requisition superior timber in unnecessarily large quantities. Furthermore, growers had little reason to raise more valuable species, and processors were discouraged from efficient milling.

Forestry was not alone in this regard, for prices of all agricultural products and many other primary commodities were similarly discriminated against relative to prices of finished goods. The price disparity was also evident relative to world prices. For example, China imported hardwoods from Southeast Asia in 1979 for about US$150 per cubic meter, while domestic timber was selling for about 150 yuan per cubic meter at the

Table 5.3. *Procurement and retail prices for timber and bamboo, 1952–85*[a]

Year	Procurement prices (yuan/m³)	Average retail prices (yuan/m³)
1952–56 (average)	20	91
1957–61 (average)	22	100
1962–66 (average)	35	100
1967–71 (average)	37	100
1972–76 (average)	39	106
1977	37	108
1978	37	109
1979	43	120
1980	56	149
1981	62	197
1982	72	216
1983	78	215
1984	—	265
1985	—	313

[a]Prices are nominal or current year prices. Procurement prices are average prices across all species and grades and represent a mix of posted procurement prices, negotiated prices, and above-quota prices, at least since 1978. Retail prices are average prices for all species and grades of timber and bamboo.

Sources: State Statistical Bureau (1984: 450, 456); *Zhongguo Tongji Nianjian* (1985: 537).

time. At the exchange rate of about 2 yuan to US$1 of that time, Chinese timber was selling at about a 50 percent discount to the world market price at Chinese ports of entry, with only part of that discount accounted for by differences in quality. When the overvaluation of the Chinese currency is considered, the disparity was in fact much greater. (Overvaluation has been partially remedied in recent years, most recently in July 1986, when the posted exchange rate was pegged at 3.71 yuan to US$1, despite the dollar's weakness.)

However, the effect on forestry was more severe than on many other raw-materials-producing sectors for several reasons. The administrative or planners' price system is calculated on the basis of allowable costs plus taxes and a predetermined profit margin. No allowance was made for capital charges, even though forests are slow-growing crops, and substantial depreciation costs are incurred while trees are on the stump. The failure to provide for interest charges may be partially explained by the fact that capital investment was financed by state appropriations rather than loans. However, shadow interest charges were still in order to reflect the full costs of production. Failure to include interest charges harmed forestry more than it did industries based on annual production cycles.

Perhaps more importantly, production cost calculations were skewed against forest management. The costs of growing timber, as contrasted with harvesting and processing it, were long ignored under the ideologically molded assumption that trees were gifts of nature lacking in labor value. Only plantation-grown timber was considered eligible for the reimbursement of growing costs, although plantation and natural forests had identical replacement costs. However, even for plantation-grown timber, procurement prices were set on the basis of unusually low, pre-1949 reference prices. Moreover, labor compensation rates were artificially low. Consequently, grower prices were also too low. For example, in the crucial state forest districts of Heilongjiang, the average forest gate price of 56.8 yuan per cubic meter in the late 1970s barely covered direct costs, providing virtually no net return to producers.

In a free market economy, growers would not sell their timber unless they were assured that the return would match or exceed their reservation price, including their production and transportation costs and interest charges. Those growers with relatively high production efficiencies would also obtain economic rents. This principle was not directly applicable in China, however, because the government had asserted title to most forests as early as 1950, and had reasserted that claim in successive state constitutions. Payments in excess of direct costs would simply consist of bookkeeping transfers between state accounts. Therefore, there was apparently no need to satisfy a grower's reservation price. The government simply had to issue production and delivery commands through the planning process in the expectation that they would be honored. In reality, this was often far from the case because producers lacked incentives.

A surrogate for stumpage prices was instituted, known as the silvicultural fee (*yulin fei*). Fees were assessed on all timber produced within the state production plan to finance forest replacement. However, the fees were orginally set at only 5 yuan per cubic meter, which failed to cover actual growing costs. Subsequent increases to 10 and later 15 yuan per cubic meter still proved insufficient. In northeastern China's state forest districts, this amounted to only 19 percent of the forest gate price for timber in 1981 despite China's crying need for additional planting and cultivation. Even then, fee receipts were restricted to financing costs incurred in the first three years after plantation, a restriction which discouraged thinning and other cultivation measures introduced during a plantation's later years. Fee receipts were also not necessarily returned to the original production areas, but might instead be allocated to other jurisdictions in need of afforestation, or actually diverted to other purposes. For a number of years, as much as 85 percent of fee receipts were treated as general government revenues rather than as funds specifically earmarked for forestry.

Moreover, the fees actually functioned more as a severance tax assessed at a flat rate per unit of output rather than as a payment for growing costs. The flat rate had the perverse effects of encouraging the logging of the most accessible and lowest quality trees and shortening the optimal investment period, since higher quality trees provided only a limited price premium that was reduced by the fee. Logging on more remote sites was also disadvantaged, discouraging dispersed logging.

The situation was somewhat different in the collective sector, where an allowance regarded as stumpage charge was made for grower costs. However, the government still set very low procurement prices, which it was able to enforce through a compulsory delivery system known as unified purchase and distribution. Beginning in the 1950s, Class I farm commodities like grain, cotton, oilseeds, and timber were subject to this system, which was intended to ensure the state's control of critical commodities for industry, defense, urban consumption, and other needs at low cost. Procurement prices were generally held virtually constant beginning in the mid-1950s for a decade or more, and even when increases were authorized the amounts tended to be modest. Small silvicultural fees failed to alleviate these problems and actually aggravated the situation by imposing an added tax burden on hard-pressed growers.

Markets and management

A second set of issues involved marketing and management. While increasing government control over priority commodities, the state planning and unified purchase and distribution systems created a variety of problems matching supplies and user needs. Handling costs were inflated because all logs had to be transshipped through timber yards operated by the Ministry of Forestry for sorting before delivery to processing plants and end users. Considerable delays arose in the handling and sorting of logs, while poor storage procedures contributed to the deterioration of product quality. Anticipating delays, handlers and end users expanded their inventories to insulate themselves against supply interruptions, but this exacerbated the inefficiency of the system as a whole. Enjoying a monopoly and responding to irrational prices, producers tended to process and deliver timber to their own rather than their customers' advantage. For example, the posted prices for such items as transmission poles and drilling platforms were attractive, so production exceeded demand by as much as a factor of five. Posted prices for other items, such as railroad ties and bridge supports, tended to be low, resulting in production shortfalls as much as 50 percent below production quotas, since it proved impossible to enforce output norms simultaneously for all product varieties (Li Kaixin 1983).

From another perspective, state control over commodities was handicapped by the limited scope of the state plan. At its peak, the planned sector accounted for no more than about one-quarter of all timber production, the remainder including not only noncommercial fuelwood but also below-grade timber and off-plan output, such as timber intended for local use. While this arrangement eased the burden on badly strained planning and official distribution systems, it also created channels for evasion and distorted the perspective of planners. For example, state forest farms were profligate in their own consumption of timber, because they enjoyed little direct benefit from additional deliveries to the state distribution system, and also tended to be careless in bucking and skidding timber in order to render it unacceptable by the state grading system. Below-grade timber could then be used for other purposes, including barter with other units, at least during periods of relaxed enforcement.

Management presented additional issues. Most of China's forests (52 percent of forest land, 61 percent of timber reserves, and about half of planned timber production) are state-owned, and most of these are administered within the Ministry of Forestry system. The principal exceptions are the holdings of the Ministry of State Farms and Land Reclamation (over 2 million hectares), the Ministry of Water Conservancy and Electric Power, and several industrial ministries like Railways, Coal Industry, and Light Industry (paper manufacturing), which have established plantations for their own use. The actual administration of the state forests is primarily a provincial and local responsibility, with 78 percent of the 4,000-odd state forest farms run by the counties, 12 percent by the prefectures, and the remaining 10 percent by the provinces. Averaging over 8,000 hectares, the farms are in turn grouped in forestry administrations or bureaus in the forest districts. The 131 bureaus administer an average area of 270,000 hectares each.

This arrangement has raised several issues. Low capitalization limited the ability of the state forest farms to expand their operations. Only 33 million hectares of state forest lands are managed by farms, of which but 22 million are treed. This situation contributes to overlogging of the more accessible forests, with 61, or nearly half, of the forestry bureaus already exhausting or in danger of exhausting their commercial resources. However, because the forest farms are located in remote areas, they have to assume some of the functions of local governments. Each bureau is on average responsible for 20,000–40,000 people, so they face major burdens creating employment for new entrants into the work force and funding hospitals, schools, and other services as well as financing the pension and health care obligations of their older employees even as their resource bases are depleted. This responsibility poses a major challenge in maintaining morale and raising the standard of living without degrading the environment.

An additional issue is the recurring rivalries between forest management and forest industry functions within the bureaus and farms. Although the two are nominally under unified authority, in practice the two operate independently for many purposes. The industry side is responsible for logging, local processing, and the like, and enjoys the bulk of state investment. By contrast, the management side is responsible for afforestation, regeneration, and cultivation, and is less well funded. This sometimes creates invidious inequalities in wages. The separation between the two functions led the industry side to disregard forest replacement costs by adopting logging procedures and slash disposal practices that handicapped regeneration, thus accelerating the exhaustion of commercial stands. However, resource exhaustion poses enormous problems for state forest bureaus because their work forces are largely permanent and immobile. As the local economic base declines, the bureaus must continue to work their stands and desperately devise improvisations to provide employment and finance services for their growing populations.

Similar problems hampered the collective or non-state sector. Beginning in the late 1950s, the unit of organization was raised in a series of abrupt steps from the household to the mutual aid team, the primary or lower level agricultural producers' cooperative, the advanced or higher level agricultural producers' cooperative, and ultimately to the commune, with its thousands of households, in 1958. The recession following the Great Leap Forward (1959–1961) resulted in a partial devolution of authority back to the constituent production teams and intermediate-level production brigades within the communes. However, the situation remained unstable and subject to considerable disarray regarding the locus of authority, with persistent tendencies to restore authority back to the commune level.

The instability itself had a damaging impact on a long-term production process like forestry. Although the production teams were the units of account, most of the country's collectively owned forests were operated by the larger communes and production brigades, which were hesitant to relinquish their control over capital assets. While centralization offered some putative economies of scale, it bred continuing resentment among neighboring collectives and collectives at different levels of authority over what they regarded as encroachment or seizure of their own property by others. This tension encouraged poaching and even preemptive liquidation of standing forests to prevent others from capturing valuable assets. It also had a debilitating effect on forest management. The production teams were reluctant to send their best workers to the collective forest farms, which numbered 240,000 by the mid-1970s, preferring instead to send fewer and less qualified workers, because any profits earned by the farms accrued to their parent units. Meanwhile, the parent collectives themselves tended to underinvest in the unprofitable, capital-intensive forest farms, which further impaired productivity and labor morale.

Countermeasures

A series of changes was instituted beginning in the late 1970s in response to the problems hampering forestry development. To Chinese forestry officials and scholars, the most pressing issue was pricing, because unrealistically low prices discouraged investment and management. Given the existing Chinese system based on administrative, cost-based pricing, the need was to formulate and win acceptance of a revised system for valuing forest resources, especially by treating silviculture as a commodity-based activity rather than as a gift of nature or as a social obligation. Then silviculture or forest management might become a profitable, self-sustaining activity with regard to timber production, although state assistance naturally would continue to be required in underforested areas like shelter-belts, where there were no immediate prospects for forestry to become self-supporting, and areas where forestry was primarily conducted for ecological purposes. This view contrasted with past practice, in which forestry was regarded merely as an extractive industry, requiring no new investments for plantations or management.

Because prices had been so low for so long, overall price increases were needed to fully incorporate all costs, particularly silvicultural costs including interest charges. Finer price distinctions among species, grades, and sizes of wood products were also needed to encourage new production and more efficient use. This would enable the state, collectives, and individuals alike to benefit from forestry.

There was also a corollary need for changes in marketing and forestry policy to make distribution more efficient and to place forestry operations on a more stable yet flexible foundation. Forestry is a long-term production process which is harmed by instability. Frequent policy changes about how much authority was to be centralized and about other issues within the collective sector had exerted a debilitating influence in the past. Every new change tended to heighten uncertainty over tenure and future earnings, particularly when the shifts involved more collectivization and fewer material incentives, which discouraged new investment and increased the likelihood of deforestation.

Recalculating administrative prices

Instituting an administrative price system is inherently difficult. The entire decision-making burden that is shared by myriads of buyers and sellers in response to changes in scarcity conditions in a market economy falls instead on the shoulders of price setters. These officials must formulate and adjust a series of equations that multiply in number and complexity as the list of goods expands. Yet price setters have little reliable information to guide their decisions in the absence of valid indicators of demand and scarcity. They must instead set prices on a cost-plus basis for various stages in the production process for both producer and consumer goods. This mechanism rewards inefficiency, particularly in an economy dominated by monopolies undisciplined by competitive bidding. Even if price setters could somehow fix rational prices at the start of the year or outset of the process, changing circumstances would soon upset the balance in a process further distorted by strategic behavior (hoarding, hidden productive capacity, and so on) thoughout the economy.

Nevertheless, because the Chinese system was structured in terms of administrative prices, adjusting the structure in the late 1970s became the first stage in a longer-term process of price reform. With regard to forestry, the critical first step was to account more fully for growing and management costs in the overall calculation of production costs and then product prices.

Although forests are valuable natural resources, the failure to appreciate the full costs of growing trees resulted in the undervaluation of forest management costs. Thus, there was no way to replace forests as they were harvested, nor were meaningful economic constraints imposed on resource exploitation. Even prime forest land proved unrewarding and was converted to other uses, while China's forest deficit remained acute.

Treating forests as gifts of nature without labor value was mistaken, if for no other reason than that true virgin forests are very few in China.

This shortage is particularly marked in southern and central China where human habitation has had a long history, so most forests are fully or partially products of human intervention through forest protection, road building, and the like, if not through actual planting and tending. Even in the more remote forests of northern and southwestern China, however, the government spends millions of yuan annually on forest protection. Thus, both natural forests and plantations deserve full-cost pricing.

More importantly, the pedigree of the forest ultimately is of no concern. Even naturally grown forests must be replaced, and the costs of replacement are no different from those for plantations, other things being equal. A country with a forest deficit, where consumption already exceeded growth by a wide margin and will continue to expand in the course of economic development, not only will inevitably have to replace most of its accessible commercial forests, but will also have to establish new plantations on a large scale for years to come.

Operating with the belief that China's ultimate goal was 20 or 30 percent forest cover, fully incorporating average costs was not sufficient. It was also considered necessary to cover the costs of growing timber on inferior land, rather than just on superior and average land. That is to say, except for land rental charges, which had been irrelevant up to the late 1970s because of the absence of markets for land, the costs of growing trees are lower, the better the land and the more favorable the climatic conditions. If the pricing system were predicated on growing costs on average or superior land, forest cultivation on inferior land would be discouraged.

On the other hand, and perhaps more importantly, producers on superior land would be deprived of economic rents if production costs were calculated according to growing costs on superior lands. Although they would still be entitled to an allowable profit margin, their net return might not exceed their gains from converting their land and labor to agricultural or other non-forestry uses. From the perspective of Chinese forestry officials, the country as a whole would suffer both economically and ecologically if superior land were removed from forestry use.

Thus, calculating stumpage prices in accordance with growing costs on inferior land promised to motivate producers with low costs to engage in forestry, which was particularly important given the urgency of expanding timber inventories. Superior land is moreover often located at lower elevations and near consumption centers, where new forest growth would lower transportation costs while providing ecological benefits. At the same time, operators of inferior land would still be able to anticipate a rate of return sufficient to keep them in production. Naturally, the government retained the option to assess taxes on the economic rents enjoyed by superior land producers in order to minimize income disparities among producers. This is an important consideration; even reform-minded political leaders agree that socialism implies less disparity between income and class groups, as well as public ownership of land.

An additional element in the price adjustments was incorporating the value of time. Time values had been disregarded in the past, when the banking and credit systems were peripheral elements in the capital formation process and depreciation charges were unrealistically low. This disregard for time values worked to the disadvantage of a long-lead-time industry like forestry. It was therefore essential that interest charges be added to production costs to discount fully for the opportunity costs of capital while timber remained on the stump.

Still another factor was the production losses from natural hazards like fire, pests, disease, wind throws, and floods. The problem was that growers were compensated on the basis of the volume of timber that they delivered. If trees were lost during the growing cycle to natural hazards, the growers were in principle ineligible for reimbursement. Because it would be impractical to maintain a continuous census of tree mortality, attention focused on developing forest insurance instead.

Forest insurance is a service purchased by the grower or forest operator (the insured) to obtain compensation for losses in the event of natural misfortune to a stand. Insurance is purchased through a contract between the forest operator and the insurer, an arrangement that enjoys the protection of law. Any grower including state forest farms, collectively owned forest farms, or specialized households engaging in forestry may purchase insurance. The provider is ordinarily the People's Insurance Company of China, although in some instances it may be a local insurance provider or pool.

Forest insurance first began to be investigated in China in 1982. By 1984, pilot projects had been organized in Guilin Municipality, Guangxi, which is within the southern collectively owned forest districts, and in Wangqing Forestry Bureau, in the Yanbian Korean Autonomous Prefecture in the lesser Xingan Mountains of Jilin. Thus, both state and collectively owned forests were studied to determine the suitability of forest insurance. By 1985, additional pilot projects were set up in Shaowu Municipality, Fujian; Benxi Municipality, Liaoning; and Yuncheng Prefecture, Shanxi.

The first problem encountered was the need for accurate appraisals of forest resources. The pilot projects all provided data on production costs and mortality to handle this problem to the satisfaction of the forestry authorities. As shown in Table 5.4 for Shaowu Municipality, the an-

Table 5.4. *Rates and costs for forest insurance, Shaowu Municipality*

Years	Chinese fir trees			Pine trees			Deciduous trees		
	Insured amount (m³/ha)	Rate (yuan/m³)	Cost (yuan/ha)	Insured amount (m³/ha)	Rate (yuan/m³)	Cost (yuan/ha)	Insured amount (m³/ha)	Rate (yuan/m³)	Cost (yuan/ha)
1–10				450	0.003	1.35	600	0.0015	0.90
1–5	900	0.0020	1.8						
6–10	1,200	0.0030	3.6						
11–20				1,500	0.003	4.50	1,650	0.0015	2.48
11–15	1,800	0.0025	4.5						
16–20	2,400	0.0025	6.0						
21 years or more	3,000	0.0020	6.0	2,250	0.003	6.75	2,400	0.0015	3.60

nualized stumpage values of the stands were used to calculate insurance coverage and premiums.

Promising results from the pilot projects have already led the forest insurance system to be used in more than 10 provinces. Forestry officials believe that as commodity production in forestry expands, coverage must be greatly extended, in accordance with the methods for calculating stumpage prices and annualized costs.

While administratively calculated prices consist primarily of production costs, it is now understood that a built-in profit margin must also be incorporated for reinvestments and incentives. The biggest change in recent years is that profit is being calculated not only for the industry side – logging, transportation, and milling – as was historically the case, but also for the silvicultural and management side as well.

Although production costs remain the primary factor in setting commodity prices, while the system of administrative prices remains in force, it is also now recognized that production costs should only be used to set minimum or floor prices for forest products. As market distribution mechanisms are developed, consumers become able to express their demand in monetary terms which can be used to determine higher prices for superior quality products. To further the goal of using prices to maximize efficiency, wider administrative price differentials are already being developed to foster conservation and discourage demand for those forest products that are harder to produce.

Administrative prices must also provide for tax payments, particularly as China's financial system shifts from primary reliance on profit remission by state-owned enterprises toward tax-based revenue sources. This change, which began in the late 1970s, provides greater incentives since enterprises have more control over their income. This trend is being extended throughout the economy.

The following formula (Equation 5.1) reflects the incorporation of these factors in calculating stumpage prices.

$$T = \frac{F(1 + L)^{n'}(1 + P)}{V(1 - S)(1 - C)}$$

$$n' = \frac{F_1(n) + F_2(n - 1) + \ldots + F_n}{F_1 + F_2 + \ldots + F_n}$$

$$= \frac{\sum_{i=1}^{n} F_i(n - i + 1)}{\sum_{i=1}^{n} F_i} \tag{5.1}$$

where

> T = stumpage price in yuan per cubic meter
> F = annual expenses in yuan per hectare
> L = interest rate
> P = profit rate
> C = tax rate
> V = volume of reserves per unit area in cubic meters per hectare
> S = tree mortality rate
> n' = weighted average year

In the above formula, stumpage prices include interest charges and forest insurance premiums. When ownership is transferred while the trees are on the stump or when timber is bought and sold, cumulative profits and taxes are calculated on the basis of annualized costs, which results in rising stumpage charges until the optimum rotation age is reached. In this way, administrative prices are synchronized with biological production cycles. However, because the stumpage charge is only a minimum price, higher prices may be obtained in accordance with scarcity conditions and quality differentials, thus providing producers with larger profits and greater incentives to reinvest in forestry. The producer with higher costs is nevertheless assured of meeting production costs and an allowable profit margin within the administrative price system.

As an example, consider a state forest farm in south China managing a 20-year-rotation plantation of Chinese fir (*Cunninghamia lanceolata*), the most common southern plantation species. Under normal operating conditions the actual annualized investment is 2,400 yuan per hectare, and average reserves are 108 cubic meters per hectare. The farm carries forest insurance with a 0.5 percent premium. The agricultural production profit rate is 20 percent of costs, the interest charge for credit is 6.3 percent, and the tax rate 5 percent. Using Equation 5.1, the stumpage charge of the plantation at age 20 is 77.9 yuan per cubic meter.

$$T = \frac{2,400(1 + 0.063)^{15}(1 + 20\%)}{108[1 - (20 \times 0.5\%)](1 - 5\%)}$$

$$= \frac{2,400 \times 2.5 \times 1.2}{108 \times 0.9 \times 0.95} = 77.9 \tag{5.2}$$

Naturally, the equation results will vary with changes in the values of the various parameters and for different species and cost functions.

Alternatively, Chinese forest economists can calculate stumpage charges through marginal cost analysis. This method gives greater attention to the producer's choice function when deciding how to use his assets. The major criterion here is the maximization of economic benefits, rather

Table 5.5. *Average and marginal production costs, including interest charges, for a Masson's pine plantation[a]*

Age of stand (years)	Aggregate costs (yuan/ha)	Timber reserves (m³/ha)	Marginal costs (yuan/m³)	Average cost (yuan/m³)[b]
8	1,519.5	13.7	43.8	110.9
12	2,301.0	35.7	24.3	64.5
16	3,154.5	58.8	24.9	53.6
20	4,282.5	95.6	37.1	44.8
24	5,770.5	127.1	65.2	45.4
28	8,022.0	155.4	76.5	51.6
32	11,047.5	170.1	89.5	64.9
36	14,701.5	216.3	120.4	68.0
40	19,525.5	247.8	158.9	78.8

[a]Costs include interest charges.
[b]Average cost = aggregate costs (including interest charges)/total timber reserves.

than the technical efficiency issues raised in purely cost-based pricing. Insofar as the timber reserves of forest farms are considered, both inputs and outputs operate under economic and technological constraints and within a specified time interval. Table 5.5 shows marginal cost analysis for a Masson's pine (*Pinus massoniana*) plantation on a southern state forest farm. This method clearly allows for variation in the average costs of trees, and therefore stumpage charges, based upon their varying growing costs and rates of growth. The marginal cost function operates in a similar way to relate the additional growing costs associated with incremental timber growth, and thus helps to identify the optimum rotation age (Figure 5.1).

Both methods reflect the fact that costs are highest in the first few years after the plantation has been established, and then taper off as trees mature. For most efficient production in the individual timber stand, the optimal rotation age or the most favorable time for producers to harvest their forests is that age at which average costs are minimal. Premature harvesting fails to maximize yield without appreciably reducing costs, while late harvesting results in little or even negative growth in yield while costs may increase. Therefore, it is important that stumpage charges reflect the sequential spacing of costs on an annualized or time-discounted basis. Annualized cost-based prices for a Masson's pine plantation in southern China are shown in Table 5.6 and illustrated in an idealized way in Figure 5.2.

Figure 5.3 illustrates such stumpage price patterns for the most popu-

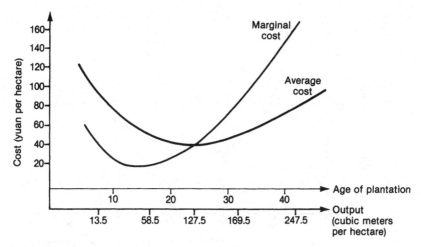

Figure 5.1.   Marginal cost relative to age and volume

Table 5.6. *Annualized stumpage charges for a Masson's pine plantation*

Year	Aggregate costs (yuan/ha)	Timber reserves (m³/ha)	Current year stumpage charges (yuan/m³)
4	973.5	0.6	2,109.3
6	1,276.5	4.5	368.8
8	1,519.5	9.0	219.5
10	1,800.0	21.0	111.4
12	2,301.0	31.0	142.4
14	2,698.0	40.5	86.6
16	3,154.5	63.0	65.1
18	3,679.5	82.5	58.0
20	4,282.5	103.5	53.8
22	4,975.5	127.5	50.7
24	5,770.5	150.0	50.0
26	6,684.0	172.5	50.4
28	8,022.0	186.5	53.1
30	9,567.0	193.5	64.3
32	11,047.5	213.0	67.4
34	12,747.0	235.5	70.4
36	14,701.5	256.5	74.5
38	16,947.0	279.0	78.9
40	19,525.5	301.5	84.2

lar plantation species in China based on the analysis of sample sites. Figure 5.3 shows how the optimal rotation period varies by species under current conditions. Although the optimal harvesting age may vary depending on changes in timber requirements, in general optimal rotation

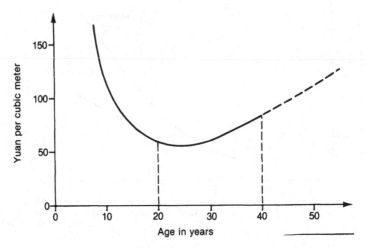

Figure 5.2.   Time sequence price curve for man-made pine forest

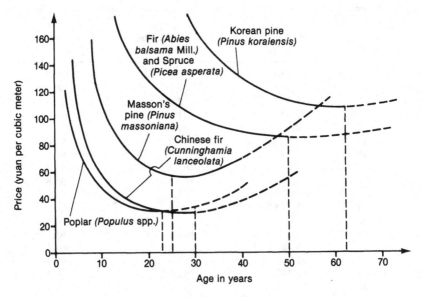

Figure 5.3.   Estimated stumpage price as a function of rotation

periods are 20 to 25 years for poplar, 30 years for Chinese fir, 25 years for Masson's pine, about 50 years for spruce and fir, and over 60 years for Korean pine.

These analytical methods make it possible to specify theoretically efficient prices for China's major forest types in all forest regions. The resulting matrix contains 30 cells for six categories of trees in five regions (Table

Table 5.7. *Calculated stumpage values at optimal rotation age, by species class and region, in yuan per cubic meter (differentials by region in parentheses)*

Species class	Differential by timber species (percent)	Central plains (80%)	South China (100%)	Southwest China (120%)	Northeast China (150%)	Northwest China (170%)
Premium class[a]	200	99.8	124.8	149.8	187.2	212.3
Class I[b]	150	74.9	93.6	112.3	140.4	159.1
Class II[c]	100	50.0	62.4	74.9	93.6	106.1
Class III[d]	80	40.0	50.0	60.0	75.0	85.0
Class IV[e]	60	35.0	43.8	52.6	65.7	74.5
Class V[f]	40	30.0	37.4	44.9	56.1	63.6

[a]Premium Class: camphor (*Cinnamonum camphora*), phoebe (*Phoebe nanmu*), Chinese rosewood (*Dalbergia hupeana* and *D. odorifera*), ginkgo (*Ginkgo biloba*), and others.

[b]Class I: Korean pine (*Pinus koraiensis*), *Phellodendron amurense*, Japanese blue oak (*Cyclobalanopsis glauca*), Manchurian ash (*Fraxinus mandshurica*), and others.

[c]Class II: Chinese fir (*Cunninghamia lanceolata*), Chinese sassafras (*Sassafras tzumu*), spruce (*Picea asperata*), fir (*Abies* spp.), Japanese larch (*Larix gmelini*), and others.

[d]Class III: Mongolian oak (*Quercus mongolica*), Masson's pine (*Pinus massoniana*), *Schima crenata*, Yunnan pine (*Pinus yunnanensis*), and others.

[e]Class IV: elm (*Ulmus* spp.), black locust (*Robinia pseudoacacia*), Chinese cryptomeria (*Cryptomeria fortunei*), Chinese sweetgum (*Liquidambar formosana*), and others.

[f]Class V: poplar (*Populus* spp.), willow (*Salix* spp.), Chinese umbrella tree (*Firmiana simplex*), Chinese wingnut (*Pterocarya stenoptera*), and others.

5.7). The categories are based on timber stumpage value. The standard or base for the classification is Class II timber in South China, calculated according to Equation 5.1.

Applications

As this extended discussion of the early phases of pricing reform has indicated, Chinese forest economists initially operated within an administrative price system. They sought to revise the calculation of production costs to incorporate silvicultural expenses, in order to justify higher producer prices and thereby promote forestry development by enhancing supply. These increases were approved by senior officials who acknowledged that artificially low prices had been a negative influence on production and encouraged wasteful consumption. The increases were effected in several stages to reduce the inflationary shock and also because it was unclear what price levels were appropriate in a dynamic context. Within the collective or nonstate forest districts of the south, procurement prices rose by an average of 30.6 percent and timber yard gate prices by an average of 20.3 percent in 1979 (Table 5.8). In 1981, procurement prices

Table 5.8. *Average forest gate prices for timber*

Year	Northeast China and Inner Mongolia state forest districts (yuan/m³)	South China collective forest districts (yuan/m³)
1952	60.3	45.1
1958	60.3	47.0
1973	56.2	66.9
1979	56.2	80.4
1980	73.5	80.4
1981	80.9	96.5
1986	143.0	—

were raised by an additional 36 percent and timber yard gate prices by another 20 percent on average. No further price increases were implemented during the following five years because budget imbalances resulted from paying higher prices to producers and increasing appropriations to consumers to cover their higher costs.

During the same period, four producer price increases were put into effect in the state forestry sector. In northeastern China, the country's principal timber production region, a 30 percent increase was implemented in 1980, one of 10 percent in 1981, yet another in 1983, and then a much larger increase of 44 percent was implemented in the fall of 1986, upon the realization that previous increases had been insufficient.

The price increases not only resulted in higher incomes for forestry producers, they also widened the price differentials for categories of timber in terms of species, grade, and size. Within the state sector, the prices for birch and larch were raised to 95 yuan from 67 yuan and 84 yuan respectively. Thus, the price differentials between the superior species (Korean pine) and ordinary hardwoods (birch) widened from 1:1.57 to 1:2.25. The margin between sawnwood and logs was narrowed from 1:2.5 to 1:1.7 to give the grower a larger share of the revenues. Meanwhile, fixed prices were lifted for small pieces of timber (less than 8 centimeters in diameter and 2 meters in length), to encourage greater use of wood fiber by making the production of small pieces of wood more attractive to the grower and logger.

Markets and management

While these changes were being implemented within the administrative price structure, other important developments were taking place outside this structure. Most basic was the relaxation of controls over the collective or nonstate forest districts, located primarily in southern China. This change began in the late 1970s with the opening of timber markets throughout the forest districts in which consumers and purchasing agents competed with one another for timber. These markets partially displaced the old unified purchase and distribution system, which was operated as a state monopoly. The markets provided producers with generally higher prices (Table 5.9) but proved controversial because they coincided with increased deforestation in parts of the country. Many foresters and conservative officials maintained that market liberalization was the cause of the increased logging, but there were also complaints from longtime consumers who suddenly had to pay sharply higher prices for their timber and found their regular supply sources interrupted.

In reaction, timber markets were closed in forest districts in 1980, and the size of the police, procuracy, and judiciary forces devoted to forestry affairs was greatly increased. After the immediate crisis was resolved partly by severe and highly publicized punishment for some violators, markets were gradually reopened. In the fall of 1984, approval was granted for markets to operate in poorer areas of the country, and then in February 1985, the compulsory procurement system was officially abolished. This change did not completely eliminate price controls. Rather, it established a three-way system in which a reduced portion of output remained subject to compulsory state procurement, though at the higher prices introduced in 1981; a second portion was subject to purchase by the state at higher, above-quota prices; and a third portion was made available for market purchase. However, market sales did not completely eliminate price controls. The markets were regulated, with buyers and sellers both subject to registration. Floating prices or price bands were established to limit upward price fluctuations. Most importantly, the number of transactions was limited in the name of conserving resources, through authorizing forestry officials to regulate the size of the annual harvest. Virtually all felling, including that of most fuelwood trees, was subject to approval to ensure that the size of the local harvest did not exceed growth on an annual basis.

State forestry units were also accorded somewhat greater freedom to market their produce, if to a lesser extent than their nongovernment-owned counterparts. Within the state sector, a dual structure was established under which the state forest bureaus were allowed to sell a portion of their output at premium prices once they had satisfied their output norms (Table 5.10).

As a result of the liberalization of the economy, by the end of 1985, 233

Table 5.9. *Urban timber market prices in China's principal consumption regions (yuan per cubic meter, except where otherwise indicated)*

Region	Standard Chinese fir logs (4–5.8 m × 14–18 cm)	Standard Chinese fir logs (5–7 m × 12–14 cm)	Standard Masson's pine logs (2–3.8 m × 20–28 cm)	Imported Douglas fir logs	Standard miscellaneous logs (2–3.5 m × 20–28 cm)	Standard warp plywood (3 ft × 6 ft × 0.5 in) (yuan/sheet)
Hangzhou	450–480	420–470	280–300	400–420	—	17.5
Nanjing	500	—	290	450	—	13.4
Xi'an	—	430	338	—	—	14.5
Beijing	—	490	—	400	—	—
Fuzhou	490	—	240	—	240	17.5
Hefei	550	480	280	450	—	—
Wuhan	500	360	215	380	—	—
Changsha	500	370	260	—	—	—
Guangzhou	500–560	—	210	400	—	—

Table 5.10. *Timber procurement and yard gate prices in Fujian (yuan per cubic meter)*

Timber product	1984[a] base price	1985[b] negotiated price	1986[b] negotiated price
A. Procurement prices			
Average log	50.42	112.76	115.10
including Chinese fir	80.75	262.28	237.56
Pine	59.84	101.86	109.99
Miscellaneous	51.90	98.35	103.15
B. Yard gate prices			
Average log	111.33	206.31	217.06
including Chinese fir	142.12	376.04	402.11
Pine	103.26	202.96	212.92
Miscellaneous	111.85	198.13	208.23

[a]Before special products tax introduced.
[b]Including special products tax.

of the 256 commodities originally subject to unified distribution by the State Planning Commission and State Commodities Supply Bureau were detached from that system. Timber, along with coal, steel, cement, and 10 other commodities remained subject to controls over distribution because of their overall economic importance as production inputs. Nevertheless, an increasing proportion even of these remaining 23 commodities was allowed to circulate through more flexible channels. With regard to timber in particular, 69.3 percent of the volume subject to unified distribution under the state plan (which, again, accounted for only one-quarter of total output) was no longer allocated at the low, preferential fixed prices. It was instead distributed at higher off-plan or floating prices, much of it through markets established in major urban areas under the auspices of the State Commodities Supply Bureau with help from the forestry departments. About 46 percent of the volume of timber consumed by local-government-run enterprises (as opposed to central-government-run enterprises which are more deeply enmeshed in the state planning system) was distributed through these markets (Zhao 1986).

These timber and forest products markets supplemented the efforts of the state forestry agencies in handling the burgeoning volume of transactions occurring as the economy developed. As an alternative to markets, state forest products exchange meetings were organized between buyers and sellers. However, the four such meetings held between 1982 and 1984 handled only 3.63 million cubic meters, less than 2 percent of the state planned timber production volume during the same period (*Dangdai Zhongguo de Linye* 1985: 424).

Organized amid rising consumer complaints and in response to the demands of prospering localities and enterprises, the markets at first were witness to very sharp price rises. In general, however, they not only provided welcome flexibility, they also enabled the authorities to supervise prices and transactions more closely. By 1986, it was reported that prices were already stabilizing owing to market adjustments, coupled with price supervision and the availability of imported wood to buffer imbalances between supply and demand (Table 5.2).

Domestic negotiated prices for timber in southern China now not only equal but actually exceed the prices of imports. This represents a marked change from the past when domestic timber was priced well below international levels, which presented a strong disincentive to growers. However, allocated timber, that is, timber distributed within the state plan at preferential prices, which still accounts for at least half of state planned production, continues to be priced below international prices.

Because of the rise in prices and the greater price differentials for timber of different quality, the markets have begun to respond more sensitively. In general, consumers seem to be more conscious of quality. Superior quality timber remains in high demand while inferior timber remains unsold. Douglas fir sells out quickly, with the supply limited by ceilings on import volume set to conserve foreign exchange. Chinese fir is also in strong demand, not only because of its quality but also because the size of the harvest is regulated in the interest of conservation. Other good quality timber like Manchurian ash (*Fraxinus mandshurica*), Korean pine, and basswood (*Tilia* spp.) is in high demand. However, small-sized miscellaneous wood has trouble finding buyers, although the large and medium-sized logs are easy to sell.

There have also been major changes in management institutions along with the changes in distribution. Especially within the collective sector, the forest farms have been under stress as a result of the sudden changes in the countryside. In all, the number of collective forest farms declined from almost 250,000 in 1975 to about 175,000 in 1982 through consolidation and abandonment. By 1984, the number of workers employed in the collective forest farms had been reduced by nearly a million, or about 40 percent (*Dangdai Zhongguo de Linye* 1985: 387, 392).

In their stead, households and other small private entities have assumed a larger share of the work load. Beginning in 1979 and extending through the end of 1985, 50 million households were involved in managing 30 million hectares of private hills and 40 million hectares of forest land under the contract responsibility system, which resembles sharecropping in its most common form. That is to say, the collective retains ownership of the land, but then contracts with households or workers to manage the land as forests or for specific forestry tasks, which are stipulated in contracts with terms as long as 15 to 20 years or for the expected life of the stand. The contracts may be inherited. In return for their efforts, the contractors receive some proportion of income from the forests or woodlots and rights to other income and commodities produced with the land. The private hills are by contrast supposed to be assigned to households more permanently, without requiring payment of a share of income back to the collective. In actuality, however, the distinction between responsibility hills and private hills is often more nominal than real.

The households' role has become increasingly important. By 1985, households were conducting over 50 percent of an expanding volume of afforestation around the country. In the arid northwest, households were assuming responsibility for soil conservation measures on erodable land, with the state subsidizing what would otherwise often be unprofitable efforts.

The most notable case involves Li Jinyao, who took over a struggling Masson's pine plantation in Xianyu County along the Fujian coast. Established back in 1969, the 70-hectare stand averaged only one meter in height when Li and three others jointly contracted to manage it. The other three soon dropped out, enabling Li to negotiate better terms. He also acquired contracts to manage a tea grove and quarry. Within only three years, he was managing 200,000 pines and had planted 70,000 Chinese firs, 70,000 eucalypts, and 5,000 tung trees. He also put in a small fruit orchard and other profitable tree cash crops. His work force grew to 30 and his trees averaged 2.3 meters in height. Although his operations ran a cash flow deficit during this period and Li encountered political criticism over the scale of his operations, he was able to continue in operation when officials intervened on his behalf and he received substantial loans totaling almost 300,000 yuan, some of which came from overseas Chinese relatives. Otherwise, he would never have been able to expand so rapidly.

Another essential ingredient in the expanded role of both household management and market-oriented forestry was the demarcation and enforcement of property rights. Beginning in 1979, authorities proceeded to define more carefully and to enforce both ownership and usage rights in forestry. This contrasted quite sharply with recent history when rights were nonenforceable, which greatly impeded forestry development. The rights themselves are inheritable and to a limited extent are alienable, it now being possible in some areas to sell timber while it is still immature. The prospect of having markets prior to the normal harvest age should relieve many potential woodlot operators of preoccupations about illiquidity, and thus favor forestry development.

Changes within the larger state sector have been less extensive, in keeping with the general pattern throughout the economy as a whole. There has nevertheless been some effort to transfer the management of specific tasks or sites to workers or dependents in groups or as individuals. There have also been some experiments in converting state forest farms into more profit-oriented enterprises, including partially worker-owned joint stock enterprises. However, these changes have proved very controversial and have yet to be widely applied or achieve demonstrated success.

Processing industries and consuming regions have been encouraged to invest in forestry operations elsewhere in the country. This arrangement ensures them timber supplies while simultaneously helping to alleviate the critical capital shortage that has greatly hampered forestry development. Capital movements across administrative boundaries have only begun in the last few years, so it is impossible to judge their success, but their potential appears to be significant.

Conclusions

As early as the 1970s, the need for urgent measures to expand forest resources and reduce China's chronic deficit in wood products was recognized. Mobilizing the population in massive annual reforestation campaigns and other episodic actions that skirted underlying structure problems was seen as insufficient, in the absence of changes in pricing, management, and marketing systems to provide stronger instititutional arrangements and material incentives for higher productivity and efficiency. Forestry needed to be economically attractive as well as socially meritorious.

The first emphasis fell on increasing price levels to provide growers with more revenues and consumers with stronger incentives to conserve wood. Prices were raised several times, and price differentials among grades and species were widened. Pending in early 1987 was a further modification of the stumpage price system. Reproduced in the appendix, the proposed "Provisional Methods for Management of the Stumpage Price System" was to be submitted to the State Council for approval in 1987. It clearly provides for full-cost stumpage prices that include interest charges and tries to regularize stumpage prices among regions.

These remarkable policy measures adjusted the level and structure of stumpage prices to reflect growing costs more adequately, with allowance for interest charges, profits, and insurance. By basing price-setting formulas on the costs of producing timber on inferior lands, Chinese authorities recognized the need to expand the forest resource base, with the goal of 20 percent forest cover by the year 2000 and one of 30 percent eventually.

At the same time, further policy changes recognized the inherent limitations of centrally administered pricing. One such limitation is the difficulty in promoting greater efficiency among growers when administered prices are set on a cost-plus basis and production is scarce relative to demand. Another is the transitional problem of ensuring appropriate rewards to producers when costs throughout the economy are rapidly adjusting to price increases in other sectors. Real income gains for growers during the period of inflationary adjustment were less than implied by nominal timber price increases, and may have been less than in competing sectors such as agriculture, in which regulations were relaxed more quickly.

A greater limitation of administrative pricing is that it inherently ignores demand conditions for wood products of various sizes, shapes, grades, and species that vary from region to region. Under conditions of extreme scarcity and excessively low prices, most output could find a buyer, but even frequently revised price formulas of considerable complexity could not efficiently match shifting market demand and supply conditions.

Therefore, further policy innovations have greatly increased the role of markets in distributing output and conveying price signals to producers and consumers. Both in the plan and off-plan segments of timber markets, increasing fractions of output have been freed for transactions at negotiated prices. At the same time, state, collective, and household producers have been given greatly increased responsibility to respond to market signals in choosing production plans. Government officials have nonetheless been forced to intervene from time to time in this process to avoid transitional problems of excessive harvesting and price rises. The prior adjustment of administered prices toward market-determined levels undoubtedly reduced such transitional problems.

Further policy adjustments can promote higher productivity in timber supply and use. State and collective forests can adjust over time to the opportunities presented by the new system, and controls over producer decisions can be further relaxed. Markets can function more efficiently as expectations stabilize, and a variety of fees imposed by local governments, freight carriers, and middlemen that capture some of the transitional rents are reduced or eliminated.

Overall, however, forestry policy in China has been improved dramatically over a relatively short period of time. Incentives have been improved. The effective participation of millions of households has been secured. The forest resource base has been expanded. A groundwork of market allocation has been established that will help rationalize production and consumption decisions. The benefits of these far-reaching changes are already being realized.

Appendix: Provisional methods for management of the stumpage price system (draft proposal submitted to the Ministry of Forestry)

Chapter I. General principles

Article 1. This system is specially formulated in accordance with the spirit of the Central Committee of the Communist Party's "Decisions on Some Problems Regarding Forestry Conservation and Forestry Development";* in order to conserve, develop, and rationally utilize forest resources; to implement the policy of "treating silviculture as the base"; and to work toward maintaining an ecological balance predicated upon "grasping both afforestation and management, integrating logging and cultivation, using the forests to nurture the forests, and sustain yield."

Article 2. The process of cultivating timber forests is a process of commodity production, the principal product of which is standing timber. Therefore, silvicultural production must practice a system of economic accounting to enable limited silvicultural capital to generate even greater economic results, to accelerate the restoration of forests, to develop forestry, and to begin a new phase in forestry construction.

Article 3. Forests are renewable resources whose cultivation requires the consumption of certain amounts of living labor and materialized labor. Therefore, forests and trees not only have use value, they also have inherent value. The monetary expression of the value of standing timber on forest land is the stumpage price. The stumpage price should be included within timber production costs, since it is an important component of timber prices and a basis for setting timber prices.

*March 8, 1981, translated in Ross and Silk (1987) [Ed.].

Chapter II. The components of stumpage price

Article 4. The theoretical stumpage price is calculated on the basis of inferior forest land, the costs and profits of silvicultural production, and taxes. As for stumpage receipts, beyond the reimbursement of silvicultural production expenses, the producers should receive a certain amount of profit and should remit to the state a certain amount of taxes.

Article 5. In accordance with the principles of investment in silvicultural production and the theory of forest reproduction, the theoretical stumpage price shall be calculated in accordance with the entirety of silvicultural production expenses on a compound interest, weighted-year basis (that is, a period of shorter duration than the forest rotation period).

Chapter III. Stumpage pricing districts and stumpage price levels

Article 6. In accordance with the economic circumstances and natural geographic conditions in forest districts, the entire country shall be divided into six stumpage pricing districts:

1. Northeast Stumpage Pricing District: all of Heilongjiang, Jilin, and Liaoning, and parts of Inner Mongolia;
2. Northwest Stumpage Pricing District: Shaanxi, Gansu, Ningxia, Qinghai, and Xinjiang;
3. Central Plain Stumpage Pricing District: all of Henan, Hebei, Shandong, and Shanxi, and parts of Inner Mongolia;
4. Southern Stumpage Pricing District: Guangdong, Guangxi, Hunan, Hubei, Jiangxi, Fujian, Zhejiang, Jiangsu, and Anhui;
5. Southwest Stumpage Pricing District: Yunnan, Sichuan, Guizhou, and Tibet;
6. Taiwan Stumpage Pricing District: Taiwan.

Article 7. There should be different stumpage price levels, that is, stumpage price lists, for different stumpage pricing districts, different species of trees, and even different age classes of the same species of trees.

Article 8. The theoretical stumpage prices should be implemented within the stumpage pricing districts. However, when difficulties are encountered in the course of implementing the theoretical stumpage prices as a consequence of China's current financial conditions or forest management capacity, transitional methods can be put into effect to attain the theoretical stumpage price levels gradually.

Article 9. Stumpage price levels for wood and bamboo consumption in bamboo, fuelwood, charcoal, and the production of forestry side-line products (such as wooden and bamboo manufactures and semi-manufactures, wood ear* and xianggu mushrooms) should be separately determined by the provinces (including the municipalities directly under the central people's government and the autonomous regions) with reference to this system.

Chapter IV. The assessment of stumpage prices and their sphere of application

Article 10. To conserve forest resources and raise the level of forest management, stumpage prices shall be calculated for all timber of any species that is felled, according to the volume felled, regardless of the purpose for the felling or the method used. In accordance with current conditions, the calculation can temporarily be made on the basis of the volume of timber that is removed. But conditions must be actively fostered for the gradual transition to calculating stumpage prices according to the actual volume of the trees.

Article 11. All units conducting capital construction in forest districts that have to fell trees must, without exception, obtain approval from the forestry departments, and pay stumpage charges in accordance with the rules. When state forests are felled, the stumpage charges shall be tendered to the forestry departments. When collectively or individually owned trees are felled, the stumpage charges shall be tendered to the holders of forestry ownership rights.

Article 12. When industrial and mining enterprises or communications and transportation departments log the specialized timber forests that they themselves have planted, they should tender stumpage charges in accordance with the standards set in the regulations governing this system, to serve as a source of capital for regeneration and afforestation.

When [government] organs, schools, and other units that have put idle land to use through tree planting and afforestation fell these trees, they must tender stumpage charges so that they afforest as they log and thereby ensure regeneration.

*Auricularia auricula judae, an edible fungus [Ed.].

Article 13. Stumpage receipts should be directly applied to forest reproduction, including seed gathering, the nursing of seedlings, afforestation, artificially assisted natural regeneration, forest cultivation, forest protection, the maintenance and construction of forestry roads, fire prevention measures, communications equipment, nursery facilities, the purchase of silvicultural implements, forestry resource surveys, scientific research, and management expenditures.

Chapter V. Stumpage price management

Article 14. Stumpage charges belong to the forestry rights holders. Stumpage charge receipts from state-owned forests belong to the central and local forestry departments and forestry enterprises, and shall be managed and utilized at the various levels. Stumpage charge receipts from collectively owned forests belong to the owners, but a specified proportion may be reserved for use as a silvicultural insurance fund. The receipts of specialized households shall be deposited in special bank accounts for specified uses, and their use shall be supervised by the forestry departments and banks, with the actual proportions to be determined by the provincial governments. With regard to the silvicultural insurance funds submitted by the collectives, the surplus may be temporarily used for loans in accordance with negotiations between the forestry departments and the forestry rights holders, to be repaid on schedule with interest.

Article 15. Stumpage receipts from cooperative afforestation and afforestation by commune members on their private hills, plots, and homesteads shall be proportionately shared by the participating units and individual owners. The ratios for apportioning receipts shall be separately stipulated by the various provinces (including municipalities directly under the central people's government, and autonomous regions).

Chapter VI. Supplementary provisions

Article 16. Following approval of the system by the state, the forestry, finance, and commodities pricing departments in charge shall jointly supervise their implementation.

Article 17. After the theoretical stumpage prices have been put into effect, the current methods for assessing silvicultural fees shall be abolished. Until stumpage prices have attained the theoretical stumpage pricing level, however, the current means for assessing silvicultural fees, consisting of capital construction and operations fees, may temporarily remain in effect without change to develop silvicultural production.

Article 18. The various provinces (including municipalities directly under the central people's government, and autonomous regions) may formulate detailed implementing regulations in accordance with this system, and report them to the Ministry of Forestry, Ministry of Finance, and State Commodities Pricing Bureau for the record.

Article 19. This system shall take effect from [date to be specified].

References

Dangdai Zhongguo de Linye (Contemporary China's Forestry). 1985. Beijing: Zhongguo Shehui Kexue Yanjiu Yuan Chubanshe.

DBC Associates, Inc. 1986. *The Market for Softwood Lumber and Plywood in the People's Republic of China.* Washington, D.C.: National Forest Products Associations.

Dowdle, Stephen. 1987. Seeking Higher Yields from Fewer Fields. *Far Eastern Economic Review,* Vol. 135, No. 12: 78–80.

Guangming Ribao (Brilliant Daily). 1986. May 27.

Jingji Ribao (Economic Daily). 1986. May 5.

Li Kaixin. 1983. Concentrate Material Strength to Guarantee the Construction of Keypoint Projects. *Honggi* (Red Flag), September 1: 16–19.

Li Ketong. 1985. Research on the Reversion of Farm Land Back to Forestry. *Exploration of Nature,* Vol. 4, No. 1: 121–24.

Regulations for Economic and Rational Application of Wood and Wood Substitutes. 1983 (January 15). Translated in DBC Associates, Inc., *The Market for Softwood Lumber and Plywood in the People's Republic of China.* Appendix G. Washington, D.C.: National Forest Products Associations, 1986.

Renmin Ribao (People's Daily). 1986. Beijing, May 25.

Ross, Lester. In press. *Environmental Policy in China.* Bloomington: Indiana University Press.

Ross, Lester, and Mitchell A. Silk. 1987. *Environmental Law and Policy in the People's Republic of China.* Maryland Studies in East Asian Law and Politics No. 8. Westport, Conn.: Quorum Books/Greenwood Press.

Schafer, Edward H. 1962. The Conservation of Nature under the T'ang Dynasty. *Journal of the Economic and Social History of the Orient,* Vol. 5, Part 3, December: 279–308.

State Council. 1982. Regulations on Soil and Conservation Work. Translated in Ross and Silk (1987).

State Statistical Bureau. 1984. *Statistical Materials for China's Commodity Trade Prices, 1952–1983.* Beijing: Zhongguo Tongji Chubanshe.

Xie Huibin. 1985. The Circumstances Surrounding the Destruction of Ecological Balance and Proposals for Its Restoration in the Yichun Forestry Region. *Beifang Huanjing*, No. 1: 62–68.

Xu Wuchuan and Chen Daping. 1982. Discussion of the Theory of Timber Price and Its Calculated Model. *Linye Kexue* (Scientiae Silvae), Vol. 18, No. 1: 71–79.

Zhao Renwei. 1986. The Dual System Problem in China's Economic Reform. *Jingji Yanjiu*, September: 12–23.

Zhongguo Tongji Nianjian (China Statistical Yearbook). 1983, 1985, 1986. Beijing: Zhongguo Tongji Chubanshe.

6 Public policy and deforestation in the Brazilian Amazon

JOHN O. BROWDER

The Brazilian Amazon's forest resource

Spanning an area of 5.5 million square kilometers, the Amazon (Figure 6.1) is the world's largest contiguous tropical moist forest. Tropical forests extend into nine South American countries, but in the Amazon the Brazilian portion (3.8 million square kilometers, or 69 percent) is the largest. The Amazon has four main types of vegetation. The dense tropical forest, *floresta densa* or "hylea" found mainly in the northern Amazon States (Amazonas, Amapá, Roraima, Pará, and Maranhão), covers 48.8 percent of the region. The less exuberant, shorter, but still continuous "transition forest," *floresta aberta* or *fina*, in the central Amazon (Acre, Rondônia, northern Mato Grosso and Goiás, and western Maranhão), covers 27 percent of the region. Farther south, mainly in Goiás and southern Mato Grosso, are savannah shrublands, *campo cerrado*, that cover 17.2 percent of the region. The fourth type, savannah grasslands, *campos naturais*, occurs mainly in the *várzea* floodplains, along the Atlantic coast in Amapá and Marajó Islands, and in northern Roraima, and covers only 6.9 percent of the "Legal Amazon" region. The tropical zone embraces about 76 percent of the Brazilian Legal Amazon region (3.8 million square kilometers).

The Brazilian Amazon region[1] alone is believed to contain some 6,000 different tree species (Correa de Lima and Mercado 1985: 152), many endemic to specific areas. The growing stock varies widely in density, from 100 to 270 cubic meters per hectare, but the natural distribution of individual tree species is sparse; an average of 84 to 90 percent of the species are represented by fewer than one individual (more than 15 cm. diameter at breast height – d.b.h.) per hectare (EMBRAPA 1981).[2] The Amazon contains 48 to 78 billion cubic meters of living timber, enough, according to one journalist, "to build every person in the world a house." In strictly monetary terms, this potential industrial roundwood, as a capital asset, would have a current (1984) market value of $1.7 trillion, making

247

Figure 6.1.   The Brazilian Amazon

Brazil one of the wealthiest natural resource owners of the net oil-importing countries.[3]

Despite the size of this timber resource, the industrial wood sector[4] plays a small but rapidly growing role in the Brazilian economy. In 1980, the most recent year for which relevant census information is available, Amazonian timber accounted for only 12.9 percent of the region's industrial output (up from 6.1 percent in 1960). Nationwide, forest products accounted for only 4.9 percent of Brazil's 1980 foreign exchange earnings (IBGE, *Anuário Estatístico* and *Censo Industrial* various years). Brazil, with 31.7 percent of the world's estimated volume of living broadleafed timber (Erfurth 1974: 86), supplies less than 10 percent of the world's consumption of tropical wood products (UNIDO 1983: 35), which is expected to

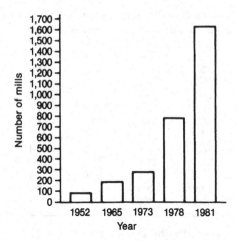

Figure 6.2. Growth in number of Amazon sawmills, 1952–81 (Browder 1986)

increase steadily in the future (Pringle 1976; FAO 1978; Myers 1981; UNIDO 1983).

In the Amazon, forestry is a leading industrial sector. Four of the region's six states and federal territories depend on wood products for more than 25 percent of their industrial output (IBGE, *Anuário Estatístico*, various years, cited in Browder 1986: 65). In Rondônia and Roraima, wood products account for more than 60 percent of industrial output. Many new urban Amazon settlements depend on local lumber industries for their only links to the national economy.

Industrial wood production in the Brazilian Amazon has expanded vigorously. The number of government-licensed mills increased more than eightfold since 1965, from 194 plants in that year to 1,639 plants in 1981 (Figure 6.2). Average annual output per mill increased from about 2,000 cubic meters of sawnwood in 1962 to 4,500 cubic meters in 1984. This rapid growth in capacity is reflected by the Amazon's increasing contribution to national roundwood production, from 14.3 percent (4.5 million cubic meters) in 1975 to 43.6 percent (17.4 million cubic meters) in 1984 (Table 6.1).

The biggest obstacle to better use of forest resources in the Brazilian Amazon is ignorance of the region's tropical hardwood resources. In 1972, only 23 of the region's estimated 1,500 different tree species accounted for 90 percent of total roundwood production (Bruce 1976: 14). In 1983, only 250 different tree species were harvested in volumes that would indicate industrial use in the Brazilian economy (IPT 1985: 6). Foreign markets for Brazilian tropical hardwoods are even narrower.

Table 6.1. *Roundwood production in Brazil, 1975–84 (million cubic meters)*

	1975	1977	1979	1981	1983	1984
Total	31.5	32.3	31.6	35.6	38.6	39.9
North (Amazônia)	4.5	6.7	8.4	13.1	16.1	17.4
Percent of total	14.3	20.7	26.6	36.8	41.7	43.6
Northeast	5.2	5.3	5.6	6.8	7.2	7.7
Percent of total	16.5	16.4	17.7	19.1	18.7	19.3
Southeast	2.2	2.0	1.2	1.6	1.7	2.2
Percent of total	7.0	6.2	3.8	4.5	4.4	5.5
South	16.9	15.3	13.4	10.9	10.2	9.0
Percent of total	53.6	47.4	42.4	30.6	26.4	22.6
Center-west	2.6	2.9	3.0	3.3	3.4	3.5
Percent of total	8.3	9.0	9.5	9.3	8.8	8.8

Note: Percentage totals may not add to 100 due to rounding.
Source: IBGE, *Anuário Estatístico* (various years).

Mercado (1980) found that of 34 species exported in 1978, five species represented 90 percent of the total (Mercado 1980: 55). Browder (1984) found that of the seven principal woods exported to the United States in 1982, mahogany accounted for 84 percent of the total. Thus, increasing production reflects not the introduction of new species, but rather more intensive cropping of traditional varieties. An important item on Brazil's agenda to enhance the market value of its forest resources, and a key to forest conservation in the Amazon, is a systematic, coordinated program of botanical identification and end-use evaluation of underutilized hardwoods.

The problem of underutilized species begins in the forest, where new woods are often given highly localized names, such as "*pau anta,*" because one logger cut an unknown tree near the banks of the river Anta Atirada in Rondônia, a geographic reference point with no national recognition. Unless the commendable identification efforts of the numerous highly qualified research institutions, such as the Instituto de Pesquisa Tecnológica de São Paulo (IPT), Instituto Nacional de Pesquisa Amazônica (INPA), Instituto Brasileiro de Desenvolvimento Florestál (IBDF), Superintendência do Desenvolvimento da Amazônia (SUDAM), and others, are coordinated and applied in the field, there is little chance that the myriad of "*pau antas*" will become the *mognos* (mahoganies) that have come to be among the Amazon's most precious forest commodities.

Notwithstanding this tremendous industrial growth potential and the equally important and fragile ecological niche of the Amazon's rain forests, the rate of deforestation in Brazil appears to be increasing exponentially with time (Fearnside 1984). Brazil's stewardship of its tropical forest patrimony has allowed, in fact promoted, its destruction.

Deforestation of the Brazilian Amazon

Considerable attention has focused on the rate of forest destruction in the Amazon, but efforts to define and measure this rate have been much criticized. The Brazilian government's Forest Cover Monitoring Program (*Programa de Monitoriamento da Cobertura Florestál do Brasíl*), based on Landsat reconnaissance, is the most widely accepted source of information. As of 1983, the program estimated that 14.8 million hectares of Amazon forests of all types had been altered, or slightly less than 3 percent of the Legal Amazon region (Table 6.2). Many scientists have argued that the actual damage has been more extensive, between 5 percent and 15 percent of the Amazon.[5]

More alarming than the actual area already altered is the apparent rate at which new areas are being cleared, a rate that appears to have been growing exponentially in some parts of the region (Figure 6.3). It has been suggested that if the deforestation rates observed in the 1970s continued unabated, then most of the Legal Amazon region would be deforested or altered by the year 2000. However, given the complexity of the social and economic causes of deforestation in the region, this extrapolation cannot be regarded as a reliable prediction.[6]

Cattle ranching has been the most important contributor to forest conversion. Deforestation due to pasture formation in the region may be roughly approximated from the region's estimated herd size of 8,937,000 head of cattle in 1980 and a widely used assumed average stocking rate of one head per hectare. Given the Landsat monitoring program's 1980 estimate of vegetation cover alteration (12,364,681 hectares), pasture formation would account for more than 72 percent of the total deforested area.

Settlement by small farmers has been the second most important cause of tropical forest destruction since 1970. Unequal land tenure regimes, the increasing mechanization of agriculture, and recurrent droughts have pushed landless farmers into the region, while government colonization and land settlement programs (e.g., the National Integration Plan and the POLONOROESTE Regional Development Program) have pulled small-scale farmers from other regions of Brazil. Massive government investments in social overhead (especially road-building), expected to exceed $6.2 billion by 1990 (Hecht 1986), have contributed to the influx. Brazil's population grew at 2.8 percent per year in the 1970s; the

Table 6.2. *Natural vegetation cover alteration in the Brazilian Amazon*

State or territory	Area (thousand hectares)	Area altered [thousand hectares (percent)]		
		By 1975	By 1978	By 1983
Amapá	14,028	15.2 (0.11)	17.1 (0.12)	17.1 (0.12)
Pará	124,804	865.4 (0.69)	2,244.5 (1.80)	4,291.4 (3.44)[b]
Roraima	23,010	5.5 (0.02)	14.4 (0.06)	14.4[c]
Maranhão[a]	25,745	294.1 (1.14)	733.4 (2.85)	1,067.1 (4.15)
Goiás[a]	28,579	350.7 (1.23)	1,028.9 (3.60)	912.0 (3.20)
Acre	15,259	116.6 (0.76)	246.5 (1.62)	462.7 (3.03)
Rondônia	24,304	121.7 (0.50)	418.5 (1.72)	1,395.5 (5.74)
Mato Grosso	88,100	1,012.4 (1.15)	2,835.5 (3.22)	6,498.0 (7.38)[d]
Amazonas	156,713	78.0 (0.05)	178.6 (0.11)	179.1[c]
Total, Legal Amazônia	500,543	2,859.5 (0.57)	7,717.2 (1.54)	14,837.3 (2.97)

[a]These states are not totally inside the area of the Legal Amazon.
[b]Updated 1983 data refer only to southern Pará.
[c]Complete data not available.
[d]Updated 1983 data refer only to the area within POLONOROESTE.
Source: Programa de Monitoriamento da Cobertura Florestál do Brasíl, IBDF, November 1985.

Amazon's corresponding rate was 6.3 percent during this period (IBGE, *Anuário Estatístico* 1983: 76). In Rondônia, the target state of the POL-ONOROESTE program, population increased by a staggering 34.2 percent per year, doubling every 2.5 years (IGBE, *Anuário Estatístico* 1983: 76). Since shifting cultivation is the main agricultural production mode among peasant migrants, rapid population increase has been an important contributing factor, responsible for an estimated 9.6 percent of total regionwide deforestation by 1980.

Other government infrastructure investments have brought additional destruction. The Tucurui hydroelectric project on the Tocantins River cost about $4 billion and alone has flooded 2,160 square kilometers of forest land (Goodland 1985: 6). Other expensive hydroelectric projects, with similar land use impacts, are under construction near Manaus (Balbina) and Porto Velho (Samuel).

Population growth

Migration to the Amazon region has been the main determinant of the region's phenomenal 6.3 percent per year population growth rate between 1970 and 1980, which was more than twice the national average.

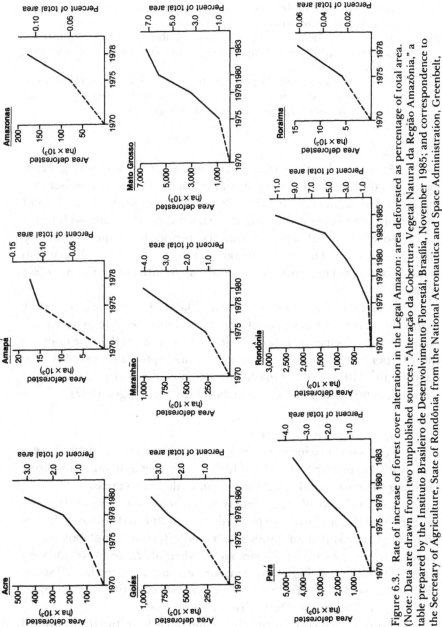

Figure 6.3. Rate of increase of forest cover alteration in the Legal Amazon: area deforested as percentage of total area. (Note: Data are drawn from two unpublished sources: "Alteração da Cobertura Vegetal Natural da Região Amazônia," a table prepared by the Instituto Brasileiro de Desenvolvimento Florestál, Brasília, November 1985; and correspondence to the Secretary of Agriculture, State of Rondônia, from the National Aeronautics and Space Administration, Greenbelt, Maryland, June 1986.)

Population growth has had three important implications for the region's forests (Fearnside 1985a):

1.   It has increased the demand for subsistence products, such as food crops, and therefore the demand for farmland.

2.   It has increased the size of the agricultural labor force, and therefore the capacity to clear forests, simply through the sheer numbers of new farmers in the region.

3.   It has increased political pressure for road-building and other social overhead investments that not only damage forests directly, but also facilitate further migration and population growth.

Inflation

Forest land in the Amazon is cheap to acquire, and with government subsidies has tended to appreciate in market value at a higher rate than inflation (until recently 200 to 300 percent per year). Amazon land cleared of forest generally has higher market value than land with forest intact. Forest removal hedges investors' profits from the devastating effects of inflation. Therefore, the rate of forest conversion is believed to be positively associated with the general rate of inflation in the Brazilian economy.[7]

The attraction of frontier land as an inflation hedge is reinforced by highly favorable tax treatment of agricultural income. Due to liberal exemptions the effective tax rate on agriculture is even less than the nominal rate of 6 percent, while non-agricultural business profits are taxed at 35 to 45 percent. The low tax adds to the demand for pasture land in the Amazon, contributing to the rise in its value (Binswanger 1987).

Displacement of small farmers

The same tax incentive makes agricultural land in settled areas more valuable to corporations and high-tax-bracket individual investors, enabling them to outbid small farmers in settled farm areas.

The consolidation of small family farms into large corporate cattle ranches or coffee and soybean plantations in southern Brazil over the last 20 to 30 years has had an important "push" effect on population movements. ("*Quando chega o boi, o homen sai*" – "when the cattle arrive, the men leave.") With the opening of the Amazon region during the last 20 years, displaced small farmers from the south have moved to the Amazon, exacerbating the population effects already mentioned.

The intraregional corollary to this pattern has been the trend toward consolidation of small family farms in the Amazon region, which also displaces small farmers who move on to clear new forest areas.

Government social overhead investments

By the end of the 1980s, the Brazilian federal government will have invested about $6.2 billion in infrastructure development in the Amazon, mostly in road-building. The notorious National Integration Program (1970–74), intended to "bring men without land to land without men" and thereby alleviate chronic poverty in the drought-stricken Brazilian Northeast, was predicated on the construction and settlement of the 2,300-km Transamazon Highway. The equally long but absurdly conceived Northern Perimeter Highway was chastised as "linking misery to nothing." And the Cuiabá-Pôrto Velho Highway, now paved, thanks, in part, to the World Bank, has opened the floodgates for many thousands of small farmers and cattle ranchers to flow into the Amazon.

Cultural attitudes

In Brazil, and elsewhere, one finds a primordial psychological aversion to dense forests, especially among recent migrants from other regions. This "fear of the forest" impedes adoption of sustainable forest land uses and pyschologically justifies forest destruction. Longstanding Portuguese tradition accords higher status to ranchers than to farmers, leading to a preference for converting forest to pasture quite apart from the expected profit (Fearnside 1985a).

International economics

Brazil finds itself in a precarious position in the international economic community. The magnitude of Brazil's foreign debt, some $105 billion, has prompted Brazilian development planners to think big in terms of remedies. Big development projects in the Amazon region, such as the $4 billion Tucurui hydroelectric project and the $4 billion Carajás mineral and timber extraction project, have had major consequences for Amazonian forests (Goodland 1985; Fearnside 1986).

Political legitimization of the former military regime

Brazil shares undefended borders with 10 other countries. All but three of these have geographic tangents in the Amazon region. The fear of foreign encroachment has provided a territorial imperative to occupy the region. It is not coincidental that most of the forest destruction visited upon the Brazilian Amazon occurred while an unpopular authoritarian government ruled, between 1964 and 1985. Anxious to legitimize its tenuous authority by placating both the urban middle class and powerful socioeconomic elites, the military government subsidized agriculture heavily to maintain low food prices while offering lucrative investment incentives to expand production. Moreover, as the spearhead of economic expansion in the Amazon, the livestock sector appeared to offer several logistic advantages over other sectors. As noted by Hecht (1985), infrastructural investments would be minimal, it was thought. A ready pool of underemployed cowboys already existed. And cattle could always walk to market if roads and bridges became impassable. Corporate entities would be preferentially treated since, it was presumed, they would have the entrepreneurial know-how and capital resources to sustain production. Moreover, North American beef producers, having widely adopted the more expensive grain feedlot system of cattle fattening by the beginning of the 1960s, were confronted with a rapidly rising demand for cheap cutter beef, for fast foods and sausages (Hecht 1985). Latin American pasture-fed cattle were viewed as an economical alternative.[8] All that was needed for Brazil to compete in foreign markets was a program to entice corporate investment to the Amazon region's land-intensive livestock sector.

Amazon regional development policies and the forestry sector

Brazil's vast Amazon region remained largely outside national political consciousness until 1946, when a new constitution was ratified, calling for a comprehensive long-term plan for the integration and development of the region. Acting on this mandate, the national Congress in 1953 established a regional planning agency, the *Superintendência do Plano de Valorização Económica da Amazônia* (SPVEA). The SPVEA was beleaguered by political problems from the start, and although a first five-year plan (1955–1960) was formulated, less than two-thirds of the resources guaranteed to the effort were forthcoming (Mahar 1979: 8–9).

By 1966, with the military regime in power, the government articulated a more assertive strategy toward the Amazon. The SPVEA was replaced by the Superintendency for the Development of the Amazon (SUDAM), which was to formulate five-year development plans to attract private investment to specific growth sectors in the region. The first of such plans included "Operation Amazônia," which had three basic objectives:

1. "The concentration of resources in areas selected in relation to their potential and existing populations . . . ."
2. "The adoption of a migration policy . . . [and] the formation of stable and self-sufficient regional population groups in the frontier zone . . . ."
3. "The [rationalization] of the exploitation of natural resources [especially forest resources]" (Lei 5.374/1967).

To attract the necessary private investment the law authorized various incentive programs, financed by an investment fund, *Fundo para Investimento Privado no Desenvolvimento da Amazônia* (FIDAM), administered by the region's development bank, the *Banco da Amazônia S.A.* (BASA). The fund was to be financed from three sources: (1) treasury transfers of 3 percent of all personal and corporate tax revenues collected by the federal government and lesser jurisdictions in the region, (2) revenues raised from the issuance of BASA stock obligations called "*Obrigações da Amazônia*," and (3) income-tax deductible deposits made by investors in support of specific SUDAM-approved development projects. With this fund, restructured as a mutual fund and renamed the *Fundo de Investimento da Amazônia* (FINAM) in 1975, three tax-based subsidy mechanisms, together referred to as the "fiscal incentive program," were made available to private investors: investment tax credit subsidies (*colaboração financeira*), personal and corporate income tax exemptions, and import duty exemptions (not addressed here).

Investment tax credit subsidies

Under present legislation private corporations in Brazil can exempt up to 50 percent of their federal income tax liabilities for investments in specific development projects in the Legal Amazon. In exchange, the corporation receives common shares of FINAM stock. The corporation may hold or sell its FINAM stock or trade it for shares of corporate stock in specific projects. Corporate stock acquired from FINAM is nontransferable for a period of four years to prevent rapid disinvestment. These tax credits may represent up to 75 percent of the total investment cost of a project. To secure the subsidy, corporations must commit their own money in an amount no less than 25 percent of the estimated total investment cost.

From January 1965 to September 1983, SUDAM disbursed $1.4 billion in tax credit subsidies to start, expand, or modify 808 existing and new private investment projects approved by its governing council (Table 6.3). Although private investment was to be the cornerstone of the incentive program and a prominent criterion in project evaluation, the projects SUDAM approved involved less private investment and reinvestment than had been anticipated. For the 808 projects approved by September 1983, only 21.6 percent of the total estimated investment had been funded by private corporations, and less than 1 percent originated in reinvested profits. In many cases the "private capital" share included land on which projects were to be located, valued at inflated appraisals. Legally, although SUDAM was permitted to finance only up to 75 percent of

Table 6.3. Distribution of SUDAM tax credit financing by sector and year (thousand U.S. dollars)[a]

Year	Livestock	Industry	Basic services	Agro-industry	Other	Total	Exchange rate[b] (cruzieros per U.S. dollar)
1965		458				458	1.896
1966	527	3,168				3,695	2.222
1967	4,057	5,960				10,017	2.669
1968	8,485	8,219	3,857	24		20,585	3.382
1969	18,001	13,094	1,555	177	179	33,005	4.076
1970	33,631	23,853	8,050	636	1,168	67,339	4.594
1971	28,337	23,390	6,345	1,312	438	59,822	5.288
1972	28,226	17,350	3,929	437	528	50,470	5.934
1973	25,789	22,279	1,639	833	1,419	51,959	6.125
1974	31,182	27,284	320	1,947	2,272	63,004	6.790
1975	52,247	55,617	390	5,034	4,557	117,850	8.127
1976	48,974	41,043	3,943	4,448	6,972	105,380	10.673
1977	53,031	33,211	5,622	1,709	4,785	98,357	14.144
1978	52,690	44,811	11,178	2,871	4,513	116,060	18.070
1979	40,594	46,806	5,813	8,382	8,747	110,340	26.945
1980	40,447	55,310	2,836	2,905	1,651	103,150	52.714
1981	37,975	65,083	3,070	8,674	3,420	118,220	93.125
1982	49,319	77,991	8,957	12,176	4,578	153,020	169.760
1983	44,202	54,160	4,151	12,632	2,418	117,560	434.020
Total	597,710	619,090	71,655	64,197	47,644	1,400,295	
Projects	469	252	31	36	20	808	

[a]Tax credit financing refers to direct tax credit subsidies as authorized by Law 5,174 of October 27, 1966.
[b]Exchange rates from World Tables (World Bank).
Source: SUDAM (1983a).

the investment costs of new projects with direct tax credit subsidies, numerous supplemental financing provisions in the law, intended to allow for such exigencies as inflation, project expansion, or diversification, reduced the real financial participation of private capital.

Industrial projects, ranging from matchstick factories to palm oil refineries, have received the largest share of tax credit financing from SUDAM: 44.2 percent of the total during this 18-year period. Fifty-nine industrial wood producers (mainly sawmills) constituted the single largest industrial beneficiary group, receiving about 35 percent of all tax credits approved by SUDAM by September 1983. Livestock projects were the second most important funding priority, representing 42.7 percent of SUDAM tax credits. Virtually all 469 livestock projects, 58 percent of all projects approved by SUDAM during this period, have been for beef cattle production (calving and fattening), with emphasis on calving.

Income tax exemptions and deductions

The second tax-based subsidy has two major elements: a total tax exemption on income from approved projects, and personal and corporate income tax deductions for the purchase of FINAM stock.

Income tax exemptions. As specified in Decreto Lei 5.174 (October 27, 1966) and revised by Decreto Lei 1.564 (July 29, 1977), companies could obtain up to 100 percent tax exemptions for up to 15 years on income derived from projects undergoing modernization, diversification, or expansion approved by the superintendent of SUDAM. Currently, this policy permits corporate income tax holidays of 10 years for projects approved before the end of fiscal 1985 (Decreto Lei 1.891, December 21, 1981). By September 1983 SUDAM had approved corporate income tax holidays for 843 projects, 39 percent of them cattle projects and 31 percent industrial wood projects.[9] Since the effective tax rate on agriculture income is very low anyway, less than the nominal rate of 6 percent, and liberal rules governing deductions of costs reduce taxable income substantially, the tax holiday has little impact. Equally important has been the ability of corporations to write off operating losses in approved Amazonian projects against other taxable income, including income earned outside the Amazon.[10]

Personal and corporate income tax deductions. As specified by Article 7 of Lei 5.174, any corporation may deduct from its taxable income up to 75 percent of the value of FINAM obligations, *Obrigações da Amazônia*, acquired from the Banco da Amazônia through an authorized

Table 6.4. *Flow of FINAM funds*

Year	Shareholders[a]	Deposits[b] (million U.S. dollars)	Disbursements[c] (million U.S. dollars)
1968	41,098	48.7	20.6
1969	67,116	63.8	33.0
1970	81,839	83.5	67.3
1971	94,906	89.8	59.8
1972	88,137	50.2	50.5
1973	72,749	62.1	52.0
1974	71,562	97.8	63.0
1975	66,836	102.5	117.8
1976	63,095	82.1	105.4
1977	48,512	153.3	98.4
1978	33,148	174.8	116.1
1979	30,925	218.7	110.4
1980	24,044	249.4	103.1
1981	n.a.	n.a.	118.2
1982	n.a.	n.a.	153.0
1983	n.a.	n.a.	117.6

[a]The number of persons and corporations declaring tax credits or tax deductions of FINAM stock.
[b]Total tax-credit deposits and FINAM stock purchases.
[c]Funds disbursed to projects.
Source: IBGE, *Anuário Estatístico* (various years).

brokerage institution. Individual taxpayers may deduct amounts equivalent to full purchase value of FINAM stocks in a given tax year, according to Article 2 of Decreto Lei 1.338 (July 23, 1974). FINAM obligations are regularly traded on the three major stock exchanges located in São Paulo, Rio de Janeiro, and Belo Horizonte, and may be exchanged for corporate stock in specific SUDAM projects that BASA offers at periodic stock sales. Corporate use of tax credit and deductions has been significant (Table 6.4). In 1980, 24,000 companies and individuals declared tax deductions as credits toward specific projects and for the acquisition of FINAM stocks worth about US$250 million. Two noteworthy patterns are evident from the flow of FINAM funds. First, in all but three years (1972, 1975, 1976), deposits have exceeded disbursements. While it may appear that FINAM shareholders oversubscribe to the Fund, Mahar (1979: 96) has shown that after 1970 "annual commitments persistently exceeded annual deposits . . . [indicating] SUDAM's failure to reduce project approvals at a rate commensurate with declining trends on supply side." Second, since 1972 the average stockholder investment in FINAM increased from US$569 to a 1980 high of US$10,375. This trend suggests the increasing concentration of corporate stock acquisition by corporations seeking to consolidate their equity interest in SUDAM-supported projects.

Rural credit and the forest sector

A second policy that has directly financed forest destruction has been Brazil's rural credit system. In 1965, the national Congress gave the National Monetary Council, the governing board of Brazil's Central Bank, wide powers to develop the institutional infrastructure to distribute subsidized rural financing to the agricultural sector. The first tier of the resulting National Rural Credit System encompasses four government banks that control various credit funds: the Banco do Brasíl, the Banco da Amazônia, the Banco do Nordeste do Brasíl, and the Banco Nacional de Crédito Cooperativo. Funding for the National Rural Credit System was to come mainly from the several existing funds in these government lending institutions, from compulsory deposits by private banks participating in rural credit programs, and from selected foreign sources. The General Fund for Agriculture and Industry (FUNAGRI) has become one of the principal sources of funds channeled by the Central Bank to rural credit programs.

The second tier of the system includes government agencies involved in planning and financing rural development activities: the former Brazilian Agrarian Reform Institute, the National Institute of Agrarian Development, and the National Economic Development Bank. The third tier includes private banks, savings and loan institutions, state banks, and cooperatives that have direct contact with rural producers.

The National Rural Credit System disburses two basic kinds of loans: credits for agriculture and for livestock. In each category are three types of financing: capital investment (*investimento*), involving fixed and semi-fixed capital investments; annual production operations (*custeio*) that are often oriented to specific crops and animals; and marketing (*comercialização*) that covers transport, storage, insurance, and tax costs of marketing farm output, and the costs of operating the guaranteed minimum price program. More than 100 specific lines of credit are included in these categories of the National Rural Credit System, including 30 different crop and animal production credit lines, and loans for electrification, grain storage, irrigation, reforestation, soil protection, fertilizer and pesticides, pasture formation, and capital equipment.

From 1973 to 1983, US$147.1 billion (in current dollars) were disbursed in rural credits, an average of US$13.4 billion per year. Benefici-

Table 6.5. *Rural credit disbursements to producers and cooperatives in the North Region, 1969–82*[a]

Year	Total disbursements (million U.S. dollars)	Disbursements by category [million U.S. dollars (percent of total)][b]		
		Investment	Production	Commercialization
1969	21.3	5.8 (27.2)	9.2 (43.2)	6.3 (29.6)
1970	20.2	6.9 (34.2)	5.7 (28.2)	7.6 (37.6)
1971	26.5	8.6 (32.4)	10.0 (37.7)	7.9 (29.8)
1972	51.1	27.5 (53.8)	14.8 (29.0)	8.8 (17.2)
1973	66.2	35.5 (53.6)	20.7 (31.3)	10.0 (15.1)
1974	72.0	30.0 (41.7)	33.2 (46.1)	8.8 (12.2)
1975	143.9	67.4 (46.8)	60.4 (42.0)	16.1 (11.2)
1976	210.3	137.3 (65.3)	54.8 (26.1)	18.2 (8.7)
1977	221.5	117.3 (53.0)	66.5 (30.0)	37.7 (17.0)
1978	297.5	173.2 (58.2)	96.5 (32.4)	27.8 (9.3)
1979	437.3	188.8 (43.2)	213.7 (48.9)	34.8 (8.0)
1980	494.8	187.4 (37.9)	267.6 (54.1)	39.8 (8.0)
1981	410.7	152.9 (37.2)	223.0 (54.3)	34.8 (8.5)
1982	350.9	114.9 (32.7)	205.4 (58.5)	30.6 (8.7)

[a]National System of Rural Credit (Banco Central).
[b]Current monetary exchange rates from *World Tables* (World Bank).
Source: Banco Central do Brasíl, *Dados Estatísticos*, "Financiamento Concedidos a Produtores e Cooperativas—Numero e Valor dos Contratos," various years.

aries in the North Region (Legal Amazônia less Mato Grosso, Goiás, and Maranhão) received 2.4 percent of these loans between 1977 and 1983. Nationwide, agricultural borrowers received an average of 80.4 percent, and livestock borrowers 19.6 percent. In the North Region, borrowing for investment purposes accounted for 44 percent of the value of credits disbursed in that region between 1969 and 1982. However, in several years, production subsidies (averaging 40 percent of total credits over the period) have exceeded investment loans (Table 6.5). In general, production credits have been available for eight-year terms with four-year grace periods. To stimulate activity in the priority regions, annual interest charges have been lower for producers in the Legal Amazon and the Northeast (12 percent) than elsewhere in the country (45 percent). Current interest charges are 3 percent per year, and after a credit reform, borrowers are now expected to pay between 70 percent and 100 percent of inflation correction costs, which were 259 percent for the 1983 calendar year as indicated by the official inflation index (Banco Central do Brasíl 1985).

Low interest rates and a six-year grace period have conferred substantial subsidies on borrowers for forest clearance. Alternative commercial

Table 6.6. *Rural credit subsidy rates, 1975–81*

	1975	1976	1977	1978	1979	1980	1981
Commercial interest rate[a]	34.6	34.4	41.1	36.4	44.8	59.4	77.6
Rural credit interest rate[b]	12.0	12.0	12.0	12.0	12.0	12.0	12.0
Effective rural interest rate[c]	5.0	5.0	5.0	5.0	5.0	5.0	5.0
Percentage of interest rate subsidized[d]	86	86	88	86	89	92	94
Interest rate subsidy relative to credit amount[e]	49	49	56	51	59	69	76

[a]Rate of return on Brazilian treasury bonds, corrected for changes in CPI and monetary correction.

[b]Interest rate of PROTERRA loans to borrowers in Legal Amazônia. See Table 6.7 for more details.

[c]Equals the internal rate of return equivalent to a credit on PROTERRA terms.

[d]Equals the difference between the commercial and effective rural interest rates, expressed as a percentage of the commercial interest rate.

[e]Defined as the present value of debt service payments on PROTERRA loans, calculated at the commercial interest rate as discount factor, expressed as a percentage of initial loan value.

Source: Commercial interest rates: International Monetary Fund (1983).

capital borrowing has become more and more costly in recent years. For instance, in 1975 the rate of return on Brazilian treasury bills was 34.6 percent while the interest on rural investment loans was 12 percent with a six-year grace period. An equivalent rate of interest on a loan with no grace period would be 4.96 percent, implying an interest subsidy of 86 percent of the commercial rate. As commercial rates rose, the implicit interest subsidy rose over this period from 49 percent to over 75 percent of the face value of the credits (Table 6.6). The availability of virtually free money from the federal government encouraged businessmen to acquire and clear more forest land than they could immediately use in pasture. Such speculative land clearance by large ranchers, using highly subsidized credit, helped to forestall squatters – who normally do not invade cleared land – and strengthened land claims (Binswanger 1987).

As part of the National Rural Credit System, the Central Bank also sponsored "special programs" offering lines of subsidized credit in support of rural development objectives. Some of the programs have been aimed at promoting production of specific commodities, such as rubber (PROBOR), sugar cane and manioc (PROALCOOL), or coffee and cacao.

Table 6.7. *Special credit programs affecting Amazon forest resources (effective 1982)*

Program	Objectives	Amortization term/grace period (years)	Annual interest charges (percent)
POLOBRASILIA	Livestock development of savannah shrublands:		Amazon: 12
	• Production	3/0	Other: 45
	• Investment		
	Fixed (e.g., pastures)	12/6	
	Semi-fixed (e.g., cattle)	8/4	
POLOCENTRO	Same as above		
POLOAMAZONIA	Land acquisition, livestock, forest, agro-industrial and mineral development in most of Amazônia:		12
	• Land acquisition	20/6	
	• Agro-industry and industry	12/3	
	• Production	1–2/0	
	• Investment		
	Fixed	12/6	
	Semi-fixed	8/4	
PROTERRA	Legal Amazônia/Northeast		12
	• Land acquisition	20/6	
	• Capital equipment	12/3	
	• Investment		
	Fixed	12/6	
	Semi-fixed	8/4	
PROASE	Increase production area for annual food crops by small producers:		Amazon: 12
			Other: 45
	• Production	2/0	
	• Investment		
	Fixed	12/?	
	Semi-fixed	5/?	
PROEXPAN	Expand area in agricultural production	8/4	Amazon: 2 Other: 45
PROPEC	Land acquisition and development for livestock:		Amazon: 12
	• Production	2–3/0	Other: 45
	• Investment	12/4	
PROALCOOL	Expand area in sugar cane production	3–8/	Amazon: 35

Source: Banco Central do Brasíl, *Manual de Normas e Instruções* (various years).

Others have been targeted to various development activities in specific regions (e.g., POLOAMAZONIA, POLOCENTRO, POLONOROESTE, POLOBRASILIA, etc.). Several of the programs are described in Table 6.7.

Forest impacts of the livestock sector

Total investment tax credits to the livestock sector and subsidized rural credit loan disbursements for pasture formation in the Amazon reached more than $730.5 million from 1966 to 1983.[11] The livestock sector clearly has been a major beneficiary of Brazilian government policies promoting Amazon development. SUDAM funding of livestock projects has contributed more to deforestation in the Amazon than any other government subsidy program. New pasture formation by SUDAM-approved cattle projects was responsible for 30 percent of the forest cover alteration detected by Landsat in Legal Amazônia between 1973 and 1983,[12] according to estimates derived from an 8.5 percent sample of such projects. Total pasture formation in the Brazilian Amazon may be roughly calculated from the region's estimated 1980 cattle herd of 8,937,000 animal units and an assumed average stocking rate of one head per hectare. Given the 1980 alteration estimate of 12,364,681 hectares, reported by the Landsat monitoring program, total pasture formation, including that resulting from SUDAM subsidies, accounted for 72.28 percent of the region's deforested area detected by satellite.

These estimates imply that nearly two-thirds of the deforestation attributable to cattle ranching has occurred without SUDAM incentives. Nevertheless, incentives give SUDAM cattle projects a staying power not enjoyed by strictly private ranches. In a survey of 40 Amazon cattle ranches that received no SUDAM incentives, the author found that only 9.2 percent of the pasture formation by such ranches occurred after 1980 (Browder 1985), while 32.8 percent of that attributed to SUDAM-subsidized ranches occurred after 1980. When funding for other lines of subsidized rural credit (especially PROTERRA) was reduced in 1980, many smaller non-SUDAM ranches lost important financial support for forest clearance and pasture formation. SUDAM-supported ranches, on the other hand, have enjoyed long-term financing from investment credits, 15-year tax holidays, and multiple investment tax credits for the same project.

Another important difference is the larger average size of the SUDAM livestock projects, and the correspondingly greater area available for future pasture expansion. The average area of the typical SUDAM-supported ranch is 23,600 hectares; non-SUDAM properties average 9,300 hectares.[13] Brazilian forestry law requires that 50 percent of the area of privately owned rural properties be left in original vegetation. By 1985, only 23.4 percent (5,500 hectares) of the area of the average SUDAM livestock project had been cleared for pasture, while 24.6 percent (2,288 hectares) of the area of the average non-SUDAM livestock property had been cleared. SUDAM projects not only have enjoyed greater financial capacity to clear forest, but also include larger forest areas still available for pasture expansion. A 1980 study of 123 beef products in the Araguaia region (southern Pará, northern Mato Grosso, and Goiás) showed that 74 percent planned to increase the pasture area on their ranches during the next two years.[14] Given these important differences in capacity, most future pasture expansion and concomitant deforestation can be expected to occur on the SUDAM-supported ranches.

The separate impacts of the National Rural Credit System and the special programs on the forest sector in the Amazon are difficult to estimate. Many borrowers also benefit from tax incentives, and it is the combined effect of tax and credit subsidies on private returns to investment that has stimulated large-scale conversions of forest to livestock operations.

In terms of forest conversion, "investment" credits have had the most important direct effect. While producers are often granted subsidized loans to bring new areas into production, usually implying forest conversion, the amount of area subsidized in this fashion is not indicated in the Central Bank's annual reports, though such data are given the Central Bank by participating local banks. In the "agriculture" credit category, estimating forest impacts is also confounded by the fact that many subsidized crops are planted together, so that estimates of forest area cleared, based on individual crop subsidies, are subject to multiple counting. In addition, many programs are vaguely defined, and the lending practices of participating local banks are not consistent nationwide. Finally, while the use of credits has been subjected to some review (fiscalicaõ), most borrowers have been free to use their subsidies without regard for program objectives and requirements. In addition, short-term "production" loans have been designated for use in pasture maintenance. However, the way Central Bank data are assembled provides little foundation for reasonable estimates of actual credit applications to these activities.

Loans have also been furnished for reforestation. Credit lines for this activity have been mainly for industrial fuelwood and pulp (eucalyptus and pine) plantings, most of which have benefited regions other than the Amazon. In 1984 approximately US$131 million of tax-credit incentives were approved for reforesting of 300,000 hectares (US$437/hectare), but only 3.0 percent of this, equivalent to 9,000 hectares, was in the North Region (IBDF 1985: 13–15). Thus, while rural credit programs have been important financial catalysts to new forest conversion in the Amazon, they have not succeeded in aiding reforestation or the economic utilization of degraded clearings.

Financial performance of subsidized Amazon cattle ranches

The government's support for livestock development in the Brazilian Amazon, through infrastructure spendings, tax credits, and subsidized rural credit, has been predicated on the prospect of profitable economic development. Only this long-term economic potential, it has been asserted,[15] could justify government policies resulting in massive forest destruction and its widely documented environmental consequences.[16]

However, financial and economic analyses of the typical SUDAM-supported ranch, based on rancher surveys administered by the author in 1984–85, demonstrate that such operations have been intrinsically uneconomic. The typical cattle ranch, without subsidies, fails to generate a positive return from livestock production. The long-term financial performance analysis demonstrates that livestock investors reap their profits only from government tax and credit subsidies, which allow a generous return to the entrepreneur's own limited financial input, despite the project's overall unprofitability. It follows that the fiscal costs to the government have been heavy, since the government has not only absorbed project losses but also provided the substantial profits reaped by private investors.

Cost-benefit structure

Five-year prospectus. The cost structure of Amazon cattle ranches comprises capital investment and operating costs. Table 6.8 presents five-year capital investment and operating cost estimates based on Browder's 1984 survey data. The typical ranch is characterized as a new undertaking, involving initial land acquisition, forest clearance, pasture formation, construction of infrastructure (fencing, roads, and mis-

Table 6.8. *Cost structure and returns on typical SUDAM beef cattle ranch (U.S. dollars per hectare during five-year development period)*

	1984
Capital investment	
1. Land cost	31.70
2. Forest clearance	
a. Manual	65.95
3. Pasture planting	26.36
4. Fencing	19.38
5. Road-building[a]	6.31
6. Miscellaneous constructions	1.25
7. Cattle acquisition[b]	90.87
Subtotal	241.82
Five-year operating costs	
1. Labor costs[c]	26.16
2. Herd maintenance[d]	21.00
3. Pasture maintenance[e]	47.34
4. Facility maintenance[f]	74.35
5. Administration[g]	4.11
Subtotal	173.00
Total costs	414.78
Total revenues[h]	112.50

[a]Road-building: Based on US$676.20/km as given by seven rancher respondents and average of 108 km/ranch divided by average pasture area (11,600 ha). The average size of the ranches in the sample was 49,500 ha.

[b]Cattle acquisition: Based on total initial herd of 4,000 animals costing US$1,054,300 on 11,600 ha of pasture.

[c]Labor costs: Based on 52 permanent employees per ranch, an average annual payroll of US $60,500, and 11,600 ha of pasture, as provided by 24 rancher respondents.

[d]Herd maintenance: Based on annual animal vaccination costing US$2.50/ha and mineral salts costing US$1.70/ha, totaling US$4.20/year.

[e]Pasture maintenance: Based on two pasture cleanings (US$10.49/ha) and one replanting (US$26.36/ha), totaling US$47.34 during the five-year period.

[f]Facility maintenance (roads, fences, corrals, buildings, etc.): Based on US$14.87/ha/year as given by 10 rancher respondents.

[g]Administration: Based on 10 percent of capital investment and five-year operating costs.

[h]Revenues: Based on observed take-off rate of 17.1 percent in 1984 (1,026 fattened steers/year), and an average 1984 price of US$254.37 per head as given by 12 rancher respondents, or a total of US$260,983 (US$112.50/ha) per five-year period, and pasture of 11,600 ha.

Source: Browder (1985).

cellaneous constructions), and cattle acquisition. Annual operating costs (labor; the maintenance of herds, pastures, and facilities; and administration) are accumulated to a total five-year cost per hectare.

The wide variation in pasture formation practices among Amazon cattle ranchers, as well as the variations in soil and vegetation types, influences forest clearance and pasture formation and maintenance costs. Based on the sample of 21 SUDAM-supported cattle ranches surveyed by the author, total investment and operating costs in the typical SUDAM-subsidized ranch over the five-year period was about $6 million (current), or $414.82 per hectare of pasture.[17] Capital investment costs were $241.82 (58.3 percent of total costs) per hectare of pasture, and five-year operating costs were $173.00 (41.7 percent) per hectare of pasture. Forest clearance, indicated by 83 percent of the sample as the single most expensive item in the ranch development program, is, in fact, the second largest investment cost ($65.95/hectare), following cattle acquisition ($90.87/ hectare).

The average return on production was $22.50 per hectare per year, or a total ranch income of $1.3 million during a typical five-year period based on the average 1984 price of $254.37 per fattened steer ($0.60/kg liveweight). Thus, cattle sales fail to cover even total operating costs in the absence of any subsidy, leaving no operating profit to justify the capital investment. However, as subsequent analysis will demonstrate, SUDAM subsidies have been generous enough to ensure ample profits on private financial investments.

Financial and economic analysis

Based on the author's survey data, the financial and economic performance of a typical SUDAM ranch was analyzed from the perspectives of the national economy and the private entrepreneur. The analysis contrasts the net benefits of Amazonian livestock investments to the national economy with the financial returns on the private investor's own equity participation, assuming the investor takes advantage of all available government incentives and subsidies. In effect, the analysis contrasts public loss and private gain.

The analysis extends over a 15-year horizon, including an initial investment and start-up period, and assumes that after 15 years all ranch assets – land, livestock, and depreciated ranch equipment – are sold at market values. Because inflation, land speculation, and credit subsidies have been central to private gains in Amazonian livestock investment, all values are recorded in nominal, rather than inflation-adjusted, terms. The general inflation rate for all prices except land values is assumed to be 25 percent per annum over the entire period, close to the average Brazilian rate during the 1970s. Land values are assumed to rise more rapidly than the inflation rate – by 2 percent annually in the base case and 5 percent in sensitivity analyses – reflecting the actual appreciation of Amazonian land in real terms during this inflationary period. Interest and amortization charges are recorded in nominal terms as well, according to the terms of various official credit programs available to SUDAM ranches. To derive net present values for economic analysis, all cost and revenue streams are discounted at a rate 6 percent above the general inflation rate.

The analysis models typical calving beef cattle operations. Initial herd size is 4,000 head, which increases to an equilibrium size of 8,750 in the fifth year of operations, at which the annual offtake of 17 percent equals the herd's natural increase. Pasture formation in the model is driven initially by the increasing herd size, but continues throughout the period because existing pasturage in use is assumed to decline in productivity over time to require replacement by the fifth year. Carrying capacity of new pasture is assumed to decline from 1.05 head per hectare to 0.65 in the fifth year of use, and to go into fallow thereafter. Thus, although total operating pasture in the model ranch stabilizes at 10,500 hectares, the total area cleared for pasture increases over the life of the operation to 27,500 hectares, some of which may represent rehabilitated fallow put back into production.

The resulting pattern of pasture formation and herd increase determines the pace of investment spending (other than initial acquisition costs for land and cattle) and operating costs, which are linked to pasture formation and herd size, respectively. Thus, for example, investments in fencing and road construction continue throughout the life of the project, as do operating outlays for herd and pasture maintenance. Revenues come from annual cattle sales that start in year two and continue throughout the project period, and from capital gains from the final sale of ranch assets, including land at appreciated prices.

Economic evaluation. The details of the base case economic and financial analyses are presented in Tables 6.9 and 6.10. Because the purpose is to compare costs and benefits to the national economy, costs and revenues are charged to the project as they are incurred. Brazilian credit and tax provisions are ignored. Comparison of operating revenues and total operating costs shows that the project generates a small operating surplus after five years. However, this surplus is too small to cover investment expenses, so that net benefits from the operation are negative in each year. The final gain from the sale of ranch assets in year 15 is far too little to offset these losses. Therefore, at a positive real interest rate of approximately 5 percent, the net present value of the project is negative $2.8 million. In other words, as the first row of Table 6.11 indicates, since the present value of total investment in the project is $5.1 million, the total loss is 55 percent of national resources invested.

Sensitivity analyses tested these conclusions under different assumptions about possible project benefits. A higher rate of appreciation in land values, at 5 percentage points above the general inflation rate annually, led to the results in the third row of Table 6.11. Even with a larger ultimate capital gain, the net present value of the project in economic terms is still negative $2.3 million, representing a loss of 45 percent of invested resources. A second departure from base case assumptions, presented in the second row, led to more favorable results. Were cattle prices assumed double those reported in the underlying survey data, implying a doubling of operating revenues in each year, the project would earn a small economic surplus, a present value of $0.5 million, which is 10 percent of invested funds. In other words, it would take a drastic improvement in project revenues to make the typical project even marginally economic at a real interest rate of 5 percent.

Financial analysis. From the private investor's perspective, the typical livestock project's profitability depends critically on the extent to which he can avoid or defer commitment of his own resources by taking advantage of government tax credits and loans, and on the extent to which he can write off costs against other tax liabilities. To show the potential impact of government incentives on private investment decisions in the Amazon, the financial analysis assumed that the private entrepreneur was able to use all available incentives: investment tax credits, tax deductions and holidays, accelerated depreciation provisions, and subsidized lending programs.

Since SUDAM-sponsored investments have been eligible for tax credits against other tax liabilities for up to 75 percent of approved investment costs, private parties have been able to finance most of the investment costs with money already due the government in taxes. Following survey findings, the analysis assumes that 54 percent of actual investment costs are met through tax credits in each year after the first. In the first year, land values are doubled in calculating the investment eligible for tax credits. This reflects the former SUDAM practice of overappraising the value of Amazon land owned by the corporate investor to ease the burden of the private investor's share in total project costs.[18]

The analysis also includes the contribution of subsidized rural invest-

Table 6.9. *Economic analysis of typical SUDAM-supported ranch (U.S. dollars)*

	Year 1	Year 2	Year 3	Year 4	Year 5	Year 6
Capital investment						
Land cost	1,578,660					
Forest clearance	251,270	123,162	168,585	230,825	315,903	606,608
Pasture planting	100,432	49,227	67,383	92,260	126,265	242,459
Fencing	73,838	36,192	49,540	67,830	92,831	178,257
Road-building	24,041	11,784	16,130	22,085	30,225	58,039
Miscellaneous construction	4,763	2,334	3,195	4,375	5,988	11,497
Cattle acquisition	1,016,000	0	0	0	0	0
Total	3,049,003	222,699	304,833	417,375	571,212	1,096,861
Operating Costs						
Labor costs	24,640	39,948	61,359	90,992	131,592	164,490
Herd maintenance	19,760	32,036	49,207	72,971	105,530	131,912
Pasture maintenance	44,560	72,243	110,965	164,554	237,976	297,470
Facility maintenance	70,000	113,488	174,316	258,501	373,840	467,300
Administration	3,840	6,226	9,563	14,181	20,508	25,635
Total	162,800	263,940	405,410	601,199	869,446	1,086,807
Total costs						
Capital costs	3,049,003	222,699	304,833	417,375	571,212	1,096,861
Operating costs	162,800	263,940	405,410	601,199	869,446	1,086,807
Loan payments	0	0	0	0	0	0
Subtotal	3,211,803	486,639	710,243	1,018,574	1,440,658	2,183,669
Tax credits	0	0	0	0	0	0
Subtotal	3,211,803	486,639	710,243	1,018,574	1,440,658	2,183,669
Tax loss or tax payment	0	0	0	0	0	0
Total	3,211,803	486,639	710,243	1,018,574	1,440,658	2,183,669
Total revenues						
Revenue from cattle sales	0	217,170	352,087	540,804	801,981	1,159,813
Revenue from land sales						
Investment loans	0	0	0	0	0	0
Operating loans	0	0	0	0	0	0
Total	0	217,170	352,087	540,804	801,981	1,159,813
Balance						
Total revenue	0	217,170	352,087	540,804	801,981	1,159,813
Total costs	(3,211,803)	(486,639)	(710,243)	(1,018,574)	(1,440,658)	(2,183,669)
Profit	(3,211,803)	(269,469)	(358,156)	(477,770)	(638,677)	(1,023,856)

Discount rate: 0
Net present value: (2,824,132)

ment (PROTERRA) credits to investment financing. Of the remaining 46 percent of project investment costs, the investor is assumed to finance half with his own funds and borrow the rest – 23 percent of total project investment costs – equally from two PROTERRA funds, the fixed investment fund and that for semi-fixed investment. The fixed investment

Table 6.9. (*continued*)

Year 7	Year 8	Year 9	Year 10	Year 11	Year 12	Year 13	Year 14
433,974	572,657	755,131	993,542	1,584,040	1,430,335	1,848,380	
173,458	228,889	301,823	397,115	633,136	571,700	738,791	
127,527	168,281	221,902	291,961	465,485	420,317	543,163	
41,522	54,791	72,250	95,061	151,559	136,852	176,850	
8,225	10,854	14,313	18,831	30,024	27,110	35,034	
0	0	0	0	0	0	0	
784,707	1,035,472	1,365,419	1,796,510	2,864,243	2,586,315	3,342,218	0
205,612	257,015	321,269	401,586	501,983	627,479	784,348	
164,890	206,113	257,641	322,051	402,564	503,205	629,007	
371,838	464,797	580,996	726,245	907,807	1,134,758	1,418,448	
584,126	730,157	912,696	1,140,870	1,426,088	1,782,610	2,228,262	
32,043	40,054	50,068	62,585	78,231	97,789	122,236	
1,358,509	1,698,136	2,122,670	2,653,338	3,316,673	4,145,841	5,182,301	0
784,707	1,035,472	1,365,419	1,796,510	2,864,243	2,586,315	3,342,218	0
1,358,509	1,698,136	2,122,670	2,653,338	3,316,673	4,145,841	5,182,301	0
0	0	0	0	0	0	0	0
2,143,216	2,733,608	3,488,089	4,449,848	6,180,916	6,732,155	8,524,519	
0	0	0	0	0	0	0	0
2,143,216	2,733,608	3,488,089	4,449,848	6,180,916	6,732,155	8,524,519	
0	0	0	0	0	0	0	0
2,143,216	2,733,608	3,488,089	4,449,848	6,180,916	6,732,155	8,524,519	
1,449,766	1,812,208	2,265,260	2,831,575	3,539,468	4,424,335	5,530,419	47,340,063
							35,296,934
0	0	0	0	0	0	0	0
0	0	0	0	0	0	0	0
1,449,766	1,812,208	2,265,260	2,831,575	3,539,468	4,424,335	5,530,419	82,636,997
1,449,766	1,812,208	2,265,260	283,157	3,539,468	4,424,335	5,530,419	82,636,997
(2,143,216)	(2,733,608)	(3,488,089)	(449,848)	(6,180,916)	(6,732,155)	(8,524,519)	82,636,997
(693,450)	(9,217,401)	(1,222,830)	(1,618,274)	(2,641,447)	(2,307,820)	(2,994,100)	82,636,997

credit carries a 12 percent nominal interest charge and is repayable in 20 annual installments after an initial six-year grace period during which no interest or amortization charges are imposed. The semi-fixed investment credit also has an interest rate of 12 percent, but is repayable over eight years after a four-year grace period. In the analysis, it is assumed that both

Table 6.10. *Financial analysis of typical SUDAM-supported ranch (U.S. dollars)*

	Year 1	Year 2	Year 3	Year 4	Year 5	Year 6
Capital investment						
Land cost	1,578,660					
Forest clearance	251,270	123,162	168,585	230,825	315,903	606,608
Pasture planting	100,432	49,227	67,383	92,260	126,265	242,459
Fencing	73,838	36,192	49,540	67,830	92,831	178,257
Road-building	24,041	11,784	16,130	22,085	30,225	58,039
Miscellaneous construction	4,763	2,334	3,195	4,375	5,988	11,497
Cattle acquisition	1,016,000	0	0	0	0	0
Total	3,049,003	222,699	304,833	417,375	571,212	1,096,861
Operating Costs						
Labor costs	24,640	39,948	61,359	90,992	131,592	164,490
Herd maintenance	19,760	32,036	49,207	72,971	105,530	131,912
Pasture maintenance	44,560	72,243	110,965	164,554	237,976	297,470
Facility maintenance	70,000	113,488	174,316	258,501	373,840	467,300
Administration	3,840	6,226	9,563	14,181	20,508	25,635
Total	162,800	263,940	405,410	601,199	869,446	1,086,807
Total costs						
Capital costs	3,049,003	222,699	304,833	417,375	571,212	1,096,861
Operating costs	162,800	263,940	405,410	601,199	869,446	1,086,807
Loan payments	(44,750)	(32,350)	(15,377)	0	111,149	111,149
Subtotal	3,167,052	454,289	694,866	1,018,574	1,551,807	2,294,817
Tax credits	(2,480,427)	(119,367)	(163,390)	(223,713)	(306,170)	(587,918)
Subtotal	686,625	334,922	531,475	794,861	1,245,637	1,706,900
Tax loss or tax payment	(163,143)	(131,577)	(154,521)	(185,175)	(252,586)	(267,368)
Total	523,482	203,344	376,954	609,687	993,050	1,439,532
Total revenues						
Revenue from cattle sales	0	217,170	352,087	540,804	801,981	1,159,813
Revenue from land sales						
Investment loans	568,575	103,333	141,442	128,143	0	0
Operating loans	81,400	0	202,705	0	434,723	0
Total	649,975	320,503	696,234	668,947	1,236,704	1,159,813
Balance						
Total revenue	649,975	320,503	696,234	668,947	1,236,704	1,159,813
Total costs	(523,482)	(203,344)	(376,954)	(609,687)	(993,050)	(1,439,532)
Profit	126,493	117,158	319,280	59,261	243,653	(279,719)

Discount rate: 0
Net present value: 1,875,436

credits were drawn in the first project year, and temporarily unused proceeds were invested by the corporation in short-term money market instruments. The private investor's own resources provide residual financing and are committed to the project only when other financial sources are used up. At the end of the project life, at the sale of the ranch, all debts are assumed paid off at their remaining net present value.

Table 6.10. (*continued*)

Year 7	Year 8	Year 9	Year 10	Year 11	Year 12	Year 13	Year 14
433,974	572,657	755,131	993,542	1,584,040	1,430,335	1,848,380	
173,458	228,889	301,823	397,115	633,136	571,700	738,791	
127,527	168,281	221,902	291,961	465,485	420,317	543,163	
41,522	54,791	72,250	95,061	151,559	136,852	176,850	
8,225	10,854	14,313	18,831	30,024	27,110	35,034	
0	0	0	0	0	0	0	
784,707	1,035,472	1,365,419	1,796,510	2,864,243	2,586,315	3,342,218	0
205,612	257,015	321,269	401,586	501,983	627,479	784,348	
164,890	206,113	257,641	322,051	402,564	503,205	629,007	
371,838	464,797	580,996	726,245	907,807	1,134,758	1,418,448	
584,126	730,157	912,696	1,140,870	1,426,088	1,782,610	2,228,262	
32,043	40,054	50,068	62,585	78,231	97,789	122,236	
1,358,509	1,698,136	2,122,670	2,653,338	3,316,673	4,145,841	5,182,301	0
784,707	1,035,472	1,365,419	1,796,510	2,864,243	2,586,315	3,342,218	0
1,358,509	1,698,136	2,122,670	2,653,338	3,316,673	4,145,841	5,182,301	0
214,977	214,977	302,488	302,488	439,224	439,224	541,725	6,368,885
2,358,193	2,948,585	3,790,577	4,752,336	6,620,139	7,171,379	9,066,244	6,368,885
(420,603)	(555,013)	(731,864)	(962,929)	(1,535,234)	(1,386,265)	(1,791,429)	0
1,937,590	2,393,572	3,058,712	3,789,406	5,084,905	5,785,114	7,274,815	6,368,885
(244,270)	(285,522)	(361,423)	(432,633)	(593,033)	(659,280)	(840,813)	33,054,799
1,693,320	2,108,050	2,697,290	3,356,773	4,491,872	5,125,834	6,434,002	39,423,684
1,449,766	1,812,208	2,265,260	2,831,575	3,539,468	4,424,335	5,530,419	47,340,063
							35,296,934
0	0	0	0	0	0	0	0
679,255	0	1,061,335	0	1,658,336	0	2,591,150	0
2,129,021	1,812,208	3,326,595	2,831,575	5,197,804	4,424,335	8,121,569	82,636,997
2,129,021	1,812,208	3,326,595	2,831,575	5,197,804	4,424,335	8,121,569	82,636,997
(1,693,320)	(2,108,050)	(2,697,290)	(3,356,773)	(4,491,872)	(5,125,834)	(6,434,002)	(39,423,684)
435,701	(295,842)	629,305	(525,198)	705,932	(701,499)	1,687,567	43,213,313

In addition to these investment incentives, the private investor is assumed able to use several operating subsidies, including tax holidays and deductibility of tax losses against nonproject income, accelerated depreciation, and subsidized intermediate-term (POLOAMAZONIA) production credits. Production credits have the same terms as semi-fixed investment credits, and are available to finance up to 50 percent of operat-

Table 6.11. *Economic and financial analysis of government assisted cattle ranches in the Brazilian Amazon*

	Net present value (U.S. dollars)	Total investment outlay (U.S. dollars)	NPV/investment outlay
I. Economic analysis			
A. Base case	−2,824,000	5,143,700	−0.55
B. Sensitivity analysis			
1. Cattle prices assumed doubled	511,380	5,143,700	+0.10
2. Land prices assumed rising 5%/year more than general inflation rate	−2,300,370	5,143,700	−0.45
II. Financial analysis			
A. Reflecting all investor incentives: tax credits, deductions, and subsidized loans	1,875,400	753,650	+2.49
B. Sensitivity analysis			
1. Interest rate subsidies eliminated	849,000	753,650	+1.13
2. Deductibility of losses against other taxable income eliminated	−658,500	753,650	−0.87

ing costs, every other year. Fixed assets other than land, under Brazilian tax codes, are depreciated by the straight-line method over a six-year period. Tax losses (operating revenues less operating costs, interest charges, and depreciation) are assumed to be fully deducted against other income taxable at the marginal corporate income tax rate of 40 percent. Since the project never generates a taxable income itself, its eligibility for income tax holiday is irrelevant.

The investor's financial analysis examines the discounted present value of all cash flows to the investor over the project life, at a real discount rate of 5 percent. Expenditures financed from outside tax liabilities, therefore, are not costs to the investor, and the availability of credit financing tied to the project at negative real interest rates adds substantially to the investment's profitability from the private investor's perspective.

In the base case, the present value of the investor's own equity input is only $0.75 million, less than 15 percent of total project investment costs, because the investor can defer his own contribution and use tax credits and government loans instead. As Table 6.11 shows, despite the project's intrinsic unprofitability, the discounted present value of net cash flows to the investor is $1.87 million. This represents a return of 2.5 times the investor's equity, despite the fact that, from a national perspective, the project loses more than half the capital invested in it. This is a strong indication of the distortions created by these incentive programs, and their effect of drawing private and – even more – public resources into uneconomic and environmentally damaging activities.

Sensitivity analyses explored how much this perverse incentive to private investors would be reduced by removing particular subsidies. First, it was assumed that interest rate subsidies were withdrawn by charging a nominal interest rate of 31 percent, 6 percent above the general inflation rate, the same rate used for discount present-value analysis. The net present value of the project to the private investor is reduced by over half, to $0.85 million, but still represents a 13 percent return on the investor's equity input. A further sensitivity test found that without deductibility of tax losses against other taxable income and without credit subsidies, the discounted present value of the investor's returns becomes a net loss of $0.65 million, nearly equal to his entire investment. Only government subsidies make such livestock investments attractive to private entrepreneurs.

The fiscal cost to the government of these subsidies and incentives is heavy, because they offset the intrinsic losses incurred in the project and provide generous returns to the private investor as well. In fact, in the base case financial analysis, the net present value of foregone tax revenues and concessionary credits is $5.6 million dollars, which is $0.5 million more than the total investment costs of the project itself. In other words, had the government invested directly in these ranches rather than stimulating private investment, it would have lost $2.8 million per ranch, the economic loss estimated in Table 6.11. Its actual loss per ranch is twice that, most accruing to private investors as profits. Such policies are fiscally burdensome as well as economically and environmentally costly.

In addition to the economic and fiscal costs of subsidized loans to cattle ranches, a complete analysis would also consider the opportunity costs of marketable roundwood destroyed in pasture formation. The author's survey found that only 18 percent of the SUDAM ranches recovered merchantable timber in clearing forests. Most ranchers simply destroyed the timber. Although it is conceivable that, given transport costs, such timber might have little stumpage value, government livestock sector development policies have given ranchers little incentive to make full use of forest resources. In contrast to the SUDAM-subsidized ranches, 42 percent of the non-SUDAM ranches surveyed had marketed timber. Two reasons for this divergence are likely. First, government subsidies substitute for private capital that would undoubtedly be raised in part from more extensive timber sales. Second, large SUDAM-subsidized land-owners clear forest quickly to demonstrate tenancy and prevent intrusions by land-grabbers and landless peasants, and to demonstrate to SUDAM inspectors that tax-credit subsidies are at work.

The opportunity costs of marketable roundwood destroyed in the process of ranch implantation may be large. On an average SUDAM ranch of 23,600 hectares (Gasques and C. Yokomizo 1985), where about 11,600 hectares could legally be converted from forest to pasture, at an average density of "merchantable" roundwood of 43.17 cubic meters per hectare (IBDF 1978), the total volume of merchantable timber that could be cut would be 500,772 cubic meters per ranch. By September 1985 there were 527 SUDAM-supported ranches, giving an estimated total marketable timber cut of 263,907,000 cubic meters. Since only 18 percent of these ranches marketed timber (generously presumed to be all marketable timber, i.e., 43.17 cubic meters per hectare), then 432 ranches marketed no timber, a potential loss of 216,333,500 cubic meters. Taking a conservative estimate of average current (1985) stumpage values for commonly extracted Amazon timber species other than mahogany, a range of $5–$10 per cubic meter as the social value of timber recovery, then the social opportunity cost of forest destruction reaches $1–2 billion on the SUDAM-supported ranches alone. This is roughly equal to the amount of SUDAM tax credits allocated to the livestock sector between 1966 and 1983.[19]

Small farmer settlement policy and the forest sector

While Brazilian Amazon development policy has emphasized the expansion of large-scale capitalist enterprises, settlement by small farmers also has been significant in regional development efforts and an important cause of tropical forest conversion. Colonization programs have been motivated by four national concerns: a growing landless peasant class, idled by drought and agro-industrial land consolidation (mainly in the Northeast and South); seasonal labor shortages in the Amazon's growing extractive industries; agricultural subsidies to stimulate domestic and export food crop production; and the military regime's desire to secure national sovereignty in a frontier region sharing undefended borders with seven neighboring nations.

The PIN Transamazon directed colonization program

The idea that the Amazon could be an agricultural frontier capable of absorbing the marginalized rural masses of the Brazilian *sertão* was embodied in Decreto Lei 1.106/1970, establishing the National Integration Program (PIN). This program's foundation was an ambitious highway-building program to integrate the Amazon with the "economic mainland" of Brazil. The east-west Transamazon Highway and north-south Cuiabá-Santarém Highway projects were planned to bisect the region. The Transamazon Highway was intended to connect the Belém-Brasília Highway with the town of Humaitá in Amazonas, a distance of 2,322 kilometers, and eventually to complete a line of roads from the Atlantic coast at Recife to the Peruvian border at Cruzeiro do Sul, a total distance of 5,560 kilometers. The areas adjacent to these roads were initially reserved for small farmers, most of whom were to be drawn from the populous, drought-beleaguered Northeast.

Colonization along the highway was to take place in a pattern of "rural urbanism," with a three-tier system of central places: agrovilas (small villages of 48 to 66 dwellings) spaced at 10-kilometer intervals, agropoli (settlements of 600 families serving 8 to 22 agrovilas with banking and postal facilities, public schools, and farm cooperatives), and ruropoli (cities of up to 20,000 with communication, medical, and administrative services and agro-industries) at 140-kilometer intervals.

The Transamazon region was divided into three Integrated Colonization Project (PIC) areas headquartered at Marabá, Altamira, and Itaituba. The ambitious plan projected the settlement of 100,000 families on 100-hectare lots by 1976. By mid-1974 only 3,700 families had received title from the National Colonization and Land Reform Institute (INCRA) (Katzman 1977). These numbers increased to 5,717 by the end of 1974 and about 7,000 by the end of 1975 (Moran 1982). By mid-1978 only 7,900 families owned titled farm lots on the Transamazon (Skillings and Tcheyan 1979). Including families with temporary land occupancy permits, no more than 12,800 families were settled through PIN in the Transamazon area (Bunker 1985). In the Marabá and Itaituba PIC areas, colonization plans were curtailed by malaria (Marabá) and poor soils (Itaituba). Because of its more fertile soils, PIC Altamira became the showcase of Transamazon colonization. The rural urbanism plan also fell short of initial objectives. Of the 66 agrovilas planned for the Altamira project, 27 were actually built and most lacked the promised amenities. Only 3 of the 15 agropoli planned for the Marabá-Itaituba segment of the highway were finally constructed and only one ruropolis was built. Maintenance of the Transamazon Highway, a continuous problem, has been minimal, leaving many stretches impassable during the rainy season, so that food often must be airlifted to settlements. Complex and often contradictory bureaucratic policies and procedures, as well as bad planning, played a large part in the failure of the program (Pompermayer 1979; Bunker 1985). Yet, underlying these shortcomings was a major policy shift away from small-farmer settlement toward a renewed emphasis on large-scale land development (mainly cattle ranching) that followed from the lobbying of the Association of Amazon Entrepreneurs, a São Paulo-based livestock interest group, and led to the POLOAMAZONIA program in 1974 (Pompermayer 1979).

The costs of the National Integration Program are difficult to measure. About $1 billion was allocated for fiscal years 1971–1974, mostly for road-building, but it is doubtful that more than $500 million was actually spent (Smith 1981). It has been estimated that highway construction costs were about $120 million.[20] The agrovilas cost about $425,000 each ($11.5 million overall), and the direct cost of relocating and settling farmers was about $13,000 per family ($103 million overall).

POLONOROESTE and semi-directed colonization in Rondônia

Colonization efforts in Rondônia have had more far-reaching social and forest sector consequences than those in the Transamazon. After nearly three decades of spontaneous settlement, the first federal initiative to bring order to the population explosion in Rondônia began in 1968. Shortly thereafter, INCRA was charged with rationalizing the distribution of land titles and planning the occupation of new frontier zones in the territory. By the end of 1980, 22,650 families had received land titles from INCRA in eight different areas of Rondônia (SEPLAN/Ro 1985). Many others were squatting on public land awaiting titles. The number of title holders increased to 24,748 by 1983. By July 1985, IN-CRA had deeded 29,944 properties to small farmers (SEPLAN/Ro 1985: 23), most of which were 100 hectares in size.

In 1981 a full-scale regional development program was established for Rondônia and western Mato Grosso. As in the PIN, the Northwest Brazil Integrated Development Program (POLONOROESTE) was predicated on massive investments in highway improvements. About $568 million (1981) was budgeted to reconstruct and pave the 1,500-kilometer Cuibá–Pôrto Velho highway. Another $520 million was budgeted for land settlement, agricultural development, and feeder roads. About $36 million was allocated for environmental protection and support of Indian communities (World Bank 1981: 1). While both PIN and POLONORESTE were based on an exaggerated conception of the importance of interregional transport, they differed in noteworthy ways. Unlike PIN in the Transamazon, INCRA's role in Rondônia has been limited to the demarcation of lots in project areas and issuing of land titles. Although the State Secretariat of Planning prepared urban plans for specific settlement sites, the rural urbanism scheme was not replicated in Rondônia. In the Transamazon, INCRA paid colonists relocation expenses and gave farmers up to eight months of salary (at about $40 per month) (Moran 1976: 18). Adding housing, local social overhead facilities, and administration, the total per capita cost of PIN was about $39,000 per colonist. In Rondônia, colonists receive no stipend; in fact, they pay nominal administrative fees for their land titles, and are expected to amortize their moving and groundbreaking costs by marketing timber from their lots. Total POLONOROESTE costs for land settlement alone come to about $10,000 per household.

A second noteworthy difference is the productivity of the soils in the two areas. In neither area are soil conditions ideal for either annual or perennial crops. However, while only about 3 percent of the Transamazon transect has agriculturally desirable soils, 33 percent of Rondônia's soils were classified as "good" for perennial agriculture (Fundação João Pinheiro 1975).

Planners of the Transamazon Highway exhibited little regard for protecting either indigenous communities or biologically rich refugia along the highway's path. While it cannot be said that POLONOROESTE planners have spared no expense to guarantee Indian land rights or conserve pristine wilderness areas in Rondônia (indeed the Brazilian government and the World Bank have been severely criticized for their sponsorship of environmental destruction in this region [Rich 1985]), it is noteworthy that in the POLONOROESTE program the government has supported initiatives to protect 46 different bounded areas (Indian reserves, biological reserves, and protected forest areas). These areas total 5.1 million hectares, or 21 percent of the total area of a state that has one of the world's richest and most diverse tropical ecosystems.

Forest sector impacts

It is virtually impossible to estimate the total forest area that has been converted by small farmer settlement in the Transamazon and Rondônia. However, the direct forest impacts of PIN in the Transamazon were probably substantial. Most of the migrants have stayed in the region, but many have sold their original lots and moved to nearby towns (Moran, in press). Assuming that by 1983 each family had cleared 50 hectares of land from their original 100-hectare lots (the maximum allowed by law), then 640,000 hectares of converted forest can be directly attributed to these settlers. This amount is equal to 14.9 percent of the area reported converted in the state of Pará (where most of the PIN settlement was centered) and only 4.3 percent of the total conversion in the Legal Amazon. Since the government effectively abandoned PIN in 1975, the program probably had no direct effect on deforestation beyond 1983.

In Rondônia, INCRA had granted 51,361 families farm lots by 1983 (SEPLAN/Ro 1985: 20).[21] Based on the author's 1985 research on forest clearance by farmers in the Rolim de Moura sector of the Gi-Paraná Settlement Project Area of Rondônia, by 1983 the typical farmer had cleared an average of 19.3 hectares. Assuming that Rolim de Moura is typical, then at the state level INCRA beneficiaries had converted 991,267 hectares, about 71.0 percent of the total area of Rondônia reported deforested by 1983. Throughout the Amazon region, where 14.8 million hectares were reported deforested by 1983, direct government colonization in Rondônia would account for only 6.7 percent of the regional total.

The direct forest conversion impacts associated with sponsored small farmer settlement in both the Transamazon and Rondônia projects totaled 11 percent of the Amazonian forest alteration detected by Landsat monitoring by 1983.

Subsidy effects

The social overhead investments in both settlement programs ($1.5 billion to $2.5 billion) tend to overshadow the substantial implicit subsidies to colonists represented by the land grants conferred in both programs. The official value of the land given to colonists was set by INCRA at about $1 per hectare. However, the market value of the land, $31.70 per hectare (based on the author's survey data, Table 6.8), indicates an implicit land subsidy of $163 million in Rondônia, or about $3,200 per colonist.

More important, land title has allowed many colonists to borrow subsidized money under various government rural credit programs. In Rondônia by 1985, an estimated 48.6 percent of the nearly 30,000 colonists with titles had borrowed money under one program or another at least once. Although the amounts, interest rates, and terms of these loans varied widely in the sample of 70 colonists in the Gi-Paraná colonization project area, most of these loans were tied to the cultivation of a certain cash crop over a specific area (e.g., seven hectares of coffee) or the purchase of livestock, either of which would involve new forest conversion. Interestingly, nearly one-fourth (23.5 percent) of the loans to these colonists were used, in part, to purchase chainsaws.

In the Transamazon, unlike Rondônia, one researcher found that "lumber operations" (i.e., forest resources) did not provide significant income to settlers and that credit programs, while contributing up to 30 percent of farmer income (in the case of upland rice subsidies), had encouraged production distortions through inappropriate crop selection and forest clearance (Smith 1981). Banks were more willing to loan for the production of specific cash crop varieties approved by INCRA and EMATER (Brazil's agricultural extension service) on the basis of experimental trials undertaken in nontropical conditions outside the Amazon. Sustaining production required expensive pesticides and fertilizers. Banks were also more willing to lend to farmers for first-year plantings in newly cleared fields, which produce higher crop yields than older fields. This practice may have encouraged farmers to cut new forest more often than necessary.

That cheap financing would encourage forest conversion is almost obvious. In Rondônia, the mean value of the area cleared by farmers by 1985 was 22.3 hectares. However, 60.9 percent of the sub-sample who were recipients of subsidized financing had cleared more than that. Farmers in Rondônia who receive rural credit tend to clear about 25 percent more forest area than those who do not receive such financing.

Small farmer settlement has been a feature of Amazon development policy since 1970 and has promoted deforestation in the Transamazon and Rondônia. Migration to the Amazon is likely to continue, even intensify, under economic conditions of austerity. Small farmer settlement is closely linked to large social overhead investments, especially in transport improvements. The regularization of land titles has enabled many titled farmers to borrow from the government's rural credit programs, further exacerbating deforestation. Although it accounts for less than half the deforestation attributable to cattle ranching, small farmer settlement clearly has been a significant cause. Finally, it should be noted that there are several private colonization projects in Amazônia, some supported by government subsidy programs. Most of these projects are relatively new and small compared to the Transamazon or Rondônia colonization programs.

Conclusions and recommendations

The Amazon is the world's largest tropical moist forest region, believed to be home to a tenth of the earth's 5 million to 10 million plant and animal species. The forests of the Brazilian Amazon alone may contain nearly a third of the world's volume of tropical broadleafed timbers: between 48 billion and 78 billion cubic meters. Yet, in spite of their enormous economic value and essential environmental functions, the rain forests of the Brazilian Amazon are being destroyed at rates that appear to be accelerating exponentially in some areas.

Nearly half the rain forest destruction in the Brazilian Amazon thus far is directly attributable to four government subsidy programs: the SUDAM program for developing the Brazilian Amazon, the Brazilian Central Bank's rural credit program, the National Integration Program in the Transamazon, and the semi-directed program of small farmer settlement in the state of Rondônia. The numbers of beneficiaries, the subsidies, and the forest impacts are summarized in Table 6.12.

Livestock production, expanded largely through government fiscal incentives, has been responsible for the largest proportion – 30 percent – of the forest conversion in the region. While it has been asserted that forest destruction may be justified if alternative uses of forest land bring large and unambiguous benefits, large-scale cattle ranching, without enormous subsidies, is economically untenable in the Amazon, its income covering only about 45 percent of costs. Regardless of the harmful environmental effects of cattle ranching on the Amazon, this activity can be discredited on economic grounds alone.

Yet the Brazilian government, during the bureaucratic authoritarian regime (1964–1985), vigorously pursued expansion of the livestock sector in the Amazon. The explanation for this apparently irrational behavior may be found in Brazil's political economy during this period. Anxious to ensure its legitimacy by appeasing powerful corporate interest groups, the government used the Amazon development program to transfer vast sums of public capital into private hands. Cattle ranching became the pretext for the appropriation of public capital by the large corporations to which the government's policies were, in the main, directed. Although ranches were inherently unprofitable as production operations, their corporate owners could nonetheless obtain large profits through government subsidies. In essence, the SUDAM livestock program has subsidized corporate profits at the considerable expense of the Brazilian taxpayer and the Amazon's forests.

Fearing social unrest in the rural backlands of the Brazilian Northeast and South, where land tenure regimes are highly unequal, the government sponsored two massive colonization programs initially directed toward small farmers with large families. The National Integration Program (PIN) of the early 1970s ambitiously sought to transplant 100,000 farmers to the Transamazon. Hastily conceived and without regard for variable soils, topographical constraints, preexisting indigenous populations, or public health problems, PIN was doomed to fail. Fewer than 15,000 farmers participated directly.

The enormously expensive Transamazon Highway, conceived partly to open up the region, is in a perpetual state of disrepair, although still in use. Most of the area along the highway cleared by small farmers is now in marginal use as pasture. Weary of trying to survive in a world they did not understand, many of the original farmers have abandoned their farms and moved to nearby towns and cities (the Amazon is the most rapidly urbanizing region of Brazil). Others, faced with tired soils or threats of intimidation by armed land-grabbers, have moved further into the Amazon to clear new forest areas.

The spontaneous settlement of Rondônia has been a different experience from that of the Transamazon, although both were predicated on the highway-based development model. In the Transamazon, colonization was directed by government planners and bureaucrats who selected the farmers to be relocated and determined the design and organization of life in the communities they would live in. In Rondônia, the government races to keep up with the droves of migrants who arrive on their own initiative. By 1985, nearly 30,000 migrants to Rondônia had received definitive land titles. Perhaps a comparable number are squatting on land in anticipation of titles. During the 1970s, as the rural population of Brazil actually declined, Rondônia's rural population exploded at an annual rate of 34 percent. The effects of this population explosion on Rondônia's forest are shocking: in 1975, only 0.5 percent of the area had been deforested; by 1980, 3.1 percent had been converted; by 1983, 5.7 percent; by 1985, over 11 percent.

Government programs to develop the Amazon are leading to its destruction. What can be done to alter this course? Numerous areas of the Amazon are known to be particularly rich in biological diversity. Other areas belong to Indians who have a moral right to live undisturbed by modern development in the lands they have conserved and cultivated for centuries. The natural integrity of such areas should be ruthlessly and tirelessly defended. However, the strictly protectionist approach to the Amazon overall is, in the author's opinion, doomed to fail. Conservation of the Amazon's rain forests must begin with an appreciation of their value as an economic asset, endowed by nature. Like any fund, with responsible stewardship, the Amazon can generate benefits in perpetuity for humans. In accordance with this "use it or lose it" philosophy, which is by no means universally shared among students of the Amazon, the following general policy recommendations are offered:

Table 6.12. *Subsidy and forest impacts of selected government programs*

Program	Total estimated subsidy (U.S. dollars)	Estimated subsidy rate (percent)	Number of direct beneficiaries	Subsidy per beneficiary (U.S. dollars)	Area deforested (hectares)	Percent of total deforested area in Brazil[a]
Livestock						
SUDAM livestock tax credits	597,710,000[b]	54[c]	469[d]	1,274,000[e]	4,432,050[f]	30.0
Rural credit for pasture formation[g]	65,072,000[h]	49[i]	3,511[j]	18,500[k]	880,000[l]	5.9
Settlement						
PIN	500,000,000[m]	n.a.	12,800	39,062[n]	640,000[o]	4.3
Rondônia						
• POLONOROESTE	520,000,000[p]	n.a.	51,361[q]	10,124[r]	991,267[s]	6.7
• Implicit land subsidy	162,800,000[t]	100	51,361	3,170[u]	—	—
• Implicit timber subsidy	150,399,000–306,779,000[v]	100	27,889[w]	550–1,100[x]	—	—

[a] Total area deforested by 1983 = 14,837,294 hectares (IBDF/PMCF).

[b] Total SUDAM tax credit assistance to the livestock sector from 1965 through September 1983 expressed in nominal U.S. dollars.

[c] Average tax credit share of total capital costs of a sample of 18 SUDAM-supported cattle ranches.

[d] Number of cattle ranches receiving SUDAM tax credit financing from 1965 through September 1983.

[e] Total estimated subsidy divided by beneficiaries.

[f] In mid-1985, the average SUDAM ranch was 11 years old and had cleared 9,450 hectares, which, multiplied by 469 ranches, is 4,432,050 hectares.

[g] "Rural credit" refers to Permanent Pasture formation loans only. Numerous other credit lines were used for deforestation as well, but these are not trackable. Hence, rural credit data given represent the minimum.

[h] $65,072,000 = total disbursements of $132.8 million from 1969 to 1982 to ranchers in the "North Region" under Permanent Pasture credit program multiplied by the nominal subsidy rate of 49 percent effective in 1975 (assumed median rate for period), as indicated in Table 6.6.

[i] 1975 subsidy rate embodied in rural credit loans as specified in Table 6.6.

[j] Complete data are not available. From 1977 through 1983, 3,511 Permanent Pasture loans were executed to cattle ranchers in the North Region.

[k] Average nominal U.S. dollar value of per capita Permanent Pasture loans, adjusted by subsidy rates given in Table 6.6, from 1977 through 1981.

[l] Derived from total Permanent Pasture loan disbursements of $132.8 million (nominal) made from 1969 through 1983 to cattle ranchers in the North Region divided by average forest clearance and pasture formation costs of $150.95 per hectare as indicated in Table 6.8.

[m] Approximate total expenditures of PIN Transamazon program.

[n] Total PIN expenditures divided by maximum number of known beneficiaries.

[o] Assumes that by 1983 each of the 12,800 PIN beneficiaries cleared a total of 50 hectares of forest, the legal limit on a 100-hectare lot (12,800 × 50).

[p] Portion of POLONOROESTE budget for 1981–86 for land settlement, agricultural development, and feeder roads.

[q] Number of families receiving permanent and provisional titles to 100-hectare lots in seven colonization areas of Rondônia by 1983 as indicated by SEPLAN/Ro (1985: 20).

[r] $520 million (land settlement, agricultural development, feeder roads) of total POLONOROESTE budget divided by 51,361 INCRA beneficiaries in Rondônia.

[s] Based on author's 1985 survey of 70 colonists in Gi-Paraná PIC of Rondônia in which the typical colonist had cleared 19.3 hectares by 1983 multiplied by total 1983 beneficiaries (51,361).

[t] Implied subsidy equivalent to $31.70 per hectare (average 1984 market price for unimproved Amazon land indicated in Table 6.8) multiplied by 100 hectares per beneficiary and 51,361 beneficiaries.

[u] $31.70 per hectare land subsidy multiplied by 100 hectares per beneficiary.

[v] Implied subsidy equivalent to $550–1,100 per beneficiary of a 100-hectare lot multiplied by 27,889 beneficiaries (54.3 percent of 51,361 total beneficiaries) who marketed an average of 110 cubic meters of timber with an estimated stumpage value of $5–10 per cubic meter through 1984 as reported by 70 colonists surveyed by author in 1985.

[w] Number of colonists who marketed timber occurring on their 100-hectare lots in Rondônia through 1984.

[x] Based on an average of 110 cubic meters marketed per beneficiary multiplied by $5–10 per cubic meter (range of stumpage values from 1978 to 1984) expressed in nominal U.S. dollars.

1. No new livestock projects should be approved for tax credit financing by SUDAM. For existing cattle projects in the Amazon, all fiscal incentives (i.e., tax credit financing and income tax deductions) should be phased out over a five-year period.

2. Similarly, the various rural credit programs should be amended by the Central Bank to prohibit new lending for fixed and semi-fixed investment on ranches in the tropical zones of the Amazon.

3. Instead, it is recommended that priority SUDAM financing be given to four categories of projects: (a) those that would reclaim and make economic use of degraded clearings in the terra firma areas; (b) large-scale agroforestry projects; (c) industrial wood projects that are based on sustained-yield cropping and selective reforestation of appropriate commercial timber species (i.e., "forest enrichment"), especially when those projects are directly linked to Brazilian-based final wood product manufacturing enterprises; and (d) projects that would promote the self-sustaining economic utilization of the Amazon's fertile *várzea* floodplains.

4. Any project that would involve the conversion of more than 50 hectares of dense or transition forest should be subjected to certain requirements for SUDAM financing that ensure the maximum possible recovery of forest resources. After the ratio of public-to-private matching shares of SUDAM projects has been determined, the private share should be adjusted upward in an amount equal to the estimated present market value and replacement cost value of the forest resources that would be destroyed by the project. Theoretically, this would provide corporations with an incentive to maximize the salvage of forest resources (timber, fuelwood) or minimize the forest area they would convert.

The development model of the Amazon region has been largely based on "growth pole" theory, which holds that investment in a leading sector propels development in other related sectors. In the Amazon, livestock was selected as the leading sector. Yet, after more than 20 years of public investment, the livestock sector has absorbed more in subsidies than it has generated in revenues and has contributed little to permanent regional employment. It has not stimulated collateral development, except in the slaughterhouse and meat-packing industries. Nor has it obviated the shortages of beef and dairy products that periodically beleaguer Brazilian consumer markets. Moreover, it has been the principal engine of destruction of the Amazon's rain forests. The growth pole approach, based on the livestock sector, has been tested in the Amazon for more than 20 years, and it has failed. Now it is time to try something new.

Acknowledgments

The author wishes to acknowledge the valuable assistance provided to him by the National Science Foundation, the Organization of

American States, the Núcleo de Altos Estudos Amazônicos–Universidade Federal do Pará, the Instituto de Florestas–Universidade Federal Rural do Rio de Janeiro, and the Instituto Brasileiro de Desenvolvimento Florestál (Brasília and Pôrto Velho), whose generous financial and logistical support made possible the field research leading to this report. The preparation of this report was made possible by the World Resources Institute.

Endnotes

1. The Brazilian Amazon region (BAR) is commonly defined in two ways. The "North Region" (defined by IBGE, the Brazilian census agency) includes the states of Pará, Amazonas, Acre, Rondônia, and Amapá and the federal territory of Roraima. The "Legal Amazon" (the definition used by SUDAM, the Superintendência do Desenvolvimento da Amazônia) includes the North Region plus Mato Grosso state and large parts of Goiás and Maranhão states. Since information in this paper is drawn from both IBGE and SUDAM, both definitions are used in the text, as necessary.

2. This estimate (84.3 percent of tree species represented by less than one individual per hectare) is based on an inventory of 36 hectares of natural "high forest" (*alta floresta*) in the Tapajós National Forest area of central Amazônia in which 134 different species of trees (15 cm. d.b.h. or more were found). Mercado (UFRRJ 1985b: 139) in an inventory of the Jamari National Forest in Rondônia (southern Amazônia) indicates that 90 percent of the species in this so-called "transition forest" (*floresta aberta*) occur in densities of less than 1 tree (10–35 cm. d.b.h.) per hectare.

3. The $1.7 trillion estimate is based on Knowles' (1966) estimate of 78.3 billion cubic meters multiplied by the average cost of roundwood production, estimated by Browder (1986: 232) at $21.87 per cubic meter. The author does not suggest that this entire stock of biomass should be auctioned off or quickly harvested to meet pressing national economic exigencies (e.g., foreign debt). Brazil's Amazon forest resource must be regarded as a capital endowment fund from which substantial annual interest income may be earned if responsible sustained-yield cropping practices are followed. In any case, the world's capacity to absorb the entirety of the region's tropical timbers in the short term (e.g., 5–10 years) is too limited for this notion to be practical.

4. "Industrial wood" includes sawnwood and pulp at various stages of processing, used as inputs to final demand manufacturing processes (excluding industrial fuelwood). In the Amazon, the sector mainly consists of lumber mills.

5. Fearnside (1985a) gives three reasons for believing that Landsat-based estimates of deforested areas are low. First, 1978–1983 data for three of the nine federal units of the Legal Amazon (Amapá, Roraima, and Amazonas), totaling nearly 40 percent of the region's area, were not included in the 1983 estimates. Second, evidently the Landsat technology has difficulty distinguishing primary (virgin) forest from secondary growth. Third, Fearnside maintains that Landsat is handicapped in detecting small forest clearings. In spite of these legitimate criticisms, most researchers, including Fearnside, use the Landsat information as the only available standard, regularly updated measure of deforestation in the Bra-

zilian Amazon. Furthermore, some of the apparent technical deficiencies in Landsat image interpretation are being resolved, according to IBDF consultants.

6. It is conceivable that as government resources become strained in the current economic crisis in Brazil, subsidies for Amazon development projects might be reduced. In this case, recent trends in deforestation rates may not continue their apparently exponential upward spiral. While general economic difficulties may constrain the pace of deforestation due to cattle ranching, it is likely to aggravate deforestation due to frontier migration and small farmer settlement as "push factors" become more intense.

7. The assertion that "deforestation is linked to longstanding economic patterns in Brazil, such as high inflation rates" and several other points made in this section of the text are abstracted from Fearnside (1985a).

8. For an excellent discussion of the preference given to the livestock sector by Amazon development policy-makers, and the role of foreign markets, see Hecht (1985).

9. The estimated value of corporate income tax exemption declared by Amazon industrial wood producers between 1970 and 1984 is $1.2 billion (1984). See Browder (1986: 139–40, fn.).

10. Personal communication from Dr. Fabio Monteiro de Barros, Senior Partner, Castro and Barros, São Paulo, Brazil, and Robert Repetto, World Resources Institute, Washington, D.C.

11. Given the proliferation of rural credit programs during the 1970s (over 100 specific credit lines), estimating with exactitude the volume of disbursements to the livestock sector in the Brazilian Amazon is difficult. The $730.5 million includes both SUDAM indirect tax credits ($598 million) and rural credit disbursements ($132.5 million). The latter is clearly a minimal estimate derived from one credit line (permanent pastures) and excludes the various special credit programs for which disbursements are not trackable. A more likely estimate of the total use of rural credit disbursements for pasture formation would be about $691 million [$147.1 billion (total disbursements) × 0.196 (share to the livestock sector nationwide) × 0.024 (share of total disbursements to North Region)]. If this amount ($691 million) was spent only on forest clearance and pasture formation, then rural credit programs to the livestock sector were responsible for conversion of 4.6 million hectares (31 percent of the total area reported deforested by Landsat) by 1983 ($691 million divided by $150.95 per hectare). These estimates seem realistic given that total rural credit disbursements to the Amazon's livestock sector between 1976 and 1983 were US$987 million in nominal terms.

12. This estimate, 29.99 percent of the total area deforested by 1983, is a direct extrapolation from the author's sample of 8.5 percent of all 469 SUDAM-subsidized cattle ranchers (by September 1983). At the microregional level, the livestock factor is a considerably more important determinant of deforestation. For instance, Tardim et al. (n.d.), in an exhaustive study of 760,000 hectares in Barra de Garcas area of northern Mato Grosso, found that SUDAM ranches were responsible for 38 percent of the area deforested.

13. Other estimates of the average area of SUDAM livestock projects range from 18,126 (SUDAM 1983) to 28,860 hectares (Pompermayer 1979). For the traditional IBGE North Region and Mato Grosso, the average size of establishment listed in livestock production (*pecuária*) has been estimated at 872 hectares (IBGE, Censo Agropecuária, 1980).

14. Personal communication from Alberto Oliveira Lima Filho in 1984, based on his 1980 market study for the Atlas Meat Processing Company.

15. The statement "Deforestation is justified only when the economic benefits to be obtained therefrom are large and unambiguous" was made by Robert F. Skillings, former chief of the Brazil Division, World Bank, and proponent of the POLONOROESTE program in Rondônia. Cited in Hemming (1985).

16. For a general review of the major environmental issues involving tropical deforestation, see Goodland (1975), Myers (1980), Lugo and Brown (1982), Guppy (1984), the Sioli and Fearnside chapters of Hemming (1985), Tangley (1986), and Buschbacher (1986). More detailed studies have focused on nutrient cycling and soil productivity (Herrera et al. 1978; Falesi 1976, 1980; Seubert et al. 1977; Alvim 1977; Serrão et al. 1979; Fearnside 1980a; Hecht 1982); soil porosity and erosion (Daubenmire 1972; Fearnside 1980b; Abreu Sa Diniz et al. 1980); hydrologic cycles and rainfall (Molion 1975; Villa Nova et al. 1976; Marques et al. 1977; Friedman 1977; Salati 1980; Gentry and Lopez-Parodi 1980); atmospheric-climatic effects (Sioli 1978; Woodwell 1978; Salati 1980; Kuklas and Gavin 1981; Woodwell et al. 1983); species extinction (Gomez-Pompa et al. 1972; Pires and Prance 1977; Prance 1982; Lovejoy and Oren 1981; Lovejoy and Salati 1983; Lovejoy et al. 1983; Myers 1985); and threats to native populations (Davis 1977; Posey 1983).

17. Land purchase price ($31.70 per hectare) applies to entire ranch (49,000 hectares), while all other capital and operating costs ($383.12 per hectare) apply only to area in pasture (11,600 hectares).

18. More recent SUDAM policy is to limit the value of the land to 10 percent of the private investor's equity contribution.

19. For an additional discussion of the social costs of timber destruction in the Brazilian Amazon, see Browder (in press) and Mahar (1979: 128–129).

20. Derived from Moran (1976: 81), who obtained a Transamazon unit road-building cost of $53,710 per kilometer.

21. This number includes 24,748 families with titles and 26,613 families who had received temporary land occupation licenses from INCRA in anticipation of definitive titles and who should be considered beneficiaries of the Rondônia settlement program. Twenty-nine percent of the colonists surveyed in the author's sample did not have definitive land titles.

References

Abreu Sa Diniz, Tatiana Deane, and Therezinha X. Bastos. 1980. Efeito do Desmatamento na Temperatura do Solo em Região Equatorial Úmida. *Boletim de Pesquisa No. 7.* Belém: EMBRAPA/CPATU.

Alvim, Paulo de Tarso. 1977. Agricultura nos Trópicos Úmidos: Potencialidades e Limitações. *In Trabalhos Apresentados no Primeiro Seminário Regional de Desenvolvimento Rural Integrado,* Vol. 2, January 24–26. Manaus: SUDAM/Fundação Getúlio Vargas.

Banco Central do Brasíl. 1985. *Manual de Normas e Instruções.* MCR No. 199, Circular 1292, Capitulo 5. Brasília.

Banco Central do Brasíl. *Dados Estatísticos* (Brasília), various years.

Binswanger, Hans P. 1987. Fiscal and Legal Incentives with Environmental

Effects on the Brazilian Amazon. Unpublished discussion paper, Agriculture and Rural Development Department, World Bank. Washington, D.C., May.

Browder, John O. 1984. Tomando Conhecimento dos Importadores Norteamericanos de Madeira Amazônica Brasileira. *Infoc Madeireiro*, Edição Especial 3 (20). Brasília: IBDF.

Browder, John O. 1985. Subsidies, Deforestation and the Forest Sector in the Brazilian Amazon. A report to the World Resources Institute. Washington, D.C.

Browder, John O. 1986. Logging the Rainforest: A Political Economy of Timber Extraction and Unequal Exchange in the Brazilian Amazon. Ph.D. diss. University of Pennsylvania.

Browder, John O. In press. The Social Costs of Rain Forest Destruction: A Critique and Economic Analysis of the Hamburger Debate. *Interciencia*.

Bruce, Richard W. 1976. Produção e Distribuição da Madeira Amazônica. Série Estudos No. 4. Brasília: IBDF.

Bunker, Stephen G. 1985. *Underdeveloping the Amazon: Extraction, Unequal Exchange, and the Failure of the Modern State*. Urbana: University of Illinois Press.

Buschbacher, Robert J. 1986. Tropical Deforestation and Pasture Development. *Bioscience*, Vol. 36, No. 1: 22–28.

Caulfield, Catherine. 1984. *In the Rainforest*. University of Chicago Press.

Correa de Lima, J. P., and R. S. Mercado. 1985. The Brazilian Amazon Region: Forestry Industry Opportunities and Aspirations. *Commonwealth Forestry Review*, Vol. 64, No. 2: 151–156.

Daubenmire, R. 1972. Some Ecological Consequences of Converting Forest to Savanna in Northwestern Costa Rica. *Tropical Ecology*, Vol. 13, No. 1: 31–51.

Davis, Sheldon H. 1977. *Victims of the Miracle*. London: Cambridge University Press.

Empresa Brasileira de Pesquisa Agropecuária (EMBRAPA). 1981. Distribuição Diamétrica de Espécies Comerciais e Potenciais em Floresta Tropical Úmida Natural na Amazônica. Boletim de Pesquisa No. 23. Prepared by João Olegário Pereira de Carvalho. Belém: EMBRAPA/CPATU.

Empresa Brasileira de Pesquisa Agropecuária (EMBRAPA). 1984. *Amazônia: Meio Ambiente e Technologia Agrícola*. Prepared by Cristo Nascimento and Alfredo Homma. Belém: EMBRAPA/CPATU.

Erfurth, Th. 1974. International Trade and Trade Flows of Tropical Forest Products. *Properties, Uses, and Marketing of Tropical Timber*, Vol. 1. Final report, FAO, Rome.

Falesi, I. C. 1976. *Ecosistema de Pastagem Cultivada na Amazônia Brasileira*. Boletim Técnico No. 1. Belém: EMBRAPA/CPATU.

Falesi, I. C. 1980. O Solo na Amazônia e sua Relação com a Definicão de Sistemas de Produção Agrícola. *Reunião do Grupo Interdisciplinar de Trabalho sobre Diretrizes de Pesquisa Agrícola para a Amazônia*, Vol. 1. May 6–10, 1974. Brasília: EMBRAPA.

FAO. *See* United Nations Food and Agriculture Organization.

Fearnside, Philip M. 1980a. Effects of Cattle Pasture on Soil Fertility in the Brazilian Amazon: Consequences for Beef Production Sustainability. *Tropical Ecology*, Vol. 21, No. 1: 125–137.

Fearnside, Philip M. 1980b. The Prediction of Soil Erosion Losses under

Various Land Uses in the Transamazon Highway Colonization Area of Brasil. In *Tropical Ecology and Development: Proceedings of the Fifth International Symposium of Tropical Ecology*, ed. J. I. Furtado, April 16–21. Kuala Lumpur: International Society of Tropical Ecology.

Fearnside, Philip M. 1984. A Floresta Pode Acabar? *Ciencia Hoje*, Vol. 2, No. 10 (Janeiro/Fevereiro): 34–41.

Fearnside, Philip M. 1985a. The Causes of Deforestation in the Brazilian Amazon. Paper presented at the United Nations University International Conference on Climatic, Biotic and Human Interactions in the Humid Tropics: Vegetation and Climate Interactions in Amazônia. February 25–March 1. São Paulo, São José dos Campos.

Fearnside, Philip M. 1985b. Environmental Change and Deforestation in the Brazilian Amazon. In *Change in the Amazon Basin: Man's Impact on Forests and Rivers*, ed. John Hemming, pp. 236–40. Manchester: Manchester University Press.

Fearnside, Philip M. 1986. Agricultural Plans for Brazil's Grande Carajas Program: Lost Opportunity for Sustainable Local Development? *World Development*, Vol. 14, No. 3: 385–409.

Friedman, I. 1977. The Amazon Basin: Another Sahel. *Science*, Vol. 197: 7.

Fundação João Pinheiro. 1975. Levantamento de Reconhecimento de Solos da Aptidão Agropastoríl, das Formações Vegetais, e do Uso da Terra em Área do Terrritório Federal de Rondônia. Belo Horizonte: MI/SUDECO/FJP.

Gasques, J. B., and C. Yokomizo. 1985. Resultados de 20 Anos de Incentivos Fiscais na Agropecuária da Amazônia. Unpublished paper, Instituto de Planejamento Economico e Social, Brazil.

Gentry, A., and J. Lopez-Parodi. 1980. Deforestation and Decreased Flooding in the Upper Amazon. *Science*, Vol. 201: 1354–1356.

Gomez-Pompa, A., C. Vasquez-Yanes, and C. Guevara. 1972. The Tropical Rainforest: A Non-Renewable Resource. *Science*, Vol. 177: 762–765.

Goodland, Robert. 1985. Brazil's Environmental Progress in Amazonian Development. In *Change in the Amazon Basin: Man's Impact on Forests and Rivers*, ed. John Hemming, pp. 5–35. Manchester: Manchester University Press.

Guppy, Nicolas. 1984. Tropical Deforestation: A Global View. *Foreign Affairs*, Spring: 928–965.

Hecht, Susanna B. 1982. Cattle Ranching Development in the Eastern Amazon: Evaluation of a Development Strategy. Ph.D. diss., University of California, Berkeley.

Hecht, Susanna B. 1985. Environment, Development and Politics: Capital Accumulation and the Livestock Sector in Eastern Amazônia. *World Development*, Vol. 13, No. 6: 663–684.

Hecht, Susanna B. 1986. Development and Deforestation in the Amazon. Unpublished report to the World Resources Institute. Washington, D.C., January.

Hecht, Susanna B. 1985. The Economics of Cattle Ranching in Eastern Amazônia. Unpublished manuscript.

Hemming, John, ed. 1985. *Change in the Amazon Basin: Man's Impact on Forests and Rivers*. Manchester: Manchester University Press.

Herrera, R., C. Jordan, H. Klinge, and E. Medina. 1978. Amazon Ecosystems:

294   J. O. Browder

Their Structure and Functioning with Particular Emphasis on Nutrients. *Interciencia,* Vol. 3: 223–231.

Instituto Brasileiro de Desenvolvimento Florestál (IBDF). 1978. *Informes Sobre a Comercialização da Madeira Amazônica.* Coleção: Desenvolvimento e Planejamento Florestál, Série Técnica 7. Brasília: COPLAN.

Instituto Brasileiro de Desenvolvimento Florestál (IBDF). 1985. *O Sector Florestál Brasileiro: 79/85.* Brasília.

Instituto Brasileiro de Geografia e Estatística (IBGE), various years. *Anuário Estatístico. Censo Agropecuária. Censo Industrial.*

Instituto de Pesquisa Tecnológica de São Paulo (IPT). 1985. Estabelecimento de Estratégia de Comercialização Nacional de Madeiras Tropicaís Brasileiras. Resumo do Relatório, No. 21.824. São Paulo.

International Monetary Fund. 1983. *Interest Rate Policies in Developing Countries.* Occasional Paper No. 22. Washington, D.C., October.

Katzman, Martin T. 1977. *Cities and Frontiers in Brazil.* Cambridge: Harvard University Press.

Knowles, O. H. 1966. Relatorio ao Governo do Brasíl Sobre a Producão e Mercado de Madeira na Amazônia. Projecto do Fundo Especial No. 52. MI/SUDAM/FAO.

Knowles, O. H. 1969. *Investment and Business Opportunities in Forest Industrial Development of the Brazilian Amazon.* SF/BR4, Technical Report 1. Rome: FAO/UNDP/FD.

Kuklas, G., and J. Gavin. 1981. Summer Ice and Carbon Dioxide. *Science,* Vol. 214: 497–503.

Lovejoy, Thomas E., and David C. Oren. 1981. The Minimum Critical Size of Ecosystems. In Ecological Studies 41: *Forest Island Dynamics in Man-Dominated Landscapes,* ed. Robert L. Burgess and David M. Sharpe. New York: Springer-Verlag.

Lovejoy, Thomas E., and Eneas Salati. 1983. Precipitating Change in Amazônia. In *The Dilemma of Amazonian Development,* ed. Emilio Moran. Boulder, Colo.: Westview Press.

Lovejoy, Thomas E., et al. 1983. Ecological Dynamics of Tropical Forest Fragments. In *Tropical Rainforests: Ecology and Management.* Oxford: Blackwell Scientific Publications.

Lugo, Ariel E., and Sandra Brown. 1982. Conversion of Tropical Moist Forests: A Critique. *Interciencia,* Vol. 7, No. 2: 89–93.

Mahar, Dennis J. 1979. *Frontier Development Policy in Brazil: A Study of Amazônia.* New York: Praeger.

Marques, J., J. M. Sanots, N. A. Villa Nova, and E. Salati. 1977. Precipitable Water and Water Vapor Flux Between Belém and Manaus. *Acta Amazônia,* Vol. 7, No. 3: 355–362.

Mercado, Roberto Samanez. 1980. A Indústria Madeireira da Amazônia: Estrutura, Produção, e Mercados. Ph.D. diss., Michigan State University.

Mitchell, J. 1984. *Workshop on the Global Effects of Carbon Dioxide from Fossil Fuels* Washington, D.C.: U.S. Department of Energy.

Molion, L.C.B. 1975. A Climatonomic Study of the Energy and Moisture Fluxes of the Amazonas Basin with Consideration of Deforestation Effects. Ph.D. diss., University of Wisconsin.

Moran, Emilio F. 1976. Agricultural Development in the Transamazon Highway. Latin American Studies Working Papers. Bloomington: Indiana University.

Moran, Emilio F. 1982. Colonization in the Transamazon and Rondônia. Paper presented at the 31st Annual Latin American Conference on Frontier Expansion in Amazônia, February 8–11, University of Florida, Gainesville.

Moran, Emilio F. In press. Resettlement in the Amazon Forests. In *People of the Rain Forest*, ed. Julie S. Denslow and Christine Padoch. Berkeley: University of California Press.

Myers, Norman. 1980. *Conversion of Tropical Moist Forests*. Washington, D.C.: National Academy of Sciences.

Myers, Norman. 1981. The Present Status and Future Prospects of Tropical Moist Forest. *Environmental Conservation*, Vol. 7, No. 2: 101–114.

Myers, Norman. 1985. The End of the Lines. *Natural History*, Vol. 2: 2–12.

Pires, J. M., and G. T. Prance. 1977. The Amazon Forest: A Natural Heritage to Be Preserved. In *Extinction Is Forever*, ed. G. T. Prance and T. Elias. Bronx: New York Botanical Gardens.

Pompermayer, Malori Jose. 1979. The State and the Frontier in Brazil: A Case Study of the Amazon. Ph.D. diss., Stanford University.

Posey, Darrell A. 1983. Indigenous Ecological Knowledge and Development of the Amazon. In *The Dilemma of Amazonian Development*, ed. Emilio Moran. Boulder, Colo.: Westview Press.

Prance, Ghillean T. 1982. *Biological Diversity in the Tropics*. New York: Columbia University Press.

Pringle, S. L. 1976. Recycling of Water in the Amazon Basin: An Isotope Study. *Water Resources Research*, Vol. 15, No. 5: 1250–1258.

Rich, Bruce M. 1985. The Multilateral Development Bank's Environmental Policy and the United States. *Ecology Law Quarterly*, Vol. 12: 681–745.

Salati, Eneas. 1980. Um Deserto no Futuro da Amazônia. Rio Branco *O Jornal*, March 31.

Secretaria de Planejamento do Rondônia (SEPLAN/Ro). 1985. *Anuário Estatístico de Rondônia, 1983*. Pôrto Velho.

Serrão, A., I. Falesi, J. B. Vega, and J. F. Teixeira. 1979. Productivity of Cultivated Pastures on Low Fertility Soils of the Brazilian Amazon. In *Pasture Production in Acid Soils of the Tropics*, ed. P. A. Sanchez and L. E. Tergas. Cali, Colombia: CIAT.

Seubert, C. E., P. A. Sanchez, and C. Valverde. 1977. Effects of Land Clearing Methods on Soil Properties and Crop Performances on a Utisol of the Amazon Jungle of Peru. *Tropical Agriculture*, Vol. 54: 307–321.

Shane, Douglas R. 1986. *Hoofprints on the Rainforest*. Philadelphia: Institute for the Study of Human Issues.

Silva, José Natalino Macedo. 1983. Influência de Duas Intensidades de Exploração no Crescimento da Floresta Residual. *Pesquisa em Andamento, No. 129*. Belém: EMBRAPA.

Sioli, Harald. 1978. Destruição da Amazônia Pode Ameaçar o Mundo. Quoted in the *Folha de São Paulo*, December 12.

Sioli, Harald. 1985. The Effects of Deforestation in Amazônia. In *Change in the*

Amazon Basin: Man's Impact on Forests and Rivers, ed. John Hemming: 58–65. Manchester: Manchester University Press.

Skillings, Robert F. 1985. Economic Development of the Brazilian Amazon: Opportunities and Constraints. In *Change in the Amazon Basin: Man's Impact on Forests and Rivers*, ed. John Hemming: 36–43. Manchester: Manchester University Press.

Skillings, Robert F., and Nils O. Tcheyan. 1979. *Economic Development Prospects of the Amazon Region of Brazil.* Monograph, School of Advanced International Studies. Washington, D.C.: Johns Hopkins University Press.

Smith, Nigel J.H. 1977. Transamazon Highway: A Cultural-Ecological Analysis of Colonization in the Humid Tropics. Ph.D. diss., University of California, Berkeley.

Smith, Nigel J.H. 1981. Colonization Lessons from a Tropical Forest. *Science*, Vol. 214, No. 4522: 744–61.

Superintendência do Desenvolvimento da Amazônia (SUDAM). 1983a. Incentivos Fiscais Liberados Pela SUDAM (Anualmente), Distribuição Setorial até o Mês de Setembro/83. Spreadsheets. Belém: SUDAM/DPO/DAI.

Superintendência do Desenvolvimento da Amazônia (SUDAM). 1983b. Relacão de Projetos Aprovados. Unpublished report. Belém.

Tangley, Laura. 1986. Saving Tropical Forests. *Bioscience*, Vol. 36, No. 1: 4–8.

Tardim, Antonio T., et al. n.d. Relatório das Atividades do Projeto. No. 1034. SUDAM/INPE.

United Nations Food and Agriculture Organization (FAO). 1978. *1977 Yearbook of Forestry Products.* Rome.

United Nations Industrial Development Organization (UNIDO). 1983. *First World-Wide Study of the Wood and Wood Processing Industries.* Sectoral Studies Series No. 2. Vienna.

United States Congress. 1984. Multilateral Development Bank Activity and the Environment. Report of the Subcommittee on International Development Institutions and Finance, Committee on Banking, Finance and Urban Affairs, U.S. House of Representatives (98th Congress, 2nd Session).

United States Department of State. 1978. Tropical Deforestation. Proceedings of the U.S. Strategy Conference Sponsored by the U.S. Department of State and the U.S. Agency for International Development, June 12–14, Washington, D.C. Washington, D.C.: U.S. Government Printing Office.

United States Interagency Task Force on Tropical Forests. 1980. *The World's Tropical Forests: A Policy Strategy and Program for the United States.* U.S. Department of State Publication No. 9117. Washington, D.C.: U.S. Government Printing Office.

Universidade Federal Rural do Rio de Janeiro (UFRRJ). 1983. *Diagnóstico da Indústria Madeireira do Estado de Rondônia.* Rio de Janeiro: MA/IBDF/DIC/UFRRJ.

Universidade Federal Rural do Rio de Janeiro (UFRRJ). 1985a. *Contribuição de Mercado Madeireiro no Desenvolvimento Regional: Rondônia.* Rio de Janeiro: MA/IBDF/DIC/UFRRJ.

Universidade Federal Rural do Rio de Janeiro (UFRRJ). 1985b. *Proposta para o*

Plano de Manejo da Floresta Nacional do Jamarí. Rio de Janeiro: MA/IBDF/DEF/UFRRJ.

Villa Nova, N. A., E. Salati, and E. Matsui. 1976. Estimativa da Evapotranspiração na Bacia Amazônico. *Acta Amazônica,* Vol. 6: 215–228.

Woodwell, George. 1978. Carbon Dioxide-Deforestation Relationships. *Proceedings of the U.S. Strategy Conference on Tropical Deforestation.* U.S. Department of State and the U.S. Agency for International Development, June 12–14, Washington, D.C.

Woodwell, George, et al. 1983. Global Deforestation: Contribution to Atmospheric Carbon Dioxide. *Science,* Vol. 222 (4628): 1081–1086.

World Bank (International Bank for Reconstruction and Development). *World Tables* (various years).

World Bank. 1981. *Brazil: Integrated Development of the Northwest Frontier.* A World Bank Country Study, June 1981. Washington, D.C.

7   West Africa: resource management policies and the tropical forest

MALCOLM GILLIS

This chapter examines the role of public policy in deforestation in each of four West African countries: Liberia, Cote d'Ivoire (hereafter, Ivory Coast), Ghana, and Gabon. We provide an overview of forest resources, deforestation, and international trade in tropical hardwoods for the entire region, and on a country-by-country basis. Patterns of property rights and foreign investment in each nation are addressed, as well as the national benefits these countries have derived from forest utilization and government capture of timber rents. We then focus upon reforestation and forest concessions policies, respectively, and examine the impact of forest-based industrialization policies. Finally, the impact of non-forestry policies on tropical forest utilization in each of the four nations is considered.

Overview

By most estimates, the rate of decline in the area of productive closed forest in West Africa has been the highest in the world since 1975, some 3.7 times the average rate for all tropical countries (Lanly and Clement 1979: 10–21). The four West African nations studied in this chapter include the Ivory Coast, with the highest rate of annual deforestation yet observed anywhere, Liberia and Ghana, in which only shreds of natural forest remain, and Gabon, where deforestation has been slow but will likely accelerate sharply before the end of the century, with the completion of the Trans-Gabon Railway.

In 1984, these four countries together accounted for nearly all of Africa's timber exports (Ivory Coast shipped 60 percent of this) but only 9.3 percent of world exports of tropical hardwood forest products (World Bank 1986: 8). West African exports of tropical hardwood products come from a relatively small resource base: less than 1.5 percent of world endowments of productive tropical forests. Clearly, the search for timber export earnings, and the associated government policies, have played a

299

major role in deforestation in the region. But as in Southeast Asia, the destructive effects of timber harvests have been magnified by the interaction of logging with institutions, property rights, and poverty. Although factors leading to rapid deforestation elsewhere have been present in West Africa, their relative roles differ from country to country. In Ivory Coast, for example, headlong promotion of commercial logging activity has outstripped all other causes of forest destruction. In Ghana, steadily worsening rural and urban poverty has clearly overshadowed all other sources of deforestation since the late 1960s. Liberia and Gabon may be special cases: Liberia, because most deforestation has been occurring in secondary rather than primary forests, and Gabon because it may yet avoid excessive depletion of its still sizeable natural forest endowment, provided forest utilization policies are reformed soon.

None of the four nations has derived many tangible or intangible benefits from forest use to offset the heavy economic, social and ecological costs of rapid deforestation. Government policies have not effectively captured rents available from the forest sector; as in Southeast Asia, governments have sold their own resources cheaply.

Forest resources and deforestation

Sustained natural forest exploitation has a much longer history in West Africa than in Southeast Asia: shipbuilders imported African oak (*iroko*) into Liverpool in sizeable quantities as early as 1823 (Adeyoju 1976: 3). By 1962, West Africa still held almost a 30 percent share of world tropical timber exports, largely by dint of its proximity to the European market. By 1984, however, West Africa's share had shrunk by more than half, with the opening in 1965 of untapped forests in Indonesia and East Malaysia, where relatively homogeneous stands allow commercial yields per hectare 4 to 5 times that of West Africa. Even so, West African countries retained a dominant position in the market for decorative hardwoods.

Liberia

It is believed that as long ago as 1680, Liberia's population substantially exceeded the 1.5 million people present in the 1970s. Widespread shifting cultivation centuries ago is thought to have destroyed the primary forest, to be replaced by secondary vegetation. Diseases imported from Europe in the 1700s, and intensification of tribal warfare and slave trading, reduced the population to less than 1 million (FAO 1986: 276).

Table 7.1. *Liberia tropical forest endowments and rates of deforestation, 1982 (million hectares)*

I. Total land area	9.63
II. Total fallow forest area	5.50
III. Tropical forest area	2.67
A. Virgin tropical forest	0.91
B. Logged-over forest	0.43
C. Other productive forest	1.33
IV. Area under timber concessions	3.82
V. Extent of deforestation (hectares per year)	
A. 1940s: 20,000	
B. 1976–80: 41,000	
C. 1981–85: 46,000	
VI. Rate of deforestation (annual deforestation as percentage of tropical forest area):	
41,000 ÷ 2,000,000 = 2.05% per year	

Source: FAO (1981: 270–272).

Over the next 200 years, the low bush in abandoned country developed into the present high secondary forest.

Liberia thus is now experiencing its second major deforestation, which has doubled in pace in the past 40 years, to 41,000 hectares annually (Table 7.1). The current measured annual deforestation rate in Liberia is roughly three times faster than the average for all nations with tropical forest endowments, but still less than one-third the rate for nearby Ivory Coast.

Relatively little forested area has been cleared for fuelwood or plantations (mostly rubber). At least one of the larger former rubber plantations (Firestone) is now used primarily as a source of commercial firewood.[1] In fact, most recent deforestation has been attributed to shifting cultivation, as would be expected in a poor country.[2] Shifting cultivation is principally of upland rice which, in 1975–76, employed 130,000 households. This crop requires clearing 1.4 hectares annually per household, and accounts for three-fourths of annual deforestation in the remaining Liberian closed forest (FAO 1981: 277).

As elsewhere, logging has indirectly contributed to deforestation. Few of the areas entered yearly by shifting cultivators are in undisturbed forest; most shifting cultivation occurs in logged-over forests. By 1982, the total area under logging concessions was well in excess of the total tropical forest area (Table 7.1, rows III and IV); some concessions include areas of unproductive vegetation. To accommodate rapid growth in con-

Table 7.2. *Ghana forest endowments, 1980 (million hectares)*

A. Total land area	23.850
B. Total "forest" area (closed plus fallow forest)	8.218
C. Virgin forest	0.000
D. Closed forest	1.718
1. Productive forest	1.167
2. Protection forest	0.397
3. Earmarked for conversion	0.154
E. Forest fallows	6.500
1. The total "forest area" occupied by shifting cultivation at the end of 1980 was estimated at 9.180 million hectares.	
2. Contains also cocoa farms, bush fallows, and food crop areas.	

Source: FAO (1981: 195).

cession areas, the network of logging roads grew rapidly after 1963. As in insular Southeast Asia, expansion of this network has converted areas once unavailable to shifting cultivation into easily accessible ones (FAO 1981: 277). The effects of opening up new forest lands with logging roads has been mitigated by historical patterns of land use in Liberia. So much of Liberia's total land area lies in forest fallows converted centuries ago that fully 90 percent of shifting cultivation takes place on such land.

Ghana

Rapid depletion of the Ghanaian tropical forest is also a long-standing phenomenon. At the beginning of this century, the natural forest estate covered 8 million hectares, or one-third of the country's area (Trum-Barima 1981: 111). By 1950, natural forest cover had dwindled to but 18 percent.

The rapid deforestation of 1900–1950 continued well into the 1970s. By 1980, virgin forest had all but vanished; closed forest area covered but 7 percent of total land area (Table 7.2). Once-forested areas, now in fallows, including degraded forest, converted forest land (cocoa), and food crop areas, now cover 27 percent of total land area. In the decade prior to 1976, average annual deforestation was 45,000 hectares per year (Table 7.3). Deforestation slowed materially in the period 1976–80 and again from 1981 to 1985, to only 22,000 hectares per year. The decline in deforestation after 1980 is not a reflection of changes in forest use policies; rather there is little left to deforest (FAO 1981: 200).

Ghana's already degraded forests are sharply threatened in the long run by the southward drift of the Sahara. But principal sources of recent

Table 7.3. *Ghana log production, recorded log export volumes, and fuelwood harvest, 1950–85 (million cubic meters)*

Year	Volume of industrial logs extracted	Volume of logs exported	Fuelwood harvests
1950	0.56	0.29	n.a.[a]
1955	1.14	n.a.	n.a.
1960	1.83	n.a.	n.a.
1965	1.59	n.a.	n.a.
1970	1.23	0.46	8.44[b]
1972	1.86	0.94	8.44[b]
1973	2.05	1.08	8.44[b]
1974	1.43	0.43	8.44[b]
1975	1.33	0.44	9.57[c]
1976	1.38	0.34	9.57[c]
1977	1.03	0.39	9.57[c]
1978	0.99	0.40	9.57[c]
1979	0.49	0.19	9.57[c]
1980	0.48	0.12	10.33
1981	0.48	0.22	10.59
1982	0.78	0.17	10.80
1983	0.68	0.21	11.15
1984	0.76	0.25	11.35
1985	0.92	0.24	11.65

[a]n.a. = not available.
[b]Average for 1970–74.
[c]Average for 1975–79.
Sources: 1950–70: John M. Page, Jr. (1974); 1972–80: Ghana Timber Marketing Board (1982); 1981–84: World Bank (1986: 36–38).

deforestation in Ghana are shifting cultivation and fuelwood harvests, both driven by rural and urban poverty. Until the mid-1960s, the second most important cause of deforestation was conversion of closed forest areas to permanent tree crops, particularly cocoa (FAO 1981: 200). While logging has not directly caused significant deforestation in recent years, new logging roads have facilitated the spread of shifting cultivation, as in Sabah, Sarawak, and Indonesia.

Tree crops and logging have been less important for deforestation since 1960, if output figures for cocoa and logs are taken as indicators of agro-conversion and logging, respectively. In the early 1960s, Ghanaian cocoa exports were nearly one-third of world exports; by 1982 they were less than 15 percent. Cocoa production plunged particularly rapidly after 1970: tonnage in 1977 was less than half that of 1970 (Roemer 1984: 205). Moreover, the volume of logs extracted in 1980 was one-fourth that of

1973, a boom period in the world timber market. Log production recovered somewhat after 1983 along with the rest of the economy, following a series of drastic economic policy reforms. Even so, 1984 production was less than half that of the early 1970s (Table 7.3).

These trends suggest that in Ghana reasons for rapid forest depletion are strikingly different from those in, say, Malaysia and Indonesia, where a battery of forestry policies (fiscal charges and concession terms) and non-forestry policies (export taxes, industrial policies) contributed significantly to a production boom that swept these two countries to dominance in the world market for tropical hardwoods in the decade after 1967.

Many of the same policies drive Ghanaian deforestation as in Southeast Asia: undertaxation of timber (Sarawak and Indonesia), the short duration of concessions (all of Malaysia and Indonesia), and ineffective reforestation programs. In Ghana, though, these policies have had decidedly secondary effects on deforestation. Further, the policies most responsible for deforestation have, ironically, limited further deforestation that might have arisen from logging alone.

Since 1963 the Ghanaian economy has undergone a clear decline. Upon winning independence in 1957, Ghana was arguably the richest country in sub-Saharan Africa (per capita income of US$500 in 1980 dollars), and its citizens were the most educated (Roemer 1984: 201). But from independence through 1983, GNP per capita declined on average 0.7 percent per year.[3] The decline has accelerated in the past two decades; the average annual rate of growth in GNP per capita was a negative 2.1 percent from 1965 to 1983 (World Bank 1985: 174).

Falling levels of economic welfare had serious consequences for the forest resource base. Between the end of the colonial period and 1980, total forest area fell by almost two-thirds. Conversion of forest land to cocoa-raising and logging clearly played a role in this process until the early 1960s, but not afterward. Forestry policies themselves, with their built-in bias toward waste, may have contributed, however marginally. But the above factors pale in comparison to the effects on the forest of impoverishment of Ghanaian citizens after 1960 (and particularly after 1970). There is little professional disagreement concerning the prime cause of sustained negative growth: the economic policies followed by successive governments. The assertion that growing poverty was primarily responsible for deforestation is supported by the rapid growth in area under shifting cultivation and by the sharp growth in fuelwood consumption after 1970 (Table 7.3) as increasing numbers of im-

Table 7.4. *Ghana: Deforestation, shifting cultivation areas, and logging areas*

	Area (hectares)
I. Deforestation (hectares/year)	
A. 1970–76	45,000
B. 1976–80	27,000
C. 1981–85	22,000
II. Area under logging	1,167,000
III. Area under shifting cultivation	9,180,000

Source: FAO (1981: 194).

poverished urban and rural families came to rely almost exclusively on fuelwood for household energy.

By 1980, areas under shifting cultivation reached 9.2 million hectares (almost 40 percent of land area) in a country of but 11 million people (Table 7.4). In Malaysia, with a 1980 population of 13.3 million people, the area affected by shifting cultivation in 1980 was less than 7.5 million hectares, 23 percent of total land area. Population growth rates from 1960 to 1983 were only marginally lower in Malaysia, where 1957 per capita income (at $570 in 1980 dollars) was only slightly higher than in Ghana.[4]

Deforestation due to fuelwood harvests accelerated sharply after the mid-1970s. By 1983, Ghanaian consumption of fuelwood was 906 cubic meters per capita, one of the world's highest, and it is still rising rapidly. Indeed, harvest of timber for fuelwood was about 12 times the harvest of logs for wood products (World Bank 1987: Annex 1,6).

The claim that impoverishment in Ghana primarily results from government policies cannot be conclusively established here. We do find that while other oil-importing middle income developing countries (a category to which Ghana belonged until 1978) had average real per capita GDP growth of 3.5 percent annually from 1965 to 1983, Ghana's GDP growth rate was 5.6 percentage points slower (World Bank 1980: Annex Table 1). Economic analysts almost unanimously attribute most of the blame to interventionist industrial, agricultural, and trade policies and to ubiquitous controls applied in a setting of chronic political instability.[5] Ironically, many of the policies that contributed to poverty in Ghana, and therefore to more widespread shifting cultivation and fuelwood harvests, also contributed to a decline in deforestation arising from logging. One of the major factors retarding both economic growth and logging export growth was the unusually deep overvaluation of the Ghanaian *cedi* in virtually every year since 1963 (see Chapter 2 for an extended discussion of effects of overvaluation). The overvaluation became most pronounced after 1976.[6] The ratio between black market and official rates for the period 1977–83 was absurdly high, ranging from a low of 4.96 in 1979 to a high of 18.18 in May of 1983.

Extreme overvaluation reduced the rate of log harvest in two principal ways. First, and most obvious, exports of any commodity were barely worthwhile unless firms were willing to bear the risks and costs of smuggling. Estimated smuggled timber exports were nearly as large as the small volume of recorded exports in the period 1977–79 (Ghana Timber Association c. 1982).

Second, deep overvaluation, in depressing export earnings, also meant a severe shortage of foreign exchange for supplies and spare parts in all industries, including logging. As a result, in the late 1970s and early 1980s even large timber firms were operating at less than 25 percent of logging capacity (*Ghana Timber News* 1979: 4).

In turn, overvaluation resulted from government attempts to maintain low nominal exchange rates despite rampant inflation arising from near complete breakdown of macroeconomic policy. Fiscal policy began to collapse in 1971; after 1976, government spending was typically nearly double government revenue levels. Agricultural policy, particularly pricing policy, was highly oppressive, with heavy controls on prices of virtually all staples. Financial policy was gutted: inflation averaged 52 percent per year for the period 1973–83 (World Bank 1985: 174).

In this setting, any effects of forestry policies on deforestation would have been decidedly secondary to the impact of deepening poverty on the extent of shifting cultivation and on fuelwood harvests. Nonetheless, forestry policies may have contributed marginally to deforestation; they contributed little to capture of timber rents, particularly after 1963.

Ivory Coast

Every comparative study of tropical forest endowments and deforestation has concluded that the Ivory Coast has experienced the most rapid deforestation rates in the world.[7] Table 7.5 shows that the forest has been receding at an extraordinarily rapid rate. At recent rates of deforestation, the forest estate in 1985 was perhaps only 17.6 percent of areas forested in 1900. Unlike Ghana (where deforestation processes were well under way before independence in 1957) deforestation was most rapid in Ivory Coast after the mid-1950s.

In the three decades from 1955 to 1985, the nation's tropical forest estate shrunk to 22 percent of its 1955 level. During that period, the

Table 7.5. *Ivory Coast: Evolution of tropical forest endowments, 1900–85, and rates of deforestation*

A. Forest cover (dense humid tropical forest):

Year	Area (million hectares)
1900	14.50
1955	11.80
1965	8.98
1973	6.20
1980	3.99
1985	2.55[a]

B. Average annual amount of deforestation (1955–85):

Period	Area deforested (hectares per year)
1956–65	280,000
1966–73	350,000
1974–80	315,000
1981–85	290,000[a]

C. Average annual rate of deforestation (percent of remaining forest per year):

Period	Annual deforestation (percent)
1956–65	2.37
1966–73	3.90
1974–80	5.08
1981–85	7.26

[a]Estimated.
Source: FAO (1981: 124–125).

average annual rate of deforestation as a percent of the remaining forest area rose from 2.37 percent in 1956–65 to 5.08 percent in 1974–80, and to an estimated 7.26 percent in 1981–85 (Table 7.5). The Ivory Coast figure is striking when compared with the estimated average rate of deforestation of 0.6 percent annually in all countries with tropical forest endowments (Steinlin 1982: 6). The implication of such rapid deforestation rates can be placed in perspective by contrasting the experience of Ivory Coast with that of Indonesia, the world's leading producer of tropical logs from 1973 until 1983. Table 7.6 shows that the annual *amount* of deforestation in Ivory Coast has been almost half that of Indonesia, a poorer country with 6 times the area and a population 16 times greater. The estimated annual deforestation *rate* in Ivory Coast after 1980 exceeded 12 times that of Indonesia.

Table 7.6. *Deforestation: Indonesia and Ivory Coast*

	Indonesia	Ivory Coast
A. Population (1985)	162.0 million	10.1 million
B. Areas deforested annually (1981–85)	600,000 ha	290,000 ha
C. Annual deforestation rate (deforestation annually as percent of forest area)	0.5%	7.26%
D. Per capita income (1985)	$530	$660
E. Area (thousand km^2)	1,919	322

Sources: Deforestation information: Table 7.5; Chapter 2, Table 2.2; population, area, and income data: World Bank (1987: Annex 1, Table 1).

The principal causes of deforestation in Ivory Coast have been shifting cultivation, logging, and agro-conversion, all in a setting in which the government has no overall land-use policy for the forest.[8] Definitive evidence on the rankings of these causes is lacking, but it is plausible that population pressure has been more important to the spread of shifting cultivation in Ivory Coast than most other countries. From 1965 to 1983, the population growth rate in Ivory Coast was, at 4.6 percent annually, nearly the highest in the world, well above the 2.45 percent average for other middle income countries (World Bank 1985: 210). Rapid population growth in Ivory Coast was a consequence both of much higher than average fertility rates (6.6 percent in Ivory Coast versus 4.6 percent in other middle-income countries) and high rates of immigration from very-low-income neighboring countries, especially Upper Volta and Mali (FAO 1981: 97–127). As a result, the area under cultivation more than doubled from 1965 to 1985 (Spears 1986: 4).

As elsewhere, the relative roles of logging and the spread of shifting cultivation in deforestation are difficult to assess. Considerable deforestation is due to the movement of shifting cultivators into areas opened up by logging. Indeed, it has been estimated that "for each 5 cubic meters of logs harvested in Ivory Coast, 1 hectare of forest disappears at the hands of the follow-on cultivator" (Myers 1983: 6).

Agro-conversion, primarily for cocoa and coffee estates, also played a significant role in deforestation, although not nearly to the extent experienced in Peninsular Malaysia from 1910 to 1970. Here again the relative roles of agro-conversion and shifting cultivation are unclear. For example, in 1966, the total Ivory Coast area formerly in "forest zones" shifted to "all cultivation" was 6.4 million hectares, of which 0.9 million was in permanent crops (cocoa, coffee, etc.), 0.5 million in food crops, and 5 million in shifting cultivation. By 1980, the total area of former forest zones

Table 7.7. *Gabon: tropical forest endowments and deforestation, 1982 (thousand hectares)*

I. Total area of country	26,767
II. Total forest area	21,861
A. Dense humid forest	20,510
1. *Okoume* species dom.	10,709
2. Other species dom.	9,801
B. Degraded forest	1,351
III. Deforestation	
A. Annual amount	15
B. Annual rate (IIIA ÷ IIA)	0.07%

Source: FAO (1981: 156–172).

converted to "all cultivation" had nearly doubled to 11.59 million hectares, but the total was not disaggregated by use (permanent crops, food crops, and shifting cultivation) (FAO 1981: 123). We thus know little about the ranking of deforestation causes in the country with the world's most rapidly receding forests.

Gabon

Gabon merits inclusion in any survey of policy issues in tropical forestry because it is atypical in so many respects.

First, Gabon was not a poor country in the 1960s and 1970s. By 1983 per capita income for the nation's 700,000 inhabitants was estimated at US$3,950 (World Bank 1985: 232). By implication, shifting cultivation practices should not have contributed much to recent deforestation in Gabon.

Second, much of Gabon's tropical forest estate remains undisturbed, in sharp contrast to Ghana, Ivory Coast, Liberia, and Malaysia. Fully 77 percent of total Gabonese land area is covered with dense forest; degraded forests constitute another 5 percent of total land area (Table 7.7). Only 6.13 million hectares or 28 percent of total forest area, are under concession agreements (Republic of Gabon, Ministère des Eaux et Forêt 1980: 31).

Third, recent deforestation rates have been among the lowest in Africa, indeed in the tropical world. At less than 0.1 percent, Gabon's deforestation rate is about 10 percent of the median rate for all countries with tropical hardwood estates, and but 1 percent of that of Ivory Coast.

Fourth, the Gabonese forest sector continues to be dominated by foreign enterprises, in contrast to Ghana, Indonesia, and most of Malaysia. Total foreign investment in the forest-based sector reached $364 million in 1981, 80 percent of total foreign investment from all sources (Gillis et al. 1983: Table 2–1).

Fifth, Gabonese taxes and forest fees have long been among the lowest in the tropical world. Relatively light timber sector taxation has partly been a consequence of the Gabonese penchant for offering income tax incentives to forest-based investment, a practice abandoned by most tropical nations by the late 1960s. In the 1970s and 1980s low timber taxes were also enabled by substantial tax revenues from the oil and uranium sectors through 1982.

Sixth, Gabon, unlike most other major tropical timber producers, has not stressed industrial policies to promote domestic log processing. Even as late as 1984, 89 percent of Gabonese log production was exported in log form, by far the highest proportion in Africa and Asia.

Finally, Gabon was the first tropical African nation to enact a reforestation tax. Enacted by the colonial government in 1957, it was earmarked for the use of the *Forestier Gabonais de Réboisement* (Gabonese Reforestation Fund) (Legault 1981: 162).

Gabon is one of six African nations for which lessons drawn from this book may be in time to be of assistance. For countries such as Ghana, Ivory Coast, and Nigeria, the lessons are at least a decade too late. But Gabon, together with Zaire, Equatorial Guinea, Benin, Congo, and the Central African Republic (CAR) all have relatively untouched forest estates (Table 7.8). The reason deforestation has been limited in these six countries has more to do with inaccessibility than with any merits of public policy, although relatively wealthy Gabon has set aside fairly large forest areas as national parks (Myers 1985: 37). But as transport infrastructure improves in Gabon, with the spread of the Trans-Gabon Railway, and as tropical forests elsewhere in Africa are further depleted, these six countries will face much the same pressures for consumption of tropical forest resources as did Ivory Coast, Ghana, and Nigeria in the past.

The fact that foreign investors remain heavily involved in the timber sector does not itself bode poorly for the future of Gabon's forest estate. Experience elsewhere suggests that the practices of large foreign firms can be even more conducive to sensible forest management, particularly if (as in Gabon) concession contracts are long enough to induce firms to consider the long-term consequences of harvest methods.[9] Moreover, existing Gabonese timber processing facilities are far more efficient than in neighboring African countries: two French firms have achieved plywood conversion ratios much higher than elsewhere in Africa or in Asia: 72 percent in one case, and 53 percent in the other.[10]

Table 7.8. *Extent and causes of deforestation in selected tropical African nations, 1982*

Country	Extent of deforestation	Cause of deforestation[a]
I. Largely undamaged or unexploited forest area		
1. Zaire	one-tenth of world total of raw forest, large areas untouched	A, L, both minimal
2. Gabon	still almost entirely forest	L, minimal
3. Equatorial Guinea	relatively untouched	A, minimal
4. Benin	three-fourths of original forest still remaining	A, minimal
5. Congo	except for south, largely untouched	L, minimal
6. Central African Republic	southern forest largely untouched	A, very minimal
II. Heavily deforested		
1. Cameroon	extensive disruption of forest	A, L, substantial
2. Ivory Coast	70 percent of primary forest in 1900 now cleared— remainder may vanish before 1995	L, A, very substantial
3. Liberia	very little remaining rain forest	A, substantial L, moderate
4. Nigeria	most area disrupted, except small area in south	P, L
5. Sierra Leone	very few undisturbed areas	P, A, substantial
6. Ghana	little rain forest remaining, no virgin forest left	P, A, L, substantial
7. Guinea	small forested area remaining	unknown
8. Cabinda	small rain forest in north	unknown
9. Madagascar	fragment of forest remains	A, substantial

[a]Principal causes of deforestation: P, population pressure, fuelwood scarcity; A, population pressure, shifting agricultural cultivation; L, logging.
Source: White (1983: 2).

Gabon does provide an extreme illustration of lost timber rents due to light timber taxation. Otherwise, there is little to be learned from Gabonese experience with public policy toward the tropical forest. The future of the Gabonese forest, however, hinges on utilization decisions taken once large new forest areas become accessible upon completion of the Trans-Gabon Railway. Gabon may yet benefit by applying lessons of experience of other countries to its still substantial forest resources.

Property rights and foreign investments in West Africa

Property rights

As elsewhere in Africa, private ownership of forest property rights is virtually absent in Liberia. The government owns the National Forest, which covers almost 1.7 million hectares, nearly all the tropical forest area (Table 7.1). The law nominally prohibits shifting cultivation in National Forest areas; the boundaries are demarcated by 7-foot-wide borders said to be patrolled by forest guards (FAO 1981: 272).

In Ghana, rights to exploit natural forest resources were originally vested in traditional groups and governed by customary law (often unwritten tribal laws). Until the early 1970s, almost all forest lands were owned by traditional communities (primarily in the Brong-Ahafo and Ashanti regions) whose titular heads were called "stools" in some areas and "skins" in others. From the time of Ghana's independence in 1957 until 1973, the lands were held in trust for the stools and skins by the central government. Applications for land use, including timber concessions, were submitted to the various stools and skins who derived significant revenues from leasing timber rights. In the early 1970s, due to political conflict, rights to these lands were transferred entirely to the central government, which now disposes of all forest rights (Peprah 1982b: 11–17). Although many African countries have recently shifted toward greater central government control of forest rights, most still preserve a degree of local communal ownership or application of customary law to forest concessions. No other central African government has expanded its jurisdiction over allocating forest harvest rights to the extent of Ghana.

As a result, the Ghanaian forest is now even more vulnerable to the "tragedy of the commons" than when property rights were vested in tribal groups. The stools and skins had at least some incentive to contain shifting cultivation to well-defined areas and limit forest damage from logging. Transferring all forest rights to the national government has left access to the remaining forest unchecked, and virtually eliminated the limited enforcement of environmental safeguards, once overseen by the stools and skins.

Tropical forest property rights are defined somewhat differently in Ivory Coast than in neighboring West African countries. The Ivory Coast Forestry Code distinguishes between forest areas reserved for the state, for individuals, and for tribal groups (Côte d'Ivoire, Loi Nr. 65–425

1965). Most forest rights are owned by the government; only a small part involves private property rights in the forest (SODEFOR 1975: 36).

Responsibility for forestry policy was in the hands of the French colonial administration until independence in 1960. Since 1965, forest use has been governed by the Forestry Code of 1965. From independence until 1974, forestry and forest industry matters were the responsibility of the State Secretariat for Reforestation, a division of the Ministry of Agriculture. In 1974, a Ministry for Forestry was created. Still, all reforestation projects and forest inventory tasks, and certain elements of forest-based industrial planning, are carried out by a semi-autonomous government agency popularly known as SODEFOR (Société pour le Développement des Plantations Forestières) (FAO 1981: 107).

Foreign investment

Liberia. Liberia's forestry sector has attracted foreign capital and technical assistance for almost four decades. In contrast to Ghana, foreign capital and ownership continued in the Liberian forest-industry sector through the early 1980s. By 1981, total foreign investment in the Liberian forest-based sector (FBS) had reached US$146.9 million, about 3 percent of the total for all the ten largest timber producers in Africa (Table 7.9). Sixteen firms with majority foreign ownership had 85 percent of Liberia's sawmill capacity by 1981 (Peprah 1982b: 17–21).

Growth in investment, foreign and otherwise, in the Liberian FBS slowed after the late 1970s, with the military takeover of the government in 1980, ending decades of relative political stability. The overthrow followed several years of stagnation in an economy that prior to 1973 had enjoyed average annual GDP growth of 5.5 percent, and a 2.8 percent population growth rate, thus allowing steadily rising per capita income (World Bank 1985: Appendix Tables 2 and 19). Since 1973, per capita income has declined at a rate of 3 percent per year (World Bank 1985: Table 2), creating further pressures for shifting cultivation and fuelwood harvests, with unfortunate consequences for forest conservation.

Ghana. Prior to independence in 1957, Ghana's timber industry was dominated by four large foreign corporations and one domestic firm that together accounted for about 80 percent of log exports.

Through 1972, multinationals had concessions granting long-term "harvesting" rights (up to 99 years) and paid nominal fees, "royalties," for the privilege. They had no legal commitment to processing. But in Octo-

Table 7.9. *Accumulated investment in the forest-based sector in selected African countries, 1981 (million U.S. dollars)*[a]

	Foreign	Domestic	Total
Liberia	124.87	22.03	146.90
Ivory Coast	558.80	145.00	703.80
Ghana	175.45	207.15	382.60
Gabon	364.46	94.07	458.53
Cameroon	365.41	335.71	701.12
Congo	157.20	28.60	185.80
Central African Republic	92.21	12.79	105.00
Zaire	95.03	54.44	149.47
Nigeria	932.16	621.44	1,553.60
Kenya	22.79	53.21	76.00
Total	2,888.38	1,574.44	4,462.82

[a]Since the estimation method involves valuing all production capacity at market prices, these data assume that the proportional age structure of all investments in all countries is the same. No recent studies have been done on the age structure of African FBS investments to allow figures to be adjusted to another vintage assumption.
Source: Gillis et al. (1983: 2–12).

ber 1973 the government directed all multinational natural resource firms to surrender 55 percent of equity to the government (Esseks 1974). After two to three years the four remaining multinational timber firms sold their interests to the government and left Ghana's timber industry. In the following years, performances of the four firms deteriorated drastically; all mills ran at a loss in 1981 (Peprah 1982a: 66). A variety of other policies have been adopted to increase Ghanaian ownership of timber firms. Preferential concession policies, frequently combined with subsidies, were used to encourage domestic control of the timber industry.[11] In the late 1960s, the government granted forest concessions and soft loans (sometimes at negative real interest rates) for capital equipment purchases to numerous Ghanaian logging contractors. As a result, the number of locally owned sawmills grew from 43 in 1962 to 61 in 1970, and plywood-manufacturing plants increased from two in the early 1960s to five in 1970.

Ivory Coast. Exploitation of Ivory Coast forests by Europeans dates from the end of the 19th century, when British and French interests began harvesting ebony logs that were then floated downstream to the coast (FAO 1981: 106). Forest sector activity grew slowly until 1951, when the port of Abidjan was opened. Tropical hardwood harvests increased spectacularly after 1958 in the east central region, spreading rapidly to the southeast by 1970. A burst of concession awards between 1965 and 1972 resulted in the allocation of more than two-thirds of the productive forest to concessionaires (Schmithusen 1977: 90–103).

The Ivory Coast was the first and largest recipient of foreign investment in the forest-based sector in French-speaking Africa. Although foreign investors established long-term interests in the Ivory Coast early in this century, not until the mid-1950s did large transnational corporations (TNCs) become substantially involved in the forest-based sector (FBS). Although the number of domestically owned firms rose sharply after independence in 1960, 91 percent of installed capacity in sawmilling remained in foreign hands by 1976. Table 7.9 shows that as late as 1981, accumulated foreign investment in the Ivory Coast's forest-based sector totalled nearly US$560 million, versus only US$145 million by domestic firms.

Independence had no substantial effects upon the forest-based sector until enactment of the new forestry code in 1965. This code was designed primarily to give foreign investors incentives by reserving large, new concession areas and longer contract terms for those firms willing to integrate to higher states of domestic processing. Together with other policies, the code resulted in one of the most laissez-faire policy regimes of any tropical nation. All operating rights were tradeable, from concessions to export quotas. Nonetheless, foreign investors continued to export in log form the most desirable types of hardwoods: *sipo, assamela, makore,* and *acajou mahogony* (Arnaud and Sournia 1980: 295). The Ministry of Waters and Forests thereupon issued new decrees, limiting the export of these species and linking new concession terms to the degree of industrialization. While these measures helped bring about new investment in processing, they did not create a viable industry as the TNCs tended to regard the investments as relatively small costs of exporting logs from the Ivory Coast. By 1984 the Ivory Coast still exported three-quarters of its production in log form, a proportion exceeded in Africa only by Liberia and Gabon.

Gabon. The first commercial export-oriented exploitation of the Gabonese forest began in 1905, when French firms first harvested ebony and the prized stems of the *okoume* species. Stands of these trees still covered over one-half the present dense forest estate in 1982 (Table 7.7). Large-scale timber exploitation dates only from 1947.[12] Harvests were initially confined largely to coastal areas, since the transport infrastructure was, and still is, poorly developed in Gabon's hilly terrain.[13] Even by 1980 Gabon had but 100 km of paved roads; logs harvested in the interior had to be floated long distances by river.

For purposes of forest exploitation, the country is divided into two zones. Lack of transport infrastructure has meant that domestic firms, with more limited capital resources, have confined their operations to Zone 1, which is nearer the coast and more easily exploited, and is now reserved to Gabonese. Zone 2, more distant, is available to foreign investors. This policy, however, has yet to prove successful in helping domestic firms. Zone 1, logged since the 1950s, has been depleted, whereas Zone 2 has only recently been opened.

Gabon is one of the world's few countries still permitting anything near 100 percent foreign equity ownership in natural resource projects. Since 1970, however, new entrants into Gabonese natural resource projects have been required to provide "free equity" to the Gabonese government, usually 10 percent of total equity. The government is also a joint venture partner through purchased equity in other timber projects.[14]

Benefits and rent capture in forest exploitation

Income, employment, exports, and tax revenues

Liberia. Value-added in Liberia's forest sector has been a rising share of relatively stagnant GDP since 1973; it doubled from 1973 to 1980 (Table 7.10), as log production expanded by 50 percent to 745,000 cubic meters. Employment in the FBS grew from 1,200 persons in 1973 to 5,900 in 1980, about 0.6 percent of the labor force (Peprah 1982b: Table 10) (Tables 7.11 and 7.12). Total value of wood and wood product exports peaked at $65 million in 1980, declining to $37 million in 1981, or about 7 percent of total exports (Table 7.13). Fiscal benefits flowing from forest sector exploitation were insubstantial through much of the 1970s, but were a relatively high proportion (20 percent) of total timber export values (Table 7.16) and also a considerable proportion of total tax revenues. In the early 1980s, forest fees and taxes accounted for slightly less than 8 percent of total government revenues, a very high percentage relative to other African countries (Table 7.14).

Ghana. The contribution of Ghana's forest sector to gross domestic product has been remarkably constant since 1965, when value-added in the forest sector was 6.2 percent of GDP (Gillis et al. 1983: 4). In the worst years of the Ghanaian economic tailspin, from 1974 until 1984, the forest sector consistently accounted for between 4.9 and 6.2 percent of the nation's declining GDP (IBRD 1986: Annex 1,2). That is to say that

Table 7.10. *Value-added in the forest sector as a percentage of GDP, selected African nations, 1965–80*

Year	Liberia	Gabon	Ghana	Ivory Coast	Kenya	Cameroon	Congo	Nigeria	Tanzania
1965			6.2	6.3		3.8	4.4		
1970							5.1		
1971			4.5		2.0	2.9		2.5	
1972				5.8	2.0			2.5	
1973	2.6				2.2			2.5	
1974	1.8				2.3			2.5	
1975	3.0	4.4			2.1			2.5	
1976	3.8	1.71		8.11	2.1		2.0		
1977	3.8	1.41		7.75	2.1			2.0	
1978	4.8	1.61		5.8	2.1				
1979	5.0	1.45		5.6	2.1				
1980	5.1	1.24	1.5	5.9	2.2	4.0		2.1	4.0

Sources: for 1967–80, Ghana, Liberia, Ivory Coast, Liberia, Tanzania, and Nigeria: national economic statistics; for 1965–67, Ghana, Cameroon, Ivory Coast, and Congo before 1965: IBRD (1976); for 1973–80, Liberia, Ghana, and Ivory Coast: Gillis et al. (1983: Table 2.1).

Table 7.11. *Evolution of employment in the forest-based sector, selected African countries, 1972–82 (thousands of workers)*

Year	Liberia[a]	Gabon[b]	Ghana[c]	Ivory Coast[d]	Kenya	Cameroon	Congo[e]	Nigeria	Zaire
1972			25.0	16.1	40.0		5.8		
1973	1.2				39.7	3.5			
1974					43.2				
1975					41.8				
1976	5.0			21.7	46.6				
1977				23.5	42.8		6.0		
1978				21.4	40.7				
1979			101.0	17.1	43.7			1,500	19.0
1980	6.0	98.0	88.5	15.1	39.0				
1981			79.0			15.0			
1982			70.0						

[a] Of total FBS employment in Liberia of 5,900, about one-third (1,980) were employed in sawmilling.

[b] From 1979 to 1981, an average of 10,500 people were employed in logging in Gabon annually.

[c] 1972 figure for Ghana is an estimate; based on information indicating 23,600 jobs in 1969, Ghanaian employment in sawmilling and plymilling was 12,500 in 1972, and had climbed to 52,000 by 1982 (Ghana Timber Marketing Board 1982).

[d] Salaried personnel only through mid-seventies, Ivorians only 40 percent of total.

[e] 1972 figure for Congo is actually for 1970.

Source: Gillis et al. (1983).

Table 7.12. *Evolution of employment in selected African forest-based sectors as a percentage of labor force, 1975–82*

Country	FBS employment (percent)							
	1975	1976	1977	1978	1979	1980	1981	1982
Liberia		0.60				0.62		
Gabon								
Logging						2.8		
Processing				•		25.2		
Total						28.0		
Ghana					1.6	1.5	1.2	1.2
Ivory Coast			0.58	0.55	0.45	0.34		
Kenya	0.59		0.58	0.57	0.60	0.51		
Nigeria								
Logging				0.11				
Total				3.56				
Cameroon								
Logging							0.3	
Total							7.0	

Source: Gillis et al. (1983).

the forest sector fared little better, or worse, than the rest of the ailing economy.

Low income growth meant low overall growth in other identifiable benefits from timber exploitation: foreign exchange earnings, employment, and tax revenue. Prior to 1965, timber exports furnished as much as 20 percent of total export earnings. But by 1972, the share of timber in total exports had fallen to 13 percent, and to but 4 percent in 1980 (Table 7.13). Taxes collected from the forest sector have also been minimal, averaging less than 1 percent of total tax revenues in the years for which figures are available.

Employment in the timber sector did, however, increase significantly from 1965 until 1979. By 1969, employment in the forestry sector had reached 23,558 (Gillis 1983: 3.3–3.10), and it peaked in 1979 at 101,000 persons, before declining to 70,000 in 1982 (Ghana Timber Marketing Board 1982), or about 1.2 percent of the labor force (Table 7.12).

Ivory Coast. The forest-based sector has played a larger role in the Ivorian economy than in any other African country, as is evident in Table 7.10. Value-added in the forest-based sector was consistently around 6 percent of GDP in the decade prior to 1981, the highest such ratio in Africa (Table 7.10).

The value of wood extracted from the forest rose from nearly US$600

Table 7.13. Forest product exports as a percentage of total exports, selected African countries (includes logs, lumber, and panels)

Year	Gabon[a]	Ghana	Ivory Coast	Liberia	Kenya	Cameroon	Congo	Nigeria[b]	Zaire
1970			22.5		1.9				1.0
1971			24.5		2.3				1.0
1972		13.3	27.2		2.7		42.0		0.3
1973			34.7	5.1	3.1			0.8	1.0
1974			22.6	4.4	3.0	14.0		0.3	2.0
1975			17.9	3.2	3.0			0.9	1.0
1976		11.3	18.9	7.1	2.2	11.2	10.7		1.0
1977	9.3	5.6	15.3	5.8	1.5	11.9	10.7		
1978	9.5	4.9	13.3	9.6	1.5	12.8	10.7	0.01	
1979	8.7	n.a.	15.9	9.3	1.9	n.a.			
1980	9.6	4.0	11.0	10.9	1.9	11.3			
1981	8.0			7.0					

[a]Until 1963, 75 percent; data not available from 1963 until 1978.
[b]5 percent during the 1960s.
Source: FAO (1981).

Table 7.14. *Forest taxes and forest fees as a percentage of total government revenues, several African and Asian countries, 1971–81*

	1971	1974	1975	1976	1977	1978	1979	1980	1981
Africa									
Ghana	0.5	0.6[a]							
Ivory Coast	6.9	8.0[a]		6.6	6.6	1.3			
Liberia	1.2			5.0	6.1	5.5	5.6	5.8	8.0
Gabon		3.1	0.7	0.7		7.6			
Cameroon	0.9	1.3[a]							
Congo	5.5	3.4[a]			3.4				
Kenya	0.2	0.2	0.2	0.1	0.2	0.2	0.2	0.3	
Nigeria	0.2	0.1	0.1	0.1	0.1				
Asia									
Indonesia		2.0	1.0	2.0	2.0	3.0	1.0	1.0	
Malaysia									
Sabah	n.a.	n.a.	n.a.	n.a.	n.a.	65.0	66.0	67.0	
Sarawak	n.a.	22.0	5.5	12.2	14.3	14.4	n.a.	n.a.	

[a]Denotes that figure is for 1973 (Ghana, Ivory Coast, Cameroon, Congo).
Sources: Africa before 1974: IBRD (1976: 27); Africa after 1973: IMF, *Government Finance Statistics Yearbook*, various issues; Asia: Chapters 2 and 3 of this volume.

Table 7.15. *Ivory Coast: Value of production and export of timber and wood products, 1977–80 (million U.S. dollars)*

	Logs	Sawnwood	Plywood	Other	Total
1977					
Production	476	90	18	8	592
Export	291	54	8	7	360
1978					
Production	473	94	26	13	606
Export	257	63	8	9	337
1979					
Production	563	114	26	18	721
Export	364	62	7	10	443
1980					
Production	730	127	27	18	902
Export	462	72	11	17	562

Source: FAO, *Forest Product Statistics*, various years.

million in 1977 to $900 million by 1980, by which time the value of logs and wood product exports had reached $562 million, or about 11 percent of total export earnings, down from the peak of 35 percent of 1973 (Table 7.15).

Notwithstanding the relatively high contribution of the forest sector to total GDP, the sector has not been a prolific source of jobs. The number of jobs in the sector reached only 24,000 in 1977 before declining to 15,000 in 1980 (Table 7.11). Forest-based employment as a percent of the labor force was particularly low in Ivory Coast relative to most other African timber producing countries (Table 7.12). By 1980, employment in the forest-based sector was less than 0.4 percent of the labor force, well below the unusually high 28 percent and 7 percent ratios in Gabon and Cameroon, respectively, and lower than even Liberia and Ghana.

In the decade prior to 1981, the Ivorian FBS was a fairly productive source of tax revenues, furnishing an average of 6 percent of government revenues annually (Table 7.14). This ratio exceeded those for other African countries save Liberia (Table 7.14), but was well below the percentages for the East Malaysian states of Sabah (67 percent) and Sarawak (14 percent).

Forest taxes and fees reached nearly one-quarter of export value in 1971. But by 1980, Ivory Coast taxes and forest fees were but 11 percent of recorded export value, below that for Liberia, and well below similar ratios in Indonesia, Sabah, and Sarawak (Table 7.16).

Table 7.16. *Taxes and forest fees as a percentage of total forest product exports, several African and Asian nations, 1971–80*

	1971	1974	1975	1976	1977	1978	1979	1980
Africa								
Ivory Coast	24.5	22.6	17.9	18.9	15.3	13.3	15.9	11.0
Liberia					21.0		19.7	
Congo	15.0	10.5		6.5	16.5			
Kenya	2.3	3.0	3.0	2.2	1.5	1.5	1.9	1.9
Cameroon	9.0	19.8						
Asia								
Indonesia		16.0	15.0	15.0	19.0	28.0	25.0	
Malaysia								
Sabah		27.0	26.0	27.0	39.0	36.0	53.0	59.0
Sarawak	14.1	16.7	12.6	11.9	13.3	13.6	13.5	18.3

Sources: Africa: IMF, *Government Finance Statistics Yearbook*, various issues; for years before 1975, IBRD (1976: Table 5.07). Asia: Chapters 2 and 3 of this volume.

The halving of the share of taxes and forest fees in total export value in Ivory Coast is difficult to explain given that export tax rates per cubic meter of logs were drastically increased from 1970 to 1979. The increase occurred in two successive steps (March 1974 and March 1979) with the result that export duties in 1979 on the high-valued woods, including *sapelli*, *sapo*, and especially *assamela*, were 900 percent of 1970 levels (Schmithusen 1977: 7). Export taxes on middle and lower valued logs (*illomba*, *fromager*) were also increased substantially, by an average of 600 percent and 250 percent respectively.

The decline in Ivory Coast forest export taxes and fees relative to total export value has been primarily due to the reduction in the share of higher-valued logs in total exports. These include *sapelli*, *sipo*, and *samba* species, all subject to the highest export tax rates. By 1978, lower-valued species constituted over half of Ivory Coast timber exports, the richer stands of *sapelli* and *sipo* having been largely depleted (Figure 7.1).

Gabon. The forest sector has played a major role in Gabon's economy since the early 1950s, accounting for 80 percent of exports as late as 1963. Then the oil and uranium sectors became the predominant sources of both domestic value-added and total exports and tax revenues, so that by 1981 the share of the forest sector in total exports had fallen to but 8 percent (Table 7.13). Gabon was still the second largest exporter after Ivory Coast of sawlogs and veneer logs in Africa as late as 1984 (World Bank 1986: Annex IX, Appendix 1,1).

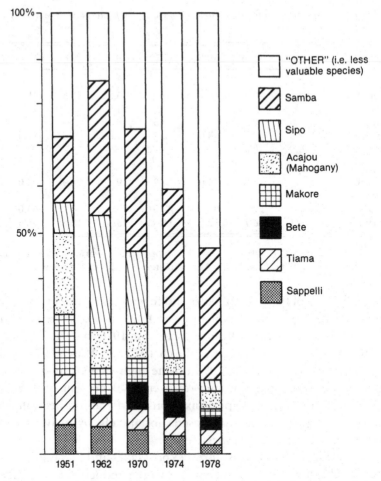

Figure 7.1. Evolution of share of precious hardwoods in the exports of the Ivory Coast, 1951–78 (Arnaud and Sournia 1980: 295)

The forest sector remains a major source of employment: government statistics indicate 28 percent of the economically active population is employed in logging and wood-processing activities. Even if the true ratio were actually half this, it would remain the highest percentage of forest-based jobs in total employment of any nation in the world (Table 7.12).

Rent capture

Liberia. In the mid-1970s, Liberia and Gabon shared the dubious distinction of having the least effective rent capture policies among Af-

Table 7.17. *Liberia and selected West African nations: Loggers' residuals after government forest tax and transport cost deductions, 1974 (selected species, U.S. dollars per cubic meter)*

	Ivory Coast	Liberia	Cameroon[a]	Congo[b]	Gabon
High-value species					
Sapelli	46	80	51/42	76/64	78
Sipo	47	98	72/63	86/74	100
Middle-value species					
Tiama	28	55	37/29	49/47	53
Kossipo	38	63	33/24	56/44	61
Iroko	26	55	27/16	50/35	49
Low-value species					
Llomba	9	25	13/5	22/11	20
Fromager	25	25	15/8	24/15	23

[a]Douala Route/Pointe Noire Route.
[b]Southern Sector (Mossendjo)/Northern Sector (Quesso).
Source: Schmithusen (1980).

rica's major timber-exporting nations. This is evident from Table 7.17, which shows, for harvest of comparable logs in different nations, the residuals remaining to loggers after subtracting forest taxes and fees, and transport costs.

By 1979, after two major increases in forest fees, Liberian policies were more strongly geared to rent capture, as suggested by Table 7.18. The table indicates that for high-valued woods, including *sapelli, acajou, tiama,* and *sipo,* the revenue system had the potential to capture about two-thirds of the total rent available in logging. For lower-valued woods (*iroko, llomba, fromager*), the tax system actively discourages harvesting, as taxes and charges exceed gross margins before tax.

Ghana. Estimating government timber rent capture after 1971 represents an impossible task. Owing to rampant smuggling, data on total (and per unit) export values are extremely suspect. Since 1971, timber tax collection totals are either unavailable or potentially erroneous, due to abysmal tax administration.

However, fairly reliable estimates of timber rent and rent capture were made for earlier years by Page, Pearson, and Leland (1976: 25–44). Like most analysts, they defined rent as the residual, after all production costs, including normal returns to capital, are subtracted from the log output value. In that study, normal returns to capital were defined as 15 percent of the total investment. This process understates rents from timber, since

Table 7.18. *Liberia: Theoretical rent capture in logging, selected species, 1979 (U.S. dollars per cubic meter)*

	(1)	(2)			(3)	(4)	(5)
		Costs					Total government
	Sales revenue	(A) Logging costs	(B) Other costs	(C) Total costs	Gross margin (Col. 1 − Col. 2C) for logger	Total taxes and forest fees (including income tax)	revenue as percentage of gross margin (Col. 4 ÷ Col. 3)
Sapelli	187	67.30	12.95	80.25	106.75	71.11	66.7
Acajou	141	67.30	12.95	80.25	60.75	48.10	79.2
Sipo	254	67.30	12.95	80.25	173.75	114.62	66.0
Iroko	100	67.30	12.95	80.25	19.75	20.50	104.0
Tiama	155	67.30	12.95	80.25	74.75	55.11	73.7
Llomba	57	67.30	12.95	80.25	negative	9.5	—
Fromager	50	67.30	12.95	80.25	negative	7.5	—

Source: Derived from Peprah (1982b: Tables 3.4 and 3.5).

Table 7.19. *Estimates of timber rents in Ghana and Indonesia, 1971–72, 1973–74 (U.S. dollars)*

	Ghana (1971–72)	Indonesia (East Kalimantan) (1973–74)
I. Rents per m³		
A. *Afformosia*	$79	n.a.
B. Mahogany (*meranti*)	$28	$45
II. Rents per hectare	$101[a]	$1,500[b]
III. Rents as percent of value of logs	26	53
IV. Percent share of rents		
A. Received by government	38	25[c]
B. Received by investors	62	75

[a]Figure is for large-scale Ghanaian firms; rents for smaller firms decline somewhat with scale of operation.

[b]Average rent per cubic meters in stand. Stands are dominated by *meranti (Shorea* sp.) but include other species as well.

[c]Note that government rent capture in Indonesia increased markedly after 1978 (Chapter 2).

Sources: Ghana: Page et al. (1976: 25–44); Indonesia: Ruzicka (1979: 73).

interest was subtracted as a cost. That being the case, normal returns to capital are correctly defined only in relation to total equity, not relative to returns to total capital (Gillis 1983).

The researchers found that available Ghanaian timber rents were 26 percent of the log output value in the early 1970s. For some particularly desirable species, rents reached 80 percent of log value. The distribution of that rent was skewed in favor of loggers: only 38 percent of total timber rent accrued to the government, primarily in the form of royalties, taxes, marketing board charges, and to a limited extent, corporate taxes. More than three-fifths of timber rent accrued to the logging industry: 35 percent went to Ghanaian investors, and 27 percent to foreign investors. Comparisons of Ghanaian and Indonesian rents are furnished in Table 7.19, prior to Indonesia's major increases in timber taxes and fees in 1978 and 1980.

The data in Table 7.19 reflect the lower incidence of rich commercial stems in Ghana than in Indonesian East Kalimantan: potential rents available per hectare were 15 times greater in Indonesia than in Ghana. However, rents per cubic meter harvested were, on average, less than twice as high in Indonesia as in Ghana because pervasive high-grading of Indonesian timber stands destroyed rent (Chapter 2).

The point is not the relative richness of Indonesian versus Ghanaian stands, but the fact that in the early 1970s both these governments performed poorly in the design and execution of rent capture policies, particularly in comparison to Sabah (Chapter 2). From the early 1970s until 1983, the Indonesian government markedly improved rent capture. However, this had not occurred in Ghana, even through 1986 (IBRD 1986: Annex V, 2–7).

Ivory Coast. Aggressive tax policy during the 1970s enabled Ivory Coast to capture a high proportion of timber rents, relative to other West African countries, as is evident from Table 7.17.

For higher-valued species, Ivory Coast loggers retained US$46 to $47 per cubic meter, after paying taxes and transport costs. This was well below comparable residuals in Liberia, Gabon, and Congo, and roughly equivalent to that of Cameroon. For middle-value species, the residual left to loggers in Ivory Coast was consistently half that of all other countries save Cameroon. For the lower-valued species, the residual was relatively high for *fromager* and relatively low for *llomba*.

These figures suggest that unless logging and skidding costs for indicated species were much lower in other West African countries, Ivory Coast forest tax policies were more strongly geared to rent capture, at least during the 1970s. In fact, reported logging and skidding costs vary little from country to country in West Africa. Ivory Coast rent capture policies thus compare well with those of neighboring timber-producing nations; albeit poorly with those of Sabah.

Gabon. We have seen that Gabon's forest sector is an important source of value-added, export earnings, and employment. It has not been, however, a major source of tax revenues. In the mid-1970s, Gabonese policy toward rent capture was, along with Liberia's, the weakest in West Africa (Table 7.17). While Liberia drastically increased timber sector taxes and fees after 1974, Gabon has yet to adjust its timber taxes or tax incentives. Clearly, Gabon's timber sector has long been among the most lightly taxed in Africa, if not the world.

Rent capture instruments in forest policy

As in Southeast Asia, the only significant forestry policy tools that West African governments have utilized for rent capture are royalties, taxes, and fees. Gabon, alone among the four nations, has attempted to capture timber rents for the domestic economy by requiring foreign firms to furnish "free equity" to local investors (primarily the government). This device has proven singularly disappointing.

Royalties and license fees: Liberia. Until recently, no royalties were charged on Liberian logs. However, firms were at least nominally subject to income taxes on any profits they reported. Finally in 1977, the government enacted a "severance fee" of $1.50 per cubic meter for all logs whether exported or processed domestically (Gray 1983: 87). This type of forest fee – the uniform specific rate royalty – creates the greatest potential for forest damage, in the form of high-grading (Chapter 2, Appendix A). It discourages loggers from selecting low-valued secondary species and defective stems of primary species. In Liberia, however, the royalty is not high enough to lead to significant damages. But the severance fee is not the only aspect of the revenue system that contributes to high-grading. The reforestation fee, set in 1980 at $3 per cubic meter, is also a uniform specific royalty, invariant with respect to species, value, or grade of timber. Table 7.18 showed that if all legal taxes and fees were actually applied, loggers in 1979 could not have afforded to harvest low-valued species nor even some middle-valued species. The widely traded *iroko* species would not have been harvested: the sum of nominal forest taxes and fees exceeded the loggers' gross margins.

Liberia's forest license fee, at US$0.25 per hectare annually, is moderate by international standards, and is too low to result in any significant effects on harvesting decisions at either the extensive or intensive margin.

Royalties and license fees: Ghana. For most of the post-independence era in Ghana, there was no standard royalty policy. Each concession agreement contained its own unique schedule of royalties. This arrangement was changed in stages from 1965 until 1976, by which time a standardized royalty system was applied to all concessionaires. The royalty now takes the unusual form of a specific charge per *tree* harvested; 39 different rates apply to 39 different species (Peprah 1982a: 19). Elsewhere in the world specific royalties are based on wood *volumes* harvested, e.g., per cubic meter or per Hoppus foot, rather than *per tree*.

One virtue of the Ghanaian system is that it is the simplest type of stumpage charge used in tropical forest revenue systems. The royalty collector merely counts stumps, without determining volumes. This system also leaves to the logger the determination of the size and species of stems harvested. In a Southeast Asian virgin forest of mixed dipterocarps the system would also have some merit. With the royalty based on stems taken rather than volume, loggers might cut oversized individual stems that they would have bypassed if the royalty were based on *volume* per tree. Bypassing large mature trees is a major problem in Asian selective cutting systems. Under the Ghanaian system more large trees would tend to be taken, providing more canopy openings for remaining young saplings. Although high-grading occurs under the Ghanaian system, in that concessionaires still leave defective trees and non-marketable species, the effects are much less pronounced than under most other royalty systems.

In addition, the Ghanaian system insures that young individual stems will not be cut: with significant and equivalent royalty charges on large and small trees, it is not worthwhile to harvest small trees. These are left to form the crop for future harvests. The Ghanaian system applied in Asia would promote forest resource conservation without the complexity of Asian systems. Minimum girth limits of eleven and seven feet, respectively, apply to valuable species and to the less valuable species, reinforcing the incentive to leave small trees.

In addition, the Ghanaian royalty system encourages greater utilization of trees harvested than Asian systems. Having paid for the whole tree, loggers will maximize utilization of stems taken (Gray 1983: 131).

Therefore, the structure, if not the level, of the Ghanaian royalty system merits consideration for logging in remaining virgin forests elsewhere. Ghana itself has no remaining virgin forest, so these comments are less relevant there. Even in Ghana's depleted forest, the system encourages conservation to a degree: small trees will tend not be taken if larger stems are available.

The royalty system has proven to be an extraordinarily weak means of raising forest revenues in recent years. With royalties expressed in cedis per tree, their dollar values (at black market rates) have decreased. In 1983, royalties (set in 1976) ranged from 54 cedis for very desirable species (US$19.60 at official exchange rates and US$1.35 at the black market rate) to 6 cedis for less desirable species (US$2.20 at the official exchange rate but US$0.15 at black market rates). At official exchange rates, the level of the Ghanaian royalty structure was second only to that of Sabah in 1983. But at black market rates, Ghanaian royalties rates were easily the lowest in the world.

This problem became clear in mid-1983, when the government began a series of devaluations that took the cedi from 2.5 cedis to 90 cedis per U.S. dollar. Even after royalty rates were sharply increased in 1985, a 1986 World Bank Mission concluded that forest royalties and fees were so low that wood taken from the forest could almost be considered a free good (IBRD 1986: 68–72). The Mission estimated stumpage value to be 20 to 60 times the royalty rates in effect in 1983, and 2 to 6 times the royalty rates proposed (but not enacted) in 1985, depending on the species. Accordingly, there remained in 1986 ample scope for at least doubling the

Table 7.20. *Ivory Coast royalty schedule, 1984 (U.S. dollars per cubic meter)*

Type of log	Exported	Locally used
High-valued species	1.50	0.75
Middle-valued species	1.00	0.50
Low-valued species	5.50	0.25

Source: Gray (1983: 87).

royalty rates over the proposed 1985 schedule. The same report also indicates extremely low fees for concession areas. Charges per hectare were termed equal to the market price of a "small stick of firewood" (IBRD 1986: Annex 5, 12).

Royalties and license fees: Ivory Coast. The timber royalty in Ivory Coast (*taxe d'abattage*) is imposed on harvested timber volumes, not on a per tree basis. Royalty rates were set in 1966 and apparently have been unchanged since that time. The royalty schedule is differentiated according to species and utilization: high-valued species bear royalties three times that of low-valued species; export logs are subject to royalty rates twice those imposed on logs used locally. The royalty is too low, relative to f.o.b. log export values, to have serious implications for forest damage from high-grading.

License and area fees too are collected from logging firms, but are also quite low. A concession license is charged at the rate of US$0.25 per hectare (*taxe d'attribution*). A more substantial fee on area is charged to finance local public works. This fee (*taxe de Gravaux d'intérêt général*) amounts to about $0.79 per hectare on richer stands and half that for poor stands. Both of these fees are one-time levies. Finally, an annual area charge (*taxe de superficie*) is levied at the rate of US$0.05 per hectare per year (Gray 1983: 67–82). These charges are collectively too small to have notable effects on harvest decisions. Very low area fees have two effects that discourage forest conservation. First, they encourage exploitation of marginal stands. Second, they provide little incentive for more intensive exploitation of more valuable stands (including the taking of so-called "secondary" species). Had these fees been increased substantially by 1970, the nation might have experienced a somewhat lower rate of deforestation than actually occurred in the 1970s and 1980s. Table 7.5 suggests, however, that increases in these charges now would have little beneficial effect.

Royalties and license fees: Gabon. The Gabonese royalty (felling tax) is but 5 percent of "posted" export values, which themselves tend to be below realized f.o.b. values. This rate is not particularly low compared to other countries (6 percent for Indonesia, around 5 percent *ad valorem* equivalent in Liberia) except Sabah. Gabon, however, collects almost no other taxes on forest products: there is no evidence that a license or area fee of any type is imposed.

Reforestation

Liberia

Reforestation activities in Liberia began on a noticeable scale only in 1971. At that time, the country's Forest Management Plan obligated concessionaires to reforest one acre for every 30,000 board feet of merchantable timber harvested, or to finance the reforestation of such an area as selected by the Forest Development Authority (FDA). The concessionaire, however, could be exempted by paying $150 into a "Reforestation Fund" for each acre he would otherwise have had to reforest. This option was cancelled when it was found that reforestation activities of firms taking advantage of the privilege were unsatisfactory. The FDA then took over the reforestation activities in 1974.[15] A further FDA Regulation (No. 7 of December 1979) imposed a reforestation fee of US$3 per cubic meter on all trees cut for commercial purposes. For 1978–1979, the amount collected totalled $2.57 million and was the second most important source of forest revenue (Schmidt 1980: 2).

Three different methods of forest regeneration are used in Liberia: natural regeneration, enrichment planting, and establishment of forest plantations. Liberian forestry policy presumes that sustained forestry yield is ensured through regeneration of the natural forest, i.e. that stocks thinned by felling would be regenerated (both the standing volume and the species composition) by the aftergrowth within 25 years.

Reforestation through enrichment planting has not been widely practiced in Liberia. Here, single plants or plant groups are cultivated in the natural forest below the canopy, supplementing the natural stock with valuable indigenous and shade-needing species. Reforestation through establishment of plantations after land-clearing (both by heavy equipment or traditional felling and wood-burning methods) has been widely employed, but not in recent years.

The nation's four major reforestation areas are located in the Bomi Hills, Glaro, Yekepa, and Cape Mount. The government had reforested a total of 4,262 hectares in these areas by the end of 1979. Companies planted another 1,366 hectares, for a cumulative total over several years of 5,628 hectares, negligible compared to annual deforestation of 40,000. The tree species *Gmelina arborea, Tectona grandis* (teak), and *Pinus* spp. cover 73 percent of the reforested area (at 37 percent, 24 percent, and 12 percent, respectively). Generally, reforestation involves planting quick-growing exotic species for pulpwood production (*Gmelina arborea, Eucalyptus* spp., *Pinus* spp. and *Albizia falcata*).

Liberia has also received substantial outside financial assistance in reforestation. For example, costs for the five-year trial phase of the Industrial Pulpwood Plantation in Cape Mount were jointly funded by the government, the World Bank, the African Development Bank, and the Federal Republic of Germany.

Ghana

Although virtually nothing remains of Ghana's virgin forest, there remains a serious deforestation threat on previously logged-over stands. Nearly one-half of the loggers were making second entry on their concessions as early as 1980.

The Forestry Department has been intermittently engaged in a long-term reforestation program, severely constrained by the country's weak fiscal performance from 1970 to 1985. Through the early 1980s, the Department received an annual governmental grant equivalent to only US$125,000 (at the black market rate). Although the Forestry Department's annual replanting target was set at 11,000 hectares, only an average of about 4,000 hectares was achieved during the 1977–78 period, a time during which deforestation from log exploitation alone reached 20,000 hectares. The government also allocated very small amounts in 1980–81 to the Forestry Department to undertake plantings in the northern regions to stem the downward drift of the Sahara Desert. Finally, firms holding concessions are nominally required to replant their land areas after harvesting. However, though incentives for compliance and penalties for noncompliance are provided by law, compliance mechanisms have been ineffective.

Reforestation fees were earmarked to fund reforestation programs. There is, however, no record indicating that amounts collected were actually used for replanting programs (*Ghana Timber News* 1979: 15). Before 1976, such fees amounted to two U.S. cents per hectare annually, and by 1986 averaged less than five U.S. cents per hectare per year.

Ivory Coast

Ivory Coast imposes no special taxes or charges on timber production for financing reforestation programs. Virtually all visible reforestation efforts have been concentrated in SODEFOR[16] and most are designed to ensure wood supply for pulp and paper projects. From 1966 until 1982, SODEFOR planted only 37,000 hectares, less than 3,000 hectares per year compared to annual deforestation of about 300,000 hectares.

The 1976–1980 Five-Year Plan did contemplate an expanded reforestation program: 10,000 hectares per year were to be planted, reaching 500,000 hectares by the year 2026 (*Afrique Industrie* 1982: 23). But between 1976 and 1980, only 13,296 hectares were reforested, about one-third of the targeted area for that period. Replanting costs incurred by SODEFOR have been high by world standards: US$2,080 per hectare in 1981 versus about US$1,000 in tropical Asia. Further, the success rate on the 13,296 hectares replanted by 1980 was estimated at only 50 percent, owing to inadequate maintenance and forest fires (FAO 1981: 157–158).

The adequacy of reforestation efforts may be assessed by comparing actual hectares reforested in 1976–80 against deforestation totals in that period: reforested hectares, at 13,296, were but 1.1 percent of the total of deforestated area.

Gabon

Gabon, as noted earlier, has collected reforestation taxes since 1957. From 1957 through 1965, they were the sole source of finance for reforestation programs. These revenues were supplemented by general budgetary funds beginning in 1965 (Gray 1983: 9), but at levels insufficient to support extensive reforestation programs. As a result, reforestation activities were virtually nonexistent from 1968 to 1975. Reforestation program areas declined from 5,000 hectares annually in the early 1960s to less than 1,000 hectares per year by 1981, and 400 hectares in 1982–83, versus average annual deforestation of 15,000 hectares per year (Legault 1981: 163).

Replanting programs received a significant boost in 1982, when reforestation taxes increased to 3.5 percent of the value of logs harvested. The primary motive, however, was to ensure future supplies of wood for a French-owned pulp and paper plant.

Concession terms

Liberia

Until 1973, all Liberian timber concession agreements were special contracts negotiated individually with each concessionaire. No provisions of general law applied to forest contracts, and no standard concession agreements were employed (Gray 1983: 195). Agreements therefore varied considerably in their terms and conditions, not only with respect to concession length but to harvest regulations and taxes and royalties charged.

Model timber concession agreements containing uniform standards for duration of contracts, logging practices, silvilcultural treatment, and taxes were finally adopted in 1973. Concession areas are large, averaging 81,000 hectares (FAO 1981: 273). Concession contract duration became 25 years, although longer terms were available at the discretion of the Forest Development Authority (FDA). While 25 years is insufficient to provide the concessionaire with clear incentives for maintaining the forest's productive value, Liberia's terms are more conducive to sensible forest management than countries with much shorter terms (Ghana, Malaysia, Ivory Coast). Harvest methods are restricted to selective cutting plans approved by the FDA. To promote natural regeneration and reduce high-grading, concessionaires are required to harvest all primary trees (except protected species) with diameters exceeding 40 cm. (d.b.h.) and to harvest 30 percent of all standing stock composed of secondary species. It is unclear, however, whether these rules, first prescribed in 1981, have actually been enforced.

Ghana

Prior to independence in 1957, timber firms in Ghana typically obtained concessions of 50 years duration, although a few were as long as 99 years (Page et al. 1976: 26). By 1976, many of the lengthier concessions had begun to expire. A number of these concessions were extremely large in area; many were held by foreign firms. For example, five of the largest foreign firms in 1971 held concessions averaging 270,000 hectares (1,036 square miles), accounting for 46 percent of the total concession area allocated at the time (Pearson and Page 1972).

While concessions granted prior to independence were of sufficient length (up to 99 years) to provide incentives for concessionaires to maintain and enhance forest estate values, the record indicates that they did not do so. This suggests that oft-repeated claims that longer concession durations are more conducive to good forest management practices (Whitmore 1984: Chapter 19) mean little without complementary measures to curtail forest depletion (silvicultural treatment, royalties that reduce high-grading, penalties for noncompliance with felling rules).

Major concessions policy changes were undertaken after 1971. First, in 1974, concession awards were restricted entirely to Ghanaians and average new concession size sharply reduced. In addition, the concession length was established at a minimum of five years and a maximum of 25 years (for concessions exceeding 800 hectares) and a maximum of 3 years for smaller concessions (FAO 1981: 196), clearly too short a period to provide incentives for forest conservation.

Ivory Coast

In Ivory Coast, as in much of French-speaking Africa, early forest concession contracts were generally of short duration, usually five years. This was consistent with the objectives of the French firms who dominated logging before independence. Richer tracts were quickly harvested, and firms moved on easily to other stands. During 1965–68, concessions of longer duration were introduced: 5 years for logging companies, 10 years for companies operating a sawmill, and 15-year permits for integrated wood-processing firms (FAO 1981: 107). However, after 1969, policy reverted to the five-year contract limit, providing forest-based firms with virtually no incentive to maintain the productive (much less the protective) value of the forest estate (World Bank 1976: 27).

Until 1972, harvest methods were not prescribed by the Ivory Coast Government. Thereafter, concessionaires were required to submit logging and road construction programs annually and provide minimal information on their past operations. However, through the 1970s, forestry officials lacked the staff and resources to administer and enforce annual allowable cuts, obligatory removal of secondary species, or verification of working programs of logging companies. Concessionaires were thus not obliged to follow any particular selection techniques or particular cutting method (World Bank 1976: 23–24). As late as 1976, the only restriction imposed on logging firms was a lower limit on girth of trees harvested (80 cm. d.b.h.), intended to protect immature trees. No other restrictions applied.

Gabon

Whereas the maximum length of logging licenses rarely reaches 30 years elsewhere in Africa and tropical Asia, logging concessions in Gabon are for 30 years at a minimum. Longer periods may be awarded for firms that invest in local processing facilities (Republic of Gabon, Ministère des Eaux et Forêt 1980: 37).

Logging concessions are 20,000 hectares at a maximum, but can be as large as 250,000 hectares for integrated wood product operations. Clearcutting is prohibited: harvesting is restricted to selective cutting, but information on permitted types of selective cutting practices is unavailable.

Forest-based industrialization policies

Liberia

Nothing exemplifies the post-1977 shift in Liberian timber sector policy better than aggressive measures then adopted to promote forest-based industrialization. From perhaps Africa's most passive stance toward timber-processing investments in the early 1970s, Liberia moved to perhaps the most aggressive posture in this policy area.

Two policy measures were introduced in 1977 to promote forest-based industrialization. The first was a major increase in the Industrialization Incentive Fee (IIF), initially applied to log exports in 1971.[17] This applies to log exports only. Second, a Forest Products Fee was imposed on exports of sawnwood.[18] Prior to 1977, there were no export levies on processed products. Plywood remained untaxed.

The IIF is the most important Liberian forest tax, both in terms of government revenues and effects on processing decisions. Sharp increases in the tax after 1977 were explicitly aimed at promoting further domestic timber-processing activities. The IIF is levied on all logs shipped from Liberia, at rates ranging from a high of $75 per cubic meter on high-valued species (*sipo*) to $35 for middle-valued species (*acajou*), and $2 to $4 on low-valued species (*llomba, azobe*).

The Forest Products Fee (FPF) also varies directly with species value and is differentiated according to the degree of domestic processing. Present rates range from $60 per cubic meter for semi-processed timber from valuable species (*sipo*) to only $1.00 for the low-valued *azobe* species. However, the FPF is a relatively minor source of forest revenue.

In spite of these policies, most Liberian exports are still in log form. Therefore, economic and political barriers to processing investments must be very strong in Liberia, since the IIF and FPF provide strong fiscal incentives for exporting processed timber. In the case of *sipo* species, incentives for domestic processing are particularly attractive. This species is commonly used for cabinet wood, and in the early 1980s constituted 25 percent of all log export volume.

F.o.b. prices for *sipo* logs in 1980 averaged $254 per cubic meter. Reckoned on the roundwood equivalent, additional domestic value-added involved in the highest degree of domestic processing (square-edged dressed on four sides) of exported timber may – in keeping with previous chapters – be reckoned at no more than 15 percent, or $38.10 per cubic meter. Therefore, for each cubic meter of *sipo* exported as processed sawn timber, the government gives up $75 in IIF revenues, and receives $15 in FPF revenues, thus bearing a $60 net tax loss to obtain additional value-added of, at most, $38.10. The implications of such heavy subsidization to processing have been fully discussed in Chapters 2 and 3 and are not repeated here.

Ghana

Ghana has employed all available devices to encourage forest-based industrialization. Before 1979, the export tax on logs was 10 percent, that on sawn timber was 5 percent, and plywood and other products were exempt. To further stimulate processed exports, all export taxes on sawn timber were abolished in 1979 (*Ghana Timber News* 1979: 7). This was largely meaningless given the deep overvaluation of the cedi in the decade prior to its substantial devaluation in 1983. Export bans on timber have also been undertaken, but not fully enforced. The first log export ban was in 1972, when *odum* and teak species export was nominally prohibited. Later in 1972, export bans were placed on 12 other major species, but not enforced. In 1976, the ban was reimposed, but suspended after three months (Peprah 1982a: 14). Again in 1979, a ban on export of fourteen important species was enacted (*Ghana Timber News* 1979: 2).

Despite announced policies, neither export taxes nor bans enacted after 1972 successfully fostered much forest-based industrialization. By 1984, the share of logs in total wood exports remained about the same as in 1970.

Two other earlier policies were used to promote forest-based industrialization: subsidized credit for indigenous investment in sawmills and plymills and income tax incentives for investment in processing plants. Both of these subsidies, substantial in dollar amounts, encouraged development of a sizeable timber-processing industry by the late 1960s, before imposition of nominal log export restrictions. By then, over three dozen sawmills, 11 of which were large by any standards, were already operating (Peprah 1982a: Table A.1,5). By 1982, there were 95 sawmills, 10 veneer and plywood plants, and 30 firms in furniture and other wood-processing activities, most in place since 1972. Nearly US$230 million in wood-processing investments were in place by 1982 (Peprah 1982a: 8,10), most of which was of pre-1970 vintage.

From 1957 through 1971, the principal means to promote forest-based industry was through subsidized timber sector loans to Ghanaian citizens operating in the timber sector. Many of these loans had terms as long as 25 years, with nominal interest rates about half that on loans offered by commercial banks. This was a substantial incentive because even commercial loan rates were consistently negative in real terms after 1963 due to rapid domestic inflation. Moreover, loans for sawmills and plymills were available through the Timber Marketing Board at zero rates of interest,[19] or at negative real rates of interest averaging about 8.0 percent from 1965 to 1973.[20]

In addition, attractive income tax incentives have been available since the mid-1960s to forest-based industries engaged in export, including manufacturing firms, sawmills, and plymills. Exporting companies were entitled, in principle, to income tax rebates as high as 50 percent (Peprah 1982a: 42).

The combination of strong tax and credit incentives offered in Ghana resulted, then, in creation of a timber-processing industry much earlier there than in Indonesia or even in Malaysia.[21] But as in Indonesia and Malaysia, many of these firms were inefficiently small operations, though government policies artificially encouraged capital-intensive techniques at all scales of production.[22] There is conflicting evidence on the social efficiency of wood-processing industries so created. Page reports that when social accounting prices were used to measure wood-processing efficiency, most exporting firms were found to be efficient (Page 1974: 121). At market prices and official exchange rates none qualified as efficient, not surprising given domestic price controls and the long history of deep overvaluation.

More recent evidence, however, indicates significant technical inefficiency in all aspects of the wood-processing industry. Recovery rates in lumber were but 70 percent of achievable levels. For plywood, recovery rates were but 67 percent of achievable rates.[23] Ghana's longtime contorted relative price structure provides ample explanation for low efficiency in Ghanaian wood processing, and therefore greater forest damage over wider areas per cubic meter of processed output.

Ivory Coast

Ivory Coast policies have also been strongly geared toward forest-based industrialization, particularly since 1972. The three most important policy instruments for this purpose have been:

1.   The export tax structure,
2.   Quotas on log exports, and
3.   Export subsidies based on the amount of value-added to manufactured products (introduced in 1985).

As in most nations exporting tropical timber, the export tax structure furnishes strong incentives for timber-processing investments. Installed

Table 7.21. *Ivory Coast: export taxes and incentives for domestic processing on selected major species*

Species	Additional domestic value-added from sawmilling (U.S. dollars per m³)	Export taxes foregone by government on sawn timber exports (U.S. dollars per m³)	Taxes foregone as percentage of increased value-added
Iroku	25.00	52.00	204
Acajou	19.20	43.00	224
Llomba	9.24	10.00	108

capacity in the processing sector did increase from 1972 to 1982, but by less than we would expect given the level of investment incentives.[24]

In Chapters 2 and 3, we saw that in Southeast Asia additional domestic value-added (per unit of log equivalent) from exporting sawn timber instead of logs rarely exceeded 10–20 percent in recent years; additional value-added in plywood has not exceeded 20–25 percent in any timber-producing country since 1979, even with the 1986 recovery of world plywood prices, to US$250–300 per cubic meter.

Ivory Coast mills are widely known for inefficiency. The average plywood conversion rate in Ivory Coast mills was estimated at only 40 percent in 1980, versus up to 53 percent and 70 percent in Gabon. There are explanations why conversion factors (logs into sawn timber or into plywood) were higher, indicating more waste, in Ivory Coast than in Southeast Asia.[25] One study found that many of the processing mills in Ivory Coast are "window-dressing industries" established solely to conform with the requirement that all log exporters invest in processing facilities. "It would seem clear that these . . . industries would not have been established without the quota system, and do not operate effectively" (Johnson 1984: 11–6). It is unlikely that additional value-added in processing, per unit of roundwood equivalent, exceeds 15 percent for sawn timber, Ivory Coast's largest category of wood product exports.

Consider the implications of the export tax structure's for investments in processing of the high-valued species *iroko* and *acajou* (mahogany). In 1980, the value of log exports per cubic meter for these two species was US$170 and $128, respectively. Additional domestic value-added from processing these logs into sawn timber may have reached $25.50 and $19.20, respectively. For each cubic meter of *iroko* exported as sawn timber the economy gained $25.50 in added value, but the government gave up $52.00 in taxes (Table 7.21). A similar pattern held for *acajou* species.

For the lower-valued species, incentives were less powerful: hence taxes foregone were only slightly greater than additional value-added generated in sawmilling.

The implications of the Ivory Coast export tax structure were strikingly similar to those for Sabah and Indonesia. In each case, the government in effect is a major provider of equity in timber-processing investment, through export taxes foregone on processed timber, but receives no equity shares in return.

Log export quotas have also been employed to induce forest-based industrialization since 1972, when log exports were restricted to 66 percent of each firm's total harvest. Log exports were scheduled for a complete ban by 1976, but this was never implemented. Instead, in 1978, rights to export Ivory Coast logs were restricted to firms with processing facilities (many of which were "window-dressing mills"). Further, in 1981, exports of high-valued logs, such as *iroko* and *kondrotti*, were banned altogether, and by 1982 quotas were applied to all companies on a case-by-case basis. Quotas were however freely traded among firms: by mid-1982 the right to export logs sold at between $4.70 and $7.10 per cubic meter.[26]

The final incentive for additional domestic log processing was implemented in 1985, when the Ivory Coast government announced export subsidies based on the amount of value-added to manufactured products (Price Waterhouse 1985: 175).

Strong incentives for forest-based industrialization in Ivory Coast had similar results as for other countries with such policies. Heavy subsidies have not only sharply reduced timber exploitation benefits to the nation, but have saddled it with highly inefficient processing capacity. Competitive pressures to improve conversion ratios, and thereby efficiency, are notably absent. The result, as elsewhere, is more rapid depletion of the natural forest endowment.

Gabon

Gabon has imposed no export quotas or bans on log exports. All concessionaires with contract areas exceeding 15,000 hectares must, however, deliver 55 percent of the harvest to the local timber-processing industry. Given the continued large share of log exports (90 percent) in total wood exports, this policy is obviously not strongly enforced.

Low subsidy/protection rates may explain the very high conversion ratios achieved in Gabonese mills, particularly plymills (Pfieffer 1980). The lack of subsidies to processing requires Gabon mills to be efficient to compete in world markets.

Non-forestry policies

Income tax incentives

Liberia. Liberia has made liberal use of tax incentives in all fields, including timber, for decades. Generalizations about typical tax incentives offered to timber concessionaires before 1973 are impossible, since each contract involved individually tailored tax terms. After adopting a standard timber contract in 1973, Liberia offered an incentive package including (a) a five-year tax holiday,[27] (b) exemption of 20 percent of income from tax after the tax holiday period, and (c) import duty exemptions on equipment, materials, and spare parts for five years (Schmithusen 1977: 131).

Other fiscal incentives available in the 1970s included log export duty exemptions, reduced severance fees, and exemption of concessionaires from national fuel and oil taxes. These fiscal incentives for timber production were generous by international standards for that period. Inasmuch as Liberian royalties and forest fees were among the world's lowest before 1977, the nation derived but very limited fiscal resources from the timber sector. This is evident in Table 7.17, which shows that loggers' residuals in Liberia (after payment of all taxes and costs) were among the highest in Africa. Since the sector provided little employment (6,000 workers), it appears that returns to Liberia from exploitation of the forest estate were exceedingly small before 1977.

The government apparently concluded the same. Beginning in 1977, royalties and reforestation fees were substantially increased and tax incentives for timber investment were sharply curtailed. Tax holidays for timber firms were ultimately abolished in 1981, and customs duty exemption periods were reduced from five to two years (Peprah 1982b: Table 3.4). Timber firms now pay corporate tax at the maximum rate of 50 percent.

Ivory Coast. The general rate of corporate income tax in Ivory Coast has been 40 percent since 1980 (Price Waterhouse 1985: 174), and timber companies are nominally required to pay an additional 10 percent tax to the National Development Fund.

However, any firm may apply for special income tax incentives that are among the most generous offered anywhere. Firms that obtain "priority" status may receive income tax holidays ranging from seven to 15 years (Price Waterhouse 1985: 175), the longest income tax holiday period available in the world. While few timber companies are eligible for the 15-

year tax holiday, they generally obtain a five-year exemption from the corporate tax. Timber firms are also eligible for a special incentive program wherein half of all amounts spent on government approved investment programs may be expensed (immediately deducted) from current income (Price Waterhouse 1985: 175), rather than depreciated over time.

Ivory Coast, like Gabon, has retained substantial income tax incentives for natural resource projects, though most other LDCs had abandoned such practices by 1970. One reason for the persistence of tax holidays and other tax incentives may be that French firms continue to dominate foreign investment in Ivory Coast. While generous income tax incentives are of little value to say, American-based firms, this is not so for French firms. The U.S. government taxes foreign income of U.S. firms and allows a credit for foreign taxes paid. With no Ivory Coast income tax liability, an American firm would have no foreign taxes to offset its U.S. tax liability. Offering tax holidays to U.S. firms merely increases their net U.S. taxes. France, however, does not generally tax the foreign income of French firms. Thus Ivory Coast tax holidays mean that the firms pay neither Ivory Coast nor French income taxes.

Generous income tax incentives available in Ivory Coast clearly made the timber sector attractive to French firms in the past. For the period 1978–80, corporate taxes from all timber firms totalled but US$6.7 million, less than 0.3 percent (one-third of one percent) of the value of log production (Ministry of Forest 1981). Clearly this government subsidy contributed significantly to expanded logging investment, and thus to the rapid rates of forest depletion observed after 1967.

Gabon. Gabon's standard corporate income tax rate is 50 percent. Exemptions, however, abound. The Gabon Investment Code allows timber companies a two-year tax holiday, plus a 50 percent tax remission in the third year. Moreover, firms are exempted from paying a variety of minor levies for 5 to 10 years.

For "privileged firms," including forestry enterprises, tax holidays may be extended to five years. Such firms also receive duty free privileges on virtually all imports for a period of 10 years.

Export taxes

Ivory Coast. The export tax (*droit unique de sortie*) is by far the most significant fiscal instrument applied in the forest sector, accounting for about 90 percent of total government revenues on log exports.

The tax is *ad valorem* and, as in Indonesia, is based on the posted price of

Table 7.22. *Ivory Coast: representative export taxes on logs and wood products, circa 1983*

Species or product	Tax rate (percent)	Taxes (U.S. dollars per m³)
Logs		
Sipo	44.6	111.50
Aboudikou	44.6	94.10
Acajou (mahogany)	44.6	60.20
Iroko	36.6	65.90
Samba	36.6	49.40
Llomba	24.6	16.00
Processed products		
Sipo sawn timber	11	
Aboudikou timber	11	
Acajou timber	6	
Iroko timber	6	
Llomba timber	2	
All plywood		
High-value species	2	
Low-value species	1	

Source: Gray (1983: 100).

log exports, in order to reduce export tax evasion through undervaluation. Posted prices have typically been slightly below actual f.o.b. prices (Gray 1983: 99). Ivory Coast's log export tax rates are nearly as high as in Sabah, and well exceed those of Indonesia and other African countries. Representative export taxes for logs ranged from 24.6 percent for the low-valued *llomba* species to a high of 44.6 percent for the prized *sipo* (Table 7.22).

Gabon. Gabonese export taxes are 20 percent of posted f.o.b. values for logs and 12 percent for sawn timber and processed wood of the predominant *okoume* species (Republic of Gabon, Ministère des Eaux et Forêt 1980: 43–85). Rates averaging but 10 to 11 percent apply to other species of log exports and processed products. Relative to other countries, the Gabonese export tax structure provides relatively small incentives for domestic processing through export taxes.

Other policies impinging upon forest use

Liberia. With one exception, non-forestry policies of the Liberian government have had little measurable impact on deforestation. Large-scale cattle ranching has been infeasible in Liberia owing to livestock diseases. No large-scale resettlement programs of the Indonesian type exist. The government has not subsidized road network construction for log transport. Liberia has no central bank; the Liberian dollar is on a par with the U.S. dollar; indeed, the U.S. currency is the principal means of payment. Thus, exchange rate policy has not been a factor except to the extent the U.S. dollar has itself been over or undervalued relative to European currencies. The U.S. dollar was in fact overvalued from 1981 to 1985, and this fact would have tended to slow Liberian timber exploitation for export purposes more than if the Liberian and U.S. dollar had been delinked.

One other policy, however, may have encouraged more rapid forest depletion. Various incentives are provided for upland rice cultivation, particularly for small (1.4-hectare) plots. According to the FAO, shifting rice cultivation practices have been the most significant cause of deforestation in recent years. The same source concludes that for the 1981–85 period, "deforestation rates should increase slightly, with increasing agricultural production, incentives given rice cultivation and increased forest planning" (FAO 1981: 277).

Ghana. The impact of forest policies on forest depletion in Ghana has been negligible, relative to that of more general macro policies: exchange rate, tax, agricultural pricing, and monetary policies. Policies governing large infrastructure projects and those promoting cattle raising have been relatively unimportant in forest depletion.

The largest infrastructure project in Ghanaian history was the Volta River Hydro-aluminum project mounted by the government and Kaiser Aluminum, begun in 1962 and completed in 1967. The dam for this project, when finished, resulted in the flooding of between 500,000 and one million hectares of land. The Volta Lake, however, submerged only a small proportion of Ghana's remaining forest cover, as it is situated well away from southeastern Ghana where the forest is concentrated. Nevertheless, the project indirectly placed additional pressure on the forest estate by creating more "shifting cultivators." Rising waters displaced many families, forcing them to turn to shifting cultivation for survival.

Forest clearing for livestock ranches has not been a significant factor in deforestation in Ghana, simply because endemic sleeping sickness (trypanosomiasis) in cattle precludes large-scale cattle ranching (Trum-Barima 1981: 115).

Ivory Coast. Several non-forestry policy instruments of the Ivory Coast government have contributed to deforestation. These include infrastructure, land-clearing, and plantation policies. Exchange rate policy has been relatively less important in forest exploitation than in Malaysia, Indonesia, or Ghana, as the CFAF franc, the currency of the Ivory Coast, is closely tied to the value of the French franc, which has basically floated against the U.S. dollar since 1971.[28]

Both the French colonial administration and later the independent government of Ivory Coast have emphasized road infrastructure investments. By 1982, Ivory Coast had more than 4,000 km of paved roads, more than any other tropical African country. Gabon by comparison had only 100 km of paved roads in 1982, and Liberia far less than Gabon. Ivory Coast initially emphasized road construction to ease tropical forest exploitation, later to facilitate transportation of coffee and cocoa produced by plantations.

The extensive road network, combined with relatively flat terrain, contributed substantially to the profitability of logging for export. The completion of the port of Abidjan in 1951 also significantly assisted log exporting (FAO 1981: 106). Since that time, two more government harbor development projects have reduced timber transport costs even further below those of other West African countries. Tolls on logging trucks utilizing the road network and harbors have been low or nonexistent.

In recent years, the government has also encouraged small-scale commercial farmers to clear forest areas for such export crops as coffee and tea by offering to transfer property rights on cleared land to the cultivator. According to Myers, small-scale farmers affected have thereby come to operate on a larger scale than shifting subsistence cultivators (Myers 1985: 161–162).

Since 1966, cocoa and coffee plantation development in once forested areas has been significant in Ivory Coast. However, official plans now call for a tree plantation program covering less than one-tenth of the nation's natural forest area. By early in the twenty-first century, these plantations are expected to supply about four million cubic meters of tropical hardwood per year, as much as present output from the natural tropical forest (Myers 1985: 359).

Gabon. As in virtually all other Francophone African nations, exchange rate policy has not been a major factor in forest extraction or processing decisions in Gabon.

The most important non-forestry policy issue affecting the future of Gabon's forest estate is the completion of the Trans-Gabon Railway. The project, with estimated costs totalling US$4 billion (in a country of 700,000 people) will, among other effects, open up over 4 million hectares of forested land (Republic of Gabon, Ministère des Eaux et Forêt 1980). Plans are already under way to award two types of concession contracts on these heretofore inaccessible areas. The basic strategy is to use low rail tariffs on log hauling to induce more intensive utilization of secondary (lower-valued) species, in the hope that they will be accepted in the world market. It is believed that this would ultimately allow heavier export taxes on such logs.

To quote:

Il ne fait aucun doute que la commercialisation regulière des "bois divers" rendue possible grace à l'existence du chemin de fer, aboutira rapidement à une valorization de ces essences avec une incidence considerable sur le produit des droits de sortie.[29]

Endnotes

1. Virtually all of the fuelwood supply for the capital city of Monrovia derives from rubberwood from old plantations. Most charcoal used in Monrovia and in rural areas is made from fallen branches in the forest, logging wastes, and wood salvaged from shifting cultivation areas (Peprah 1982b).

2. Liberia's 1983 per capita income has been estimated at US$470 (1985 dollars), well below that of Ivory Coast ($660), just above nearby Senegal's ($370), and well above Ghana ($380) (World Bank 1987: Appendix Table 1).

3. Roemer (1984) places the decline at 0.7 percent per year from 1957 to 1981. Since 1981, the decline has been even more rapid.

4. Recall that Ghana has no remaining virgin forest upon which shifting cultivation can intrude and that in Malaysia, particularly East Malaysia, upwards of 5 million hectares of undisturbed forest remain.

5. See for example Roemer (1984), Killick (1978), and DeWilde (1980).

6. Typically, a black market exchange rate that is a multiple of the official rate indicates not only overvaluation, but substantial capital flight as well. When, as in Ghana, black market rates are consistently several times higher than official rates over several years, we can be sure that severe overvaluation is present.

7. See, for example, Simon and Kahn (1984: 160–161). Simon and Kahn cite three separate sources indicating that not only has Ivory Coast ranked first among all nations in deforestation rates, but that this rate has been nine or ten times that of the median LDC in a sample of 48.

8. As late as 1986, a leading World Bank specialist concluded that an overall land-use policy for the forestry zone was nonexistent (Spears 1986: 4).

9. Concessions are awarded for a period "equal to or superior to 30 years" (Republic of Gabon, Ministère des Eaux et Forêt 1980: 37).

10. Conversion ratios for plywood tend to be 40 percent in Ivory Coast and in Indonesia. The two firms were Compagnie France/Gabon and Société de Haute Mondah (Pfieffer 1980: 40–42).

11. This policy resulted in the proliferation of many small Ghanaian firms working at less than minimum efficient scales. Default rates on the government

loans were high. The soft loans made firms choose inappropriate capital intensity (Peprah 1982a: 24).

12. Even by 1983, virtually all foreign timber firms were French-based (Gillis et al. 1983).

13. There remains, however, the project to complete the Trans-Gabon Railway, so as to "open up" the interior of the country.

14. The Gabonese government owns 10 percent of Société Forestière SA, 25 percent of Compagnie Forestière Gabon, 51 percent of Société Nationale du Bois de Gabon, and 62 percent of a pulp and paper mill (Gillis et al. 1983: 4–8).

15. Other objectives of Liberia's reforestation program are to provide employment for the unemployed and underemployed rural populace and to bring the rural people into contact with organizational-framework institutions and agricultural and forestry techniques that they are at present generally unfamiliar with.

16. SODEFOR activities have in recent years been largely financed by World Bank and French AID funds: 40 percent of SODEFOR support came from the Ivory Coast government in 1976–80.

17. Details on the Industrialization Incentive Fee are provided in Schmidt (1980: 4–8).

18. Details on the Forest Products Fee are available in Peprah (1982b: Chapter 4) and Gray (1983: 98–99).

19. Personal observation of the author, in Accra, Ghana, June 1970; see also Page (1974).

20. Computed from inflation figures provided in World Bank (1985: 174).

21. Also, attempts were made several times during the period from 1960 to 1972 to compel firms to invest in sawmilling capacity as a prerequisite to obtaining or extending timber holdings. The policy was not consistently applied, however, and only five firms in the sample survey conducted by Page in 1974 noted that concessions policy had provided a major incentive to invest in wood processing facilities (Page 1974: 106).

22. Negative real rates of interest were the chief factor in inducing greater capital intensity; tax exemptions and overvaluation of the exchange rate also contributed.

23. The "achievable" recovery rate is the level of recovery that can be expected from the best-managed operations (World Bank 1986: Annex VIII, 8).

24. Ivory Coast capacity in sawmilling increased by 100 percent (60 plants to 80 plants; plywood capacity also doubled (2 plants to 5 plants). But capacity in veneer and chipboard actually declined.

25. The 40 percent figure for conversion in Ivory Coast is from Pfeiffer (1980: 40–42).

26. Report by Michelline Mescher to the author from interviews; Abidjan, Ivory Coast, September 1983.

27. The tax holiday may be extended to seven years for investments in excess of $2 million (Price Waterhouse 1985: 203).

28. Values for the CFAF Franc vis-à-vis the dollar in selected recent years: 1970, 276; 1976, 239; 1978, 225; 1980, 211; 1984, 485.

29. Official policy, as quoted in Republic of Gabon, Ministère des Eaux et Forêt (1980: 55).

References

Adeyoju, Kolode. 1976. *A Study on Forest Administration in Six African Nations.* Rome: FAO.

Afrique Industrie. 1982. May, p. 23.

Arnaud, Jean Claude, and Gerard Sournia. 1980. Les Forêts de Côte d'Ivoire, Une Richesse Naturrelle en Voie de Disparision *Les Cahiers d'Outre Mer,* p. 295.

Bates, R. H., and M. F. Lofnic, eds. 1980. *Agricultural Development in Africa.* New York: Praeger.

Côte d'Ivoire, Code Forestier. 1965. Loi Nr. 65–425. Abidjan, Ivory Coast, December 20.

DeWilde, John C. 1980. Case Studies: Kenya, Tanzania and Ghana. In *Agricultural Development in Africa,* ed. R. H. Bates and M. F. Lofnic. New York: Praeger.

Esseks, John D. 1974. The Nationalization of the Ownership of Resources in Ghana. Paper prepared for the Colloquium on Ghana, Department of State, Washington, D.C., October 11.

FAO. 1981. *Forest Resources of Tropical Africa.* Rome: FAO.

FAO. 1986. *Tropical Forest Assessment Project.* Rome: FAO.

Ghana Timber Association. Circa 1982. Internal Memorandum.

Ghana Timber Marketing Board. 1982. Personal Communication.

Ghana Timber News. 1979. Vol. 7, No. 3.

Gillis, Malcolm. 1983. Evolution of Natural Resources Taxes in Developing Countries. *Natural Resource Journal,* Vol. 23, No. 2.

Gillis, Malcolm, Ignatius Peprah, and Michelline Mescher. 1983. Foreign Investment in the Forest-Based Sector in Africa. Unpublished. Cambridge, Mass.: Harvard University.

Gray, John. 1983. *Forest Revenue Systems in Developing Countries.* Rome: FAO.

IBRD. 1976. *West Africa Forestry Sector Study.* Washington, D.C.: World Bank.

IBRD. 1986. *Ghana: Forestry Sector Review.* Washington, D.C.: World Bank.

Harberger, A. C. 1984. *World Economic Growth.* San Francisco: Institute for Contemporary Studies.

Johnson, Brian. 1984. *Lumbering in Forest Ecosystems: Easing Pressures of the Tropical Timber Trade on Forest Lands.* London: IIED.

Killick, Tony. 1978. *Development Economics in Action: A Study of Economic Policies in Ghana.* New York: Columbia University Press.

Lanly, J. P., and L. Clement. 1979. *Present and Future Forest and Plantation Areas in the Tropics.* FAO Miscellaneous papers. Rome: FAO.

Legault, Faustin. 1981. Le Réboisement en La République Gabonaise. In *Where Have All the Forests Gone,* ed. V. H. Sutlive, N. Altshuler, and M. D. Zamura. Williamsburg, Va.: College of William and Mary.

Mescher, Michelline. 1983. Report to the author from interviews, Abidjan, Ivory Coast, September.

Ministry of Forests, Ivory Coast. 1981. *Centrale de Bidans, 1978–80.* Abidjan: Government of Ivory Coast.

Myers, Norman. 1983. The Tropical Forest Issue. In *Progress in Resource Management and Environmental Planning,* ed. T. O'Riordan and R. K. Turner. New York: John Wiley and Sons.

Myers, Norman. 1985. *The Primary Source: Tropical Forests and Our Future*. New York: W. W. Norton.

O'Riordan, T., and R. K. Turner, eds. 1983. *Progress in Resource Management and Environmental Planning*. New York: John Wiley and Sons.

Page, John. 1974. The Timber Industry and Ghanaian Development. In *Commodity Exports and African Economic Development*, ed. Scott Pearson and John Cownie. Lexington, Mass.: D.C. Heath.

Page, John. 1976. The Social Efficiency of the Timber Industries in Ghana. In *Using Shadow Prices*, ed. I.M.D. Little and M.F.G. Scott. London: Heinemann Educational Books.

Page, John M., Scott R. Pearson, and Hayne E. Leland. 1976. Capturing Economic Rents from Ghanian Timber. *Food Research Institute Studies*, Vol. 15, No. 1.

Pearson, Scott, and John M. Page. 1972. *Development Effects of Ghana's Forest Products Industry*. Accra: USAID.

Peprah, Ignatius. 1982a. Foreign Investment in Forest-based Sector of Ghana. Unpublished. Cambridge, Mass.: Harvard University.

Peprah, Ignatius. 1982b. Foreign Investment in Forest-based Sector of Liberia. Unpublished. Cambridge, Mass.: Harvard University.

Pfieffer, Ernst. 1980. *Promotion of Marketing Trade and Value-Added*. Addis-Ababa, Ethiopia: FAD, August.

Price Waterhouse. 1985. *Corporate Taxes: A Worldwide Summary*. New York: Price Waterhouse.

Republic of Gabon, Ministère des Eaux et Forêt. 1980. *La Forêt Gabonaise*. Libreville, Gabon.

Roemer, Michael. 1984. Ghana 1950–80: Missed Opportunities. In *World Economic Growth*, ed. A. C. Harberger. San Francisco: Institute for Contemporary Studies.

Ruzicka, I. 1979. Indonesian Logging. *Bulletin of Indonesian Economic Studies*, Vol. 15, No. 2 (July).

Schmidt, Gerald W. 1980. The Forest Fees and Taxes of the Republic of Liberia. *FAO/World Bank Report*. Monrovia, Liberia, February.

Schmithusen, Franz. 1977. *Forest Utilization Contracts on Public Land*. Rome: FAO.

Schmithusen, Franz. 1980. Forest Utilization Contracts: A Key Issue in Forest Policy. United Nations Centre for Transnational Corporations, Workshop on Tropical Hardwoods, Pattaya, Thailand, September 5.

Simon, Julian, and Herman Kahn, eds. 1984. *The Resourceful Earth*. New York: Basil Blackwell Inc.

Spears, John. 1986. *Key Forest Policy Issues for the Coming Decade in the Rain Forest Zone*. Washington, D.C.: World Bank.

Steinlin, Hans Jung. 1982. Monitoring the World's Tropical Forest. *Unasylva*, Vol. 34, No. 137: 6.

SODEFOR. 1985. La Forêt Dense Humide en Côte d'Ivoire. Abidjan, Ivory Coast.

Sutlive, V. H., N. Altshuler, and M. D. Zamura. 1981. *Where Have All the Forests Gone*. Williamsburg, Va.: College of William and Mary.

Trum-Barima, K. 1981. Forests of Ghana – A Diminishing Asset. In *Where*

Have All the Forests Gone, ed. V. H. Sutlive, N. Altshuler, and M. D. Zamura. Williamsburg, Va.: College of William and Mary.

White, Peter T. 1983. Tropical Rain Forests: Nature's Dwindling Treasure. *National Geographic,* Vol. 163, No. 1 (January).

Whitmore, T. C. 1984. *Tropical Rain Forests of the Far East,* 2nd edition. Oxford: Clarendon Press.

World Bank. 1976, 1979, 1980, 1985, 1986, 1987. *World Development Report.* Washington, D.C.: World Bank.

8 Subsidized timber sales from national forest lands in the United States

ROBERT REPETTO

Background

The U.S. National Forest system includes 191 million acres of forest land, more than a quarter of the national total. Because timber production in the past drew more heavily on accessible, higher quality stands owned by private industry, the national forests contain a much higher fraction of standing sawtimber stocks, almost one-half. Much of this is mature, old-growth timber or grows on relatively low productivity sites, so despite a large increase in harvesting between 1940 and 1965, the national forests' contribution to annual net timber *growth* and harvest is more in line with their share in forest area. The national forests are also prime recreation areas, attracting more than 225 million visitor days of recreational use each year, almost ten times as many as in 1950. In addition, national forest lands have drawn increasing attention as potential repositories of significant mineral and energy resources.

These increasing and potentially conflicting demands have ensured controversies over forest management policies (Wilkinson and Anderson 1985). Often, these have arisen over attempts to restrict the multiple-use management principles employed by the U.S. Forest Service, or to reorient the application of those principles in favor of one or another specific use. A long-standing and sharp controversy of this nature has pitted recreational and conservationist interests against the Forest Service over timber harvesting and associated road-building in areas allegedly unsuitable for commercial timber production due to inaccessibility and relatively high growing costs. Many of the areas in question are either roadless tracts potentially eligible for restrictive wilderness designation, or in regions of recreational value. While the Forest Service defends its timber operations under broader measures of multiple-use benefits and costs, critics charge that they fall far short of recovering even their direct costs and should be curtailed. This chapter assesses the controversy over these "below-cost"

353

timber sales to illustrate multiple-use forest policy problems in high-income countries.

Until the 1870s there was no provision for managing timber in federal forests, which were rapidly being alienated to private owners. Effective conservation began with the creation of forest reserves in the 1890s, and legal authorization for the sale of "dead, matured, or large growths of trees" at not less than appraised value. Yet, so much timberland was by then privately owned that demands on the reserves were small. In fact, only since World War II have the national forests provided a significant fraction of timber supplies, now about one-third of the softwood harvest. About half the volume of timber ever sold from national forest land, and over two-thirds of the value, has been marketed since 1965.

When forest reserves were transferred to the Department of Agriculture and the Forest Service created in 1905, the main goal was to develop the West. Although Congress later sanctioned purchases of forest land in the East, so little public land remained in the East that the national forests have always been mainly a Western resource. The Forest Service has promoted development by selling timber, with little thought for its own profit. For example, on Forest Service lands, appraised stumpage values are estimated to cover the *buyer's* costs of extracting, processing, and selling the wood with a reasonable profit (Beuter 1985). Although it is required to consider economic efficiency in its operations, the governing laws do not require the Forest Service to cover its own costs of growing and marketing timber. The minimum bid price specified in advertised sales does not, even in principle, reflect Forest Service supply costs. In fact, despite several urgings by Congress, as yet the Forest Service has not installed an adequate accounting framework for assessing those supply costs, an omission that has prolonged and confused the debate.

By 1908 the provision that 25 percent of timber receipts were to be transferred to local governments was in place, and by 1913 10 percent of receipts had been earmarked for constructing and maintaining roads and trails. In 1930 the Knutson-Vanderberg (K-V) Act set aside "deposits" from timber purchasers for reforestation and brush disposal, later expanded to cover conservation and wildlife protection activities. Later legislative revisions sanctioned credits against timber payments for purchaser built roads, and raised payments to local communities to 25 percent of gross receipts. These laws constructed the present system that links support for local communities, road construction, and future forestry activities to timber sales. These links, and the interests of timber purchasers, create budgetary and political incentives for continued timber operations even on marginal sites (Johnson 1985).

In 1960 the Multiple Use Sustained Yield Act codified the policy of pursuing sustained yields of the several kinds of benefits obtained from the national forests. The legislative guidelines for multiple-use management were so general that they, in effect, ratified the considerable administrative discretion with which the Forest Service managed the domain under its charge. In the 1960s and 1970s conflicts over multiple use of national forests intensified around the Wilderness Act of 1964, which confirmed – contrary to the spirit of multiple-use management – previous administrative wilderness designations and included large additional areas from the timber base as potential restricted use areas. Subsequent reviews of roadless areas for potential wilderness designation (RARE I and II) have provided fuel for the "below cost" timber sale issue, particularly with respect to the 11 million acres of national forest land remaining undesignated pending further study. Defenders of Forest Service timber operations believe that conservationists employ invalid economic arguments for restricting forest use to their favored purposes. Conservationists, on the other hand, accuse the Forest Service of conducting uneconomic sales to justify accelerated road construction in roadless areas, thereby forestalling future possible wilderness designation (Jackson 1986: 11–15).

The National Environmental Protection Act (NEPA), which facilitated public challenges of such ecologically damaging Forest Service harvesting practices as extensive clear-cutting, also helped influence Congress to mandate a comprehensive planning process for the use of national forests and grazing lands, including targets for timber production, in the Forest and Rangeland Renewable Resources Planning Act (RPA) of 1974, and to provide new authorization for timber sales in the National Forest Management Act of 1976. In this law, provision 6(k) requires the Forest Service to "identify lands within the management area which are unsuitable for timber production, considering physical, economic, and other pertinent factors to the extent feasible, as determined by the Secretary, and shall assure that, except for salvage sales or sales necessitated to protect other multiple-use values, no timber harvesting shall occur on those lands for a period of ten years." This was an attempt to address the problem of uneconomic timber operations by directing the Forest Service to demarcate those regions economically or otherwise unfit for sustained timber production. The guiding criterion was that forest land economically or physically unsuitable for profitable timber investments should be segregated before management plans are formulated. Thus far, however, the Forest Service has not fully complied with that directive. Although it has withdrawn a substantial fraction of the national forests from the timber base because of physical unsuitability and environmental risk, it has sought to continue managing all forests physically suitable for timber production under flexible multiple-use criteria, deriving harvest plans for each area through administrative planning driven by timber harvest targets. At present, production targets thus determine the areas deemed economically suitable for harvesting, not the reverse. According to a recent review, ". . . a court might well conclude that the current regulations do not carry out the intent of Congress as to the separate timberlands evaluation of section 6(k)" (Gorte and Baldwin 1986).

Rapidly rising timber prices in the 1970s, rising imports, and the perception of supply limitations on private forests, led to forecasts of continuing price rises and interest in expanding national forest harvests to contain inflation. In June 1979, a directive from President Carter urged accelerated harvest of mature timber, and the pro-development philosophy of the Reagan administration reinforced this policy. The allocation of planned timber harvests among the national forests has tended to follow the distribution of the allowable cut, despite a number of studies arguing that it would be economically rational to concentrate the harvests further on the old-growth forests of the Pacific Northwest where stands tend to be richer, values higher, and costs lower (Clawson 1976). Harvesting this old-growth timber is retarded by the Forest Service's policy and by mandating legislation stipulating that harvest schedules produce the maximum volume of timber that can be sustained over the long run. Under this policy the annual cut is based on long-run growth potential, and precludes accelerated harvest of existing overaged stands, even those with negligible or negative annual growth (Nelson 1982). Were capital charges assessed on the inventory of standing timber, the losses involved in holding old-growth stands far past the economically optimal rotational age would be obvious (Dowdle and Hanke 1985).

There is a certain symmetry in these economic evaluations of public timber management: too little management in productive national forests, too much timber management in unproductive forests – both stemming in part from the lack of an adequate framework for economic analysis of timber operations. The Forest Service has been challenged by several studies over the past decade charging that even existing levels of harvest in many unproductive forests are uneconomic, because direct costs greatly exceed the value of the output. The claims made by these studies were strengthened by the precipitous drop in stumpage prices in the recession of 1982, and their slow recovery in subsequent years. Congress has repeatedly directed the Forest Service to consider economic efficiency in its timber operations and to develop a system of cost accounting able to identify uneconomic timber operations. In its response, the Forest Service has insisted that although an accounting system can be developed to show the effects of timber sales on cash flows to the federal government, economic evaluation of its timber operations requires analysis of multiple-use benefits (USDA, Forest Service 1986).

Evaluating below-cost timber sales

Growing timber as a renewable resource is a process that extends across decades or centuries. Revenues come periodically as trees reach harvest size. Costs also occur at intervals, as reforestation, thinning, and harvesting are needed. Assessing economic returns inevitably requires comparing costs and returns over many years, and applying an appropriate interest charge to equate values in different years is essential. Although Forest Service planning models calculate discounted present values in analyzing future benefits and costs, not charging interest is a fundamental deficiency in Forest Service cost accounting. Not only does it understate costs, it also seriously distorts investment and management decisions.

Forest Service accounts are not set up to compare costs and revenues over time. Past expenses involved in bringing particular stands of timber to harvest are not accumulated, with interest, and charged against the sales receipts, nor are future benefits of timber management activities discounted back to a common base year. Rather, costs are charged to various functional accounts by administrative district in each accounting period. Accounts for a particular year and national forest will include receipts from sales that have already taken place, and costs incurred on sales that will take place in the future.

There are more fundamental complexities. National forests yield various benefits, and some expenditures serve several purposes: roads may provide access to both logging trucks and recreational vehicles, and timber "salvage" harvests may yield timber and reduce risks of forest fires or insect damage. Timber harvests may raise water yields in downstream watersheds, but also increase sediment loadings. "Goods" and "bads" from the national forests are often jointly produced.

There is no foolproof way of allocating such joint costs, and the Forest Service argues that timber operations that apparently lose money can be economic when non-timber benefits are taken into account. In this respect, it correctly insists that the existence of important joint costs requires evaluation of total non-timber and timber benefits and costs. An optimal forest management plan is one that maximizes total net benefits over time. An efficient management plan is one that produces the maximum timber benefit, given specified levels of non-timber benefits and budgetary and other resource constraints. It is conceivable that an efficient plan could result in continuing revenue losses on timber operations, if harvests were mainly in support of non-timber objectives. It is likely that within an efficient plan some individual sales would fail to cover their direct (separable) costs (Scheuter and Jones 1985).

This optimization principle provides little practical help, however, either for forest managers or for outsiders concerned about management of the public forests. The impacts of timber operations on wildlife habitat, hunting, recreation of various kinds, soil erosion, and water quality are not readily ascertained and are often sharply debated. Nor are these public benefits fully priced through user fees, and surrogate estimates yield a substantial range of values (Sog and Loomis 1984). Therefore, instead of asking whether forest plans attain overall optimal patterns of forest use, evaluations of forest operations have concentrated on tests of incremental efficiency, asking whether alternatives to proposed harvest plans that *maintained* non-timber benefits and *reduced* timber harvests would reduce costs more than revenues. This incremental efficiency test focuses attention on the marginal returns to timber sales and the separable costs of timber operations (Krutilla and Bowes, in press).

Ideally, an incremental efficiency test would allocate the joint costs of durable timber roads across current *and future* harvests and sales, rather than charging such costs solely to current sales, and calculate the net present value of changes in timber sales over the useful lifetimes of the roads. Studies that have done this have been limited in geographical scope, because they rely on individual forest plan projections rather than on current forest-wide or region-wide statistical accounts. Projecting future net revenues involves uncertain assumptions, especially about future costs and prices.[1] However, such studies have been carried out in the San Juan National Forest in Colorado, the Tongass National Forest in Alaska (Hyde and Krutilla 1979), the Hoosier National Forest in Indiana (Minckler 1985), the Clearwater National Forest in Idaho (Helfand 1983), the Lolo and Kootenai National Forests in Montana, and others (Hyde 1981; Hyde and Krutilla 1979; Helfand 1983; Scheuter and Jones 1985; Minckler 1985; U.S. General Accounting Office 1984). With the exception of the Forest Service study in Montana, these analyses found that future harvests and sales would be unprofitable, despite previous investments in road construction. Once existing mature stands of timber are harvested, the present value of future rotations on low-productivity sites tends to be unfavorable. Therefore, including future benefits and costs would not seem generally to improve the estimated efficiency of current timber operations in marginal areas.[2]

The 1985 Congressional Research Service estimates

A recent Congressional Research Service (CRS) study of all timber sales in six western Forest Service regions during 1981 and 1982 examined both the net cash flows to the U.S. Treasury and a measure of net economic benefits (Beuter 1985). The main limitation of studies that examine sales results for one or two years is that the timber market conditions of the particular years greatly affect the results. Since stumpage markets were severely depressed in 1982, below-cost sales were prevalent. Partially offsetting this bias, the winning bid amounts were used instead of eventual sales receipts, which are usually somewhat lower because of price adjustment clauses written into many contracts and other reasons.

The cost accounting assumptions underlying this study had to be adjusted to approximate a test of incremental net revenues. Payments to counties were redefined as a revenue transfer rather than as a cost to the federal government. Road construction financed from purchaser credits was treated as a logging cost met from timber receipts, and netted out of the accounts. Only roads financed from appropriated funds were treated as an incremental harvest cost, because road expenditures have been virtually constant in real terms for at least a decade, so that within entire regions current outlays approximately equal the sum of amortization charges for past construction expenditures. Expenditures under K-V deposits for reforestation and purchased land management services were similarly netted out against timber sale deposits and receipts.

Adjusting the CRS estimates to compare net bid receipts and remaining resource costs results in the figures presented in Table 8.1. Two conclusions stand out. First, of the six western regions examined, region 5, the Pacific Southwest, and region 6, Northwest, account for most of the revenues and profits. The results for the Southwestern and Northern regions also show positive net receipts for both comparison years, although by narrower margins. Second, on balance, timber operations in the Rocky Mountain and Intermountain belt, regions 2 and 4, clearly do not cover direct costs. Even if road costs are ignored altogether, net receipts fall far short of the costs of sale preparation, administration, and support for the regions as a whole, despite the fact that these aggregate statistics combine below-cost and above-cost operations. This finding lends credibility to the contention that many timber operations in these regions do, in fact, fail a test of incremental economic efficiency.

The rejoinder to this conclusion is that both receipts and costs are

Table 8.1. *Estimated costs and returns from western timber sales, 1981 and 1982 (million dollars)*

	Region					
	(1) Northern	(2) Rocky Mountain	(3) Southwestern	(4) Intermountain	(5) Pacific Southwest	(6) Pacific Northwest
1981						
Cash bids	75.5	4.8	43.5	4.8	319.4	1,335.0
Total direct costs	28.4	13.5	8.8	13.8	41.6	97.7
Road cost to government	3.1	2.4	0.8	6.8	2.9	2.3
Sale preparation and administration	18.5	7.5	6.2	4.7	26.5	54.3
Pre-sale and support costs	6.8	3.6	1.8	2.3	12.2	41.1
Net revenues	47.1	−8.7	34.7	−9.0	277.8	1,237.3
1982						
Cash bids	43.5	2.1	14.8	3.2	99.8	414.7
Total direct costs	33.8	12.7	8.8	12.4	53.1	108.9
Road cost to government	6.8	1.0	1.4	4.9	7.5	11.5
Sale preparation and administration	16.4	6.3	4.9	4.0	25.2	41.8
Pre-sale and support costs	10.6	5.4	2.5	3.5	20.4	55.6
Net revenues	9.7	−10.6	6.0	−9.2	46.7	305.8

Source: Beuter (1985: 93).

affected by logging practices that serve non-timber objectives. Restrictions on harvesting operations to protect or enhance recreational, wildlife, and other values raise loggers' costs, and reduce the amounts that can be bid for timber. Similarly, management for non-timber benefits raises the Forest Service costs. A recent study, based on an evaluation of hundreds of sales, shows that the effects on net timber revenues can be substantial (Benson and Niccoluci 1985).

On closer examination, this fact is seen as largely beside the point. Most of the extra costs incurred are actually to mitigate the effects of logging and road construction on recreation, wildlife, and ecological stability, and to protect those values against losses due to timber operations. The relevant test of efficiency is whether the timber benefits cover the incremental costs of timber operations, *for the same level of non-timber benefits*. The extra costs incurred to maintain non-timber benefits unchanged when timber is harvested are part of the incremental costs of logging, and should be included in incremental benefit-cost evaluations. Just as the non-timber benefits of logging should be included in evaluations, the non-timber costs of timber operations also must enter the calculation.

The 1984 General Accounting Office estimates

A study carried out by the General Accounting Office (GAO) in 1984 also examined data on over 3,000 individual timber sales in 1981 and 1982, for the four regions 1, 2, 4, and 6 (U.S. GAO 1984). The methodology of this study departed further from that implied by an incremental efficiency test. By comparing sales receipts, rather than winning bids, with sales costs within accounting studies, the GAO study implicitly matched revenue data with cost data from different sales, unlike the CRS study. Although the Forest Service data the GAO examined did not identify the costs associated with individual sales, forest-wide figures were used to estimate unit costs for each sale. Although the GAO methodology inappropriately treated payments to counties as costs, and included purchaser credits for roads with government returns, these biases are offsetting. The GAO report did not contain the information to adjust the estimates to conform to Table 8.2, but a sensitivity analysis in the report found that excluding the payment to counties from sales costs would turn less than 10 percent of the below-cost sales to above-cost sales (Table 8.2). Despite these limitations, the results reproduced in Table 8.2 bear out the earlier finding: over 90 percent of sales in regions 2 and 4 generated revenues insufficient to recover direct costs, while sales in region 6 generate large net revenues, and those in region 1 are mixed.

Table 8.2. *National forest timber sales, regions 1, 2, 4, and 6: Summary of gains and losses for fiscal years 1981 and 1982 (thousand dollars)*

	Sales showing gains		Sales showing losses	
	Number of sales	Amount of gain	Number of sales	Amount of loss
1981				
Region 1	135	$ 12,955	132	$19,016
Region 2	5	51	75	14,117
Region 4	8	86	62	13,450
Region 6 (pine)	211	106,539	108	12,332
Region 6 (Douglas fir)	838	597,624	56	5,097
Total	1,197	$717,255	433	$64,012
1982				
Region 1	74	$ 3,691	169	$26,220
Region 2	1	3	73	13,860
Region 4	3	3	73	10,422
Region 6 (pine)	142	26,976	145	20,634
Region 6 (Douglas fir)	717	121,237	217	21,639
Total	937	$151,910	677	$92,775

Source: U.S. GAO (1984: 10).

Moreover, this study provided detail on the pattern of above- and below-cost sales in individual forests.

The GAO report examined a number of individual sales to explore the reasons for losses. GAO found that the principal explanation was timbering low productivity stands of low-valued species on difficult terrain that raised harvesting costs. The Forest Service explanation, that losses were incurred to open up areas for logging by building roads, or to replant areas with higher valued, high-productivity stands, was not supported by the data: the GAO found that subsequent sales would not yield positive economic returns either. Therefore, despite its flaws, this study contributes to an understanding of the pattern of efficient and inefficient operations.

The 1984 Congressional Research Service estimates

Another Congressional Research Service study provided a longer time perspective on timber revenues by examining forest-by-forest receipts between 1973 and 1983, averaging out year-by-year fluctuations (Wolf 1980). This approach partially rectifies the deficiency of studies discussed above that relied on revenues estimated only from sales in one or two years. Only cost figures for 1982 were used, because breakdowns for earlier years were unavailable. However, this study provided alternative estimates of separable costs – a minimum estimate comprising only timber sale administration and resource support, and a broader estimate also including costs of reforestation and timber stand improvement and timber roads financed by appropriated funds. This procedure allows finer resolution of the distinction between separable and possible joint costs. These four cost categories made up about two-thirds of recorded timber costs. When average revenues were inflated into 1982 prices and compared to these costs, the same pattern emerged as in other studies. Timbering in the Pacific states accounts for most Forest Service revenues and profits. Operations in the Deep South also show positive net revenues. On the other hand, most states in the Rocky Mountain, Intermountain, and Eastern regions, as well as Alaska and some states in the Northern and Southwest regions, realized revenues well below costs. There were 23 states in all in which average revenues between 1973 and 1983 fell short of direct costs. In 21 of them, revenues were below costs over the decade even when the narrowest concept of separable cost was adopted, counting only the most direct resource support and sale administration expenses. This study therefore strongly supports the evidence reviewed above.

Natural Resources Defense Council and Wilderness Society estimates

A drawback in using figures for whole regions or states to estimate below-cost timber sales is that such aggregate estimates combine sales with positive and negative net returns, and so underestimate the extent of below-cost sales. Forest Service spokesmen sometimes carry this to the national level, netting revenue-generating sales in the Pacific Northwest against below-cost sales in other regions. Studies by the Natural Resources Defense Council (NRDC) and the Wilderness Society examined costs and revenues for each national forest (Barlow et al. 1980; Sample 1984). The NRDC analysis compared receipts and costs for the period 1974–1978 by national forest. Receipts and costs can refer to different sales, under this "cash flow" approach. Although several variants were computed, the estimates presented below included all road costs except those financed by effective purchaser credits, because current road costs on a forest-by-forest basis approximately equal the sum of amortization charges for past construction outlays. However, as in the CRS study cited above, most of the forests characterized as failing to meet efficiency tests do so when road charges are reduced or eliminated altogether. The possible jointness of road costs is not a decisive issue for most

Table 8.3. *National forests that consistently experience below-cost timber sales, by region*

Region 1 (Northern)	Region 2 (Rocky Mountain)	Region 3 (South-western)	Region 4 (Inter-mountain)	Region 5 (Pacific Southwest)	Region 8 (Southern)	Region 9 (Eastern)	Region 10 (Alaska)
Beaverhead	Bighorn–Shoshone	Carson	Ashley	Angeles	Daniel Boone	Chequamegon	Tongass
Bitterroot	Black Hills	Cibola	Boise	Cleveland	Chattahoochee–Oconee	Chippewa	Chugach
Clearwater	Grand Mesa–Uncompahgre–Gunnison	Coronado	Bridger-Teton	Inyo	Cherokee	Huron–Manistee	
Custer	Medicine Bow	Gila	Caribou	Los Padres	George Washington	NFS in Missouri	
Deerlodge	Nebraska	Lincoln	Challis	San Bernardino	Ozark–St. Francis	Mark Twain	
Gallatin	Rio Grande	Prescott	Dixie		NFS North Carolina	Nicolet	
Helena	Arapahoe–Roosevelt	Santa Fe	Fishlake		Croatan	Ottawa	
Lewis and Clark	Routt	Tonto	Humboldt		Uwharrie	Shawnee	
Lolo	Pike–San Isabel		Manti–La Sal		Pisgah	Superior	
Nezperce	San Juan		Salmon		Nantahala	Hiawatha	
	White River.		Sawtooth		Jefferson	Wayne–Hoosier	
			Targhee		Caribbean	Green Mountain	
			Toiyabe			Monongahela	
			Uinta			White Mountain	
			Wasatch-Cache				

Table 8.4. *Acreage of national forests that consistently record below-cost sales, and total national forest acreage, by region*[a]

Region	(A) Regional total of below-cost acres	(B) Regional total of national forest system	Col. A ÷ Col. B (percent)
1	16,635,405	24,017,177	69
2	19,802,858	19,918,565	100
3	14,415,525	20,427,109	71
4	28,773,082	31,087,119	93
5	5,345,722	19,769,805	27
6	—	24,356,651	0
8	6,149,641	12,494,241	49
9	10,907,512	11,412,905	96
10	23,043,437	23,043,437	100
Total	125,073,182	186,527,009	67

[a]See table 8.3 for a listing of national forests in each region.

marginal timber operations. The Wilderness Society's study of forest-by-forest losses in 1982 adopted a similar definition of costs, land receipts, but used bid values instead of actual receipts as the measure of revenues.

Table 8.3 presents a list of national forests by region that lost money over the six years 1974–78 and 1982, and Table 8.4 shows the percent of total acreage they represent. Although there were negative cash flows for individual years in other national forests, only those that showed consistent overall losses were listed. The results duplicate earlier findings. All national forests in Alaska, the Rocky Mountain, and Intermountain regions show losses over the whole period. None of those in the Pacific Northwest do. In the East, almost all national forests consistently lose money on timber sales. In other regions – California, the South, the Southwest, and the North – some forests consistently lose money; others do not. Overall, 125 million acres of national forests, 67 percent of the total system, show a consistent pattern. According to a variety of studies, timber operations in these forests chronically generate less revenues than the direct costs of timber operations, even when conservative estimates of separable costs of operations are adopted.

When taken as a whole, this body of evidence is difficult to discount. The claim that sales are generally economic when viewed as part of a longer-term sequence of timber operations has been repeatedly examined and rejected. The argument that below-cost sales are an artifact of particular years of depressed market demand is also untenable. The defense that below-cost sales are only incidental to efficient timber management in larger areas also fails, because timber operations in entire forests and even regions consistently fail to cover direct costs. The justification that net revenues to government are substantially reduced by restrictions and requirements that serve non-timber objectives is mistaken, because appropriate measures of incremental timber costs *include* the costs of protecting non-timber outputs. The evidence is strong that in many national forests timber operations do fail tests of economic efficiency.

Upward and downward biases in the estimates

Exact tests of the efficiency of Forest Service timber in marginal areas are impossible to carry out, because the accounting and statistical base is far from adequate for that difficult task. Yet, since the underlying issues of forest management are important and pressing, it is also impossible to defer judgments. Continuing current patterns of national forest management until all uncertainties are removed would itself be an important policy decision. Judgments must be made on the basis of imperfect information, such as the studies reviewed above. This involves weighing the likely effects of upward and downward biases in the methodologies and estimates.

There are several sources of bias that may underestimate the net incremental benefits of timber operations. The first is allocating costs to timber accounts that actually *enhance*, not merely protect, non-timber objectives. Forest Service representatives report that this is likely not only because of intrinsic joint costs but also because non-timber expenditures are sometimes budgeted under timber accounts, which are more readily funded. The empirical importance of this bias is difficult to assess, but is significant that in many Rocky Mountain, Intermountain, and Alaskan forests, timber operations fail efficiency tests by large margins when minimal definitions of separable costs are adopted, including only direct sale preparation and administration expenses and excluding all road costs. A second is treating the cost of a road as a current expense, rather than amortizing it over timber operations during its useful life. Again, it is significant that when road costs are omitted altogether, many forests fail incremental efficiency tests. Also, since road expenses in constant dollars have been nearly constant for over a decade, and are projected nearly constant over the next planning period, total annual amortization charges on a forest-wide basis must approximate current capital spending.

A less obvious potential bias is buried in the appraisal system. The

Forest Service may be collecting substantially less than the true stumpage value of timber in the large proportion of its sales that has a single bidder. This negatively distorts efficiency tests based on comparisons of Forest Service timber revenues and direct timber costs. In fact, there may be greater residual value in such timber sold, but that value is transferred to buyers to the extent that winning bids are less than competitive market prices. Each year a substantial fraction of actual sales have only one bidder. Between 1973 and 1979, years for which published data are available, 25 percent of total sales had one bidder, but in regions where most below-cost sales occur – those other than regions 5, 6, and 8 – over 40 percent of timber sales had only one bidder. Naturally, single bidders rarely go above the advertised price. If appraisals underestimate current market values, single bidders in non-competitive sales get a real bargain. In fact, Forest Service statistical equations based on sale characteristics and past competitive bids show that appraised values are consistently low. In implementing its residual value appraisals, the Forest Service bases cost estimates on operations of average efficiency, and often relies on cost and price data that are outdated and inaccurate. Although a balance must be struck, since appraised values that are too close to estimated market prices would result in too many unsuccessful no-bidder sales, Forest Service regions that use alternative, transactions-based appraisals typically approximate market values more closely. Noncompetitive bidding based on appraised values that are well below market values contributes to Forest Service losses, and also underestimates the net benefits of timber operations.

There are also biases in the estimates that overestimate the efficiency of timber operations. The omission of capital charges on timber expenditures has already been discussed. Other biases are less obvious. For example, just as national aggregate Forest Service accounts obscure losses by combining negative returns in some regions and positive returns in others, so *individual* sales often combine timber that can profitably be sold and timber that could not be sold if offered by itself. Each forest contains stands of low-value timber on steep, inaccessible slopes that could not be sold at the Forest Service's legal minimum price (essentially the cost of reforestation plus one dollar per million board feet). These stands are often combined in a single sale with more valuable, more accessible, timber that has commercial value. The Forest Service adjusts the appraised value of these profitable stands downward, so that it can raise the appraised value of the unprofitable timber to the legal base rates. This allows bidders to meet or exceed appraised values and still make a profit on the sale as a whole. The larger sale justifies extending the road network and

Table 8.5. *Volume of national forest timber appraised at negative rates and estimated revenue loss by region*[a]

Region	Total volume (million board feet)	Volume losing[b] (million board feet)	Percentage of volume losing	Lost receipts[c]
1	928,544	690,166	74.3	− 11,787,290
2	268,427	108,092	40.3	− 31,570
3	355,999	59,881	16.8	− 64,480
4	218,231	122,376	56.1	− 718,510
5	1,710,576	908,017	53.1	− 9,897,930
6	4,642,863	1,416,292	30.5	− 17,209,630
10	84,751	53,256	62.8	− 2,621,970
Total	8,209,391	3,358,080	40.9	− 42,331,380

[a]See table 8.3 for a listing of national forests in each region.
[b]Volumes whose indicated advertised rates are less than zero.
[c]Loss of revenue for selling timber for less than indicated rates, equal to indicated rates minus high bid.
Source: O'Toole (1984: 16–17).

also provides greater retained deposits on a larger timber volume. However, the total sales revenue may be less than it would have been had only the high-valued timber been sold at its unadjusted appraised value. According to one observer, it is largely for this reason that the average price of timber sold from national forests in Montana is only half that from state forests (Jackson 1986: 14). The total costs of the sale become much higher, and the returns to the treasury much lower. In some cases, the Forest Service procedure amounts to paying the successful bidder to haul away unsalable timber. According to a Forest Service examination of the reasons and possible remedies for below-cost sales, "In many cases sales are larger than the size that would maximize bidding competition and revenue" (USDA, Forest Service 1985: I–4).

An analysis of this cross-subsidization, based on all 1983 timber sales, showed that no less than 40 percent of the timber sold in the western regions in 1983 was appraised to have a negative stumpage value. That is, according to Forest Service estimates, efficient operators could not harvest it at a profit (O'Toole 1984). Although Forest Service appraisals are often conservative, much of this noncommercial timber could be marketed only by combining it in the same sales with more valuable stands.

A minimum estimate of the resulting losses on such sales compares total sales revenues with what the Forest Service estimated the more valuable species would bring if sold alone. It is a minimum estimate of the revenue loss because winning bids for valuable species often exceed Forest Service appraisals. Also, such a comparison by species does not capture cross-subsidization across the same species at more and less favorable locations. Results are given by region in Table 8.5. They suggest an annual revenue loss of $42 million dollars from cross-subsidization within timber sales. These figures imply that timber harvests could be reduced by perhaps as much as 40 percent by excluding the most uneconomic timber from sales, and federal *revenues* would actually rise by at least $40 million per year. *Net* returns would rise by much more, of course, because the most un-economical sales add disproportionately to costs. Combining these losses with sales that produce net revenues reduces apparent inefficiencies.

Non-timber benefits and below-cost timber sales

In discussing the issue of below-cost sales, the Forest Service dis-counts any attempt to devise incremental efficiency tests and emphasizes that timber operations must be evaluated as part of an overall forest management plan. The Forest Service cites contributions to recreational access; control of fire, disease, and insects; improvement of wildlife hab-itat, forage, and fuelwood production; water management; and commu-nity stability as benefits that more than offset any revenue losses. The Forest Service maintains that both in the aggregate and in individual forests, in carrying out forest plans, below-cost timber sales are economic if total benefits are compared with total costs. The implied efficiency test is that no other set of management activities would yield a greater surplus of *total* benefits over *total* costs.

Table 8.6, reproduced from the Forest Service Annual Report for 1985, illustrates this point. It suggests that total benefits of cutting timber exceed direct costs, excluding roads, in all regions. Roads are regarded by the Forest Service as capital assets, and their costs not written off as a recurrent expense. What is remarkable about this table is the size and variation of non-timber benefits associated with the timber program rela-tive to the value of the timber itself. They range from around 36 percent of timber value in the Pacific and Southern states, where all analyses of below-cost sales show timbering to be profitable, to 174 percent, 300 percent, and 688 percent of timber value in the Intermountain, Rocky Mountain, and Alaskan regions, where all studies show timbering to be unprofitable. The Forest Service claims not only that there are no trade-offs between timber and other outputs from the national forests, but also that the non-timber benefits from logging exceed the timber benefits by large margins in entire regions.[3]

Table 8.6. *Values, costs, and associated outputs for fiscal year 1984 timber sale program (million dollars), by region[a]*

	Region									
	(1) Northern	(2) Rocky Mountain	(3) South-western	(4) Inter-mountain	(5) Pacific Southwest	(6) Pacific Northwest	(8) Southern	(9) Eastern	(10) Alaska	Total
Value of products sold[b]	24.0	4.7	8.5	5.8	88.7	333.7	76.9	19.2	1.7	563.2
Associated non-timber values[c]	28.5	14.1	13.3	10.1	31.9	124.8	27.1	19.7	11.7	281.2
Wildlife and fish	15.1	6.8	6.4	4.7	14.9	59.6	12.5	8.2	5.7	133.9
Recreation	13.1	6.9	6.2	5.0	16.6	64.1	13.8	11.3	5.9	142.9
Range	0.2	0.1	0.1	0.1	0.2	0.8	0.1	0.1	—	1.7
Fuelwood	0.1	0.3	0.6	0.3	0.2	0.3	0.7	0.1	0.1	2.7
Production costs[d]	35.1	12.3	12.3	14.2	62.0	122.1	39.4	21.8	11.7	330.9
Net (value less cost)	17.4	6.5	9.5	1.7	58.6	336.4	64.6	17.1	1.7	513.5
Roads[e]	45.4	19.8	13.3	14.2	48.4	101.9	90.6	24.5	25.1	333.2

[a]Data are for National Forests and Grasslands only. They do not include regional office or Washington office costs.

[b]This is the value of sawtimber, pulp, poles, and miscellaneous products such as posts, fuelwood, and Christmas trees. It does not include road values (purchaser credit or purchaser elected roads) or brush disposal, but does include K-V and salvage sale fund collections. The total value sold includes nonconvertible product value (approx. $1.6 million) and the value of the long-term sale volume released (approx. $1.1 million). These values are not included in the tables.

[c]These represent total quantities of selected outputs associated with the annual timber program, based on constant per million board feet relationships in the 1985 RPA data base, current management alternative. These are the best estimates of field managers. Values per unit output are based on those published in Table F.2, adjusted to 1984 terms, of the draft environmental impact statement for the 1985–2030 Resource Planning Act Program (EIS No. 840007, filed Jan. 11, 1984), except free-use fuelwood which is estimated annually by field managers. A Forest Service task force is currently studying the assignment of such associated outputs, costs, and benefits to the timber program.

[d]These are National Forest costs of producing sawtimber, pulp, poles, and miscellaneous products. These include timber management planning, silvicultural examination, sale preparation, harvest administration, salvage sale activities, resource support to timber and K-V reforestation, and TSI. Not included are general administration, timber management support to other resources, and road costs.

[e]Roads are considered capital assets that have a cost and a value. Included are Forest Service appropriated, purchaser credit, and purchaser elected road construction, and all engineering support expenditures.

Source: U.S. Forest Service Annual Report (1985: 82).

There is considerably more irony in these data than in most government statistics. They imply that the Forest Service, under fire from environmentalists for its inattention to strict business principles, justifies its timber sales by their substantial recreational and environmental benefits. The supposed beneficiaries, including not only the major environmental groups but also fish and game management agencies in many of the states where below-cost sales occur, are vocal in opposition to the benefits they are allegedly receiving, and in fact are suing the Forest Service to stop it from providing them. For example, the Idaho Department of Fish and Game testified as follows before Congress in June 1985:

Because the more profitable timber sites were logged first, many "sales below cost" are in severe sites, on steep slopes, on unstable soils, at higher elevations, and require significant road construction. These characteristics increase the chances of damage to wildlife, retard the rate of recovery of treated areas, and result in the loss of security areas for wildlife. Therefore, adverse impacts to wildlife are usually greater than they are on the better timber sites (Idaho Department of Fish and Game 1985).

In the same hearing, the Wyoming Game and Fish Department testified:

We have serious concerns that center around recent decreases in quality wildlife habitat due to timber harvests, the potential for further reductions due to the proposed increase in timber harvests presented in recent draft forest plans, the methods used to establish timber harvest targets, and the long-term habitat changes that will result from this level of harvest in Wyoming (Wyoming Game and Fish Department 1985).

A spokesman for the National Audubon Society, one of the largest wildlife recreation interest groups, disputed the recreational benefits of timber operations in Congressional testimony:

Now, there are enough roads within national forest boundaries to drive back and forth across the nation fifty times, with two more round trips being added every year. New roads simply shift road-oriented recreation from one place to the other, but do not increase it. And they certainly diminish the experience for every other type of recreationist (Evans 1985).

These quotations illustrate the disbelief of environmental groups and wildlife management agencies in the benefits claimed by the Forest Service from its logging activities. Four general points are raised, and many examples are adduced in support. First, while the Forest Service assumes its road program will result in recreational benefits by improving access, opponents maintain that national forests are already adequately roaded relative to projected recreational needs. For example, the Forest Service management plan for the George Washington National Forest claimed that road construction would raise recreational use, even though present capacity exceeds projected demand by a factor of three, most of the new roads would remain closed to the public, and the roadless areas available for scarcer primitive recreational use would shrink.

Second, the Forest Service claims its timber activities will benefit fish and wildlife and those who hunt or observe them, both by providing open browsing areas and by improving visitors' access. Opponents claim that the loss of undisturbed old-growth forests and tree-sheltered winter range will reduce the available habitat for many species, stream siltation and changes in stream flow will impair fisheries, and the road program will increase hunting pressure on remaining habitats. In the Bitterroot National Forest in Montana, for example, the state Department of Fish and Wildlife argued that further road construction would reduce elk herds significantly.

Third, the Forest Service claims that timbering will raise water runoff and yields from many watersheds and increase the availability of water to downstream users. Opponents argue that soil erosion and increased sediment loading will decrease water quality, and the increase in seasonal variation in stream flows will impose additional costs on downstream users. Moreover, the increased water runoff typically occurs in the early spring, when its marginal value in agriculture is low. There is no downstream storage capacity below much of the Shoshone National Forest. Yet, the Forest Service assumed that increased runoff would be worth $45 per acre-foot, an estimate of the value of *delivered* irrigation water on the farm.

Fourth, the Forest Service argues that road construction associated with timber harvests lowers the costs of forest protection activities, and below-cost salvage and rehabilitation harvests prevent the loss of valuable timber from pest infestation and fire. Opponents claim that the timber lost in marginal sites has no economic value, and if harvested at maturity, would actually result in economic losses. Moreover, they adduce data that fire protection costs are higher in roaded areas due to greater risks of larger fires caused by human visitors. They also cite research findings that pests and disease often perform a natural thinning and culling function, with no significant adverse effect on stand growth or value.

These arguments depend on quantitative environmental assessments and are difficult to resolve in general (LeMaster et al. 1987). Reviews of dozens of proposed forest management plans by groups representing supposed beneficiaries have repeatedly challenged the non-timber benefits claimed by the Forest Service. Their concern is not whether the non-timber benefits are sufficiently large to justify the considerable net losses from timber operations in marginal regions, but whether impacts, on balance, are beneficial or damaging. Given the opposition of supposed beneficiaries, it is difficult to believe that the net benefits are as large as Forest Service data indicate.

Contributions to community stability

Another repeated justification for below-cost sales is economic support for local industry and the stability of communities dependent on timbering activities in national forests. Sawmills and manufacturing plants have been established and jobs created on the basis of a continued flow of timber from national forests. Reduction or cessation of below-cost sales would drive many of these industries out of business and destabilize the local economy, it is maintained. A special concern for small firms leads the Forest Service to set aside sufficient sales in each market area to ensure that small firms (defined as independent logging or milling companies with fewer than 500 employees) attain their historical share of purchases.

A 1983 joint study by the Forest Service and the Small Business Administration examined this set-aside program and the participation of large and small firms in all Forest Service timber sales from 1975 to 1979 (U.S. Forest Service and Small Business Administration, 1983). Table 8.7 presents some summary results. It shows that large firms, those that employ more than 9,500 workers, bought more than half the timber sold by the Forest Service over the years 1973–1979, and harvested 55 percent. If the Pacific and Southern regions (regions 5, 6, and 8), where below-cost sales are less prevalent, are excluded, the predominance of large firms is greater. In the remaining regions, characterized by chronic below-cost timber sales, large firms purchased and also harvested 65 percent of the timber. So most of the economic subsidy goes to large firms. In fact, the largest firms, those with more than 1,000 employees, bought 42 percent of all the timber sold between 1973 and 1979; the smallest, with fewer than 100 employees, bought only 18 percent.

The jobs supported by below-cost timber sales are surely vitally important to those who hold them, their families, and local communities. Yet the federal government has not preserved jobs in other industries against market forces, and federal subsidies are no guarantee of job stability. There have been substantial shifts in employment in the timber industry in the last decade. They resulted both from broad economic factors like recession, high interest rates, and an overvalued exchange rate that made Canadian imports cheap, and from regional shifts in the industry to the South.

From a state-wide perspective, these employment shifts are small. In 1983, as Table 8.8 shows, in the 23 states where timber sales are consis-

Table 8.7. *Volume and percentage of sawn timber harvested and purchased by small and large business, during study period, by region (million board feet)*

	Small business	Percent	Large business	Percent	Total
Harvested FY 1975–79					
Region 1	1,309	32	2,844	68	4,153
Region 2	358	43	480	57	838
Region 3	327	29	811	71	1,138
Region 4	519	34	1,008	66	1,527
Region 5	2,756	35	5,056	65	7,812
Region 6	9,310	52	8,577	48	17,887
Region 8	1,486	64	829	36	2,315
Region 9	395	82	84	18	479
Region 10	114	20	456	80	570
Total	16,574	45	20,145	55	36,719
Purchased FY 1973–79					
Region 1	1,919	33	3,963	67	5,882
Region 2	790	51	761	49	1,551
Region 3	441	28	1,154	72	1,595
Region 4	818	39	1,287	61	2,105
Region 5	4,313	37	7,405	63	11,718
Region 6	16,130	57	12,136	43	28,266
Region 8	2,190	65	1,195	35	3,385
Region 9	882	80	223	20	1,105
Region 10	238	37	409	63	647
Total	27,721	49	28,533	51	56,254

Source: U.S. Forest Service and Small Business Administration (1983).

tently below cost, employment in logging and sawmilling exceeded 1 percent of total state private sector employment only in Idaho, where it was 4 percent, and in Maine, where it was 2.1 percent. Similarly, past studies have not found that payments to counties are essential to their economic stability. Like timber revenues themselves, revenue-sharing payments to local communities are heavily concentrated in the Pacific and Southern states. Most counties in the states listed in Table 8.8 receive timber revenues less than the $0.75 per acre ceiling on federal payments in lieu of taxes on federal lands to local governments. Since timber revenue-sharing is offset against these payments, for these counties a drop in revenues is fully made up by higher payments in lieu of taxes and total receipts are unaffected.

In any case, an analysis of the distribution of timber revenue-sharing indicates that these contributions go predominantly to communities that are better off. Table 8.9, from a 1978 study by the Advisory Commission

Table 8.8. *Timber industry and total private employment in below-cost sale states*

	(A) Employment in logging and sawmills	(B) Total private employment	Col. A. ÷ Col. B (percent)
Alaska	1,099	149,860	0.733
Colorado	797	1,079,087	0.074
Idaho	9,828	244,131	4.026
Illinois	471	3,717,194	0.013
Kentucky	2,730	903,733	0.302
Maine	7,178	339,526	2.114
Michigan	2,728	2,591,941	0.105
Minnesota	1,292	1,405,677	0.092
Missouri	97	1,578,421	0.006
Montana	1,660	204,174	0.813
Nevada	13	342,830	0.004
New Hampshire	445	348,102	0.128
New Mexico	1,237	351,994	0.351
New York	2,897	5,980,849	0.048
North Carolina	11,867	1,985,206	0.598
South Dakota	659	172,270	0.383
Tennessee	4,006	1,392,223	0.288
Utah	14	918,615	0.002
Vermont	1,351	165,557	0.816
Virginia	7,352	1,665,628	0.441
Wisconsin	3,734	1,523,754	0.245
Wyoming	940	148,488	0.633
West Virginia	3,177	437,835	0.726
Total	65,572	27,647,095	0.237

Source: U.S. Department of Labor (1983).

on Intergovernmental Relations, shows the distribution of payments among recipient counties. The 235 counties with the highest median family income, 36 percent of all recipient counties, received 80 percent of all payments in 1976. Using 1983 data and concentrating only on the states listed in Table 8.8, there was still a positive correlation between per capita income and per capita timber payments across all the counties that share in the revenues.

In a state-by-state comparison of counties that received timber revenues with those that did not, Table 8.10 shows that in all the Western states where below-cost sales are chronic, median family income, on average, is higher in the counties that receive timber revenues than in those that do not. At best, below-cost sales are contributing to the welfare and stability of communities that are already better off. In the eastern part of

Table 8.9. *Distribution of national forest receipt sharing payments (1976) to counties grouped by median family income, 1970*

	Median family income					
	$6,000 or less	$6,000–7,000	$7,000–8,000	$8,000–9,000	$9,000 or more	Total[a]
Number of counties	165	112	140	153	82	652
Actual national forest payment, 1976 (thousand dollars)	5,159	3,026	12,125	49,994	30,355	100,658
Forest payments as a percent of income within counties grouped by family income levels	0.17	0.08	0.18	0.33	0.06	0.13
Payments as a percent of total national forest payments	5.13	3.01	12.05	49.67	30.16	100

[a]Total may vary slightly due to rounding.
Source: Advisory Commission on Intergovernmental Relations (1978).

the country, and in California, the findings are reversed. Counties that do not receive timber revenues are generally more urban and industrial and have higher incomes than those that do receive them. But, in these states as well, any loss in timber revenues would be fully offset for almost all counties by higher payments in lieu of taxes.

Fiscal impacts

Within the broader framework for evaluation that the Forest Service recommends, the evidence in support of the large claimed non-timber benefits associated with below-cost timber sales is weak. That is not to say that such benefits are unlikely to exist. Undoubtedly, some below-cost timber sales do contribute to overall forest management and to the well-being of dependent communities. Rather, the evidence fails to support the proposition that such benefits are sufficiently large and systematic to offset the widespread and significant economic losses from timber operations.

This section attempts a rough quantification of the magnitude of such losses, subject to the shortcomings in the information base discussed above. The results provide a conservative estimate of the net non-timber

Table 8.10. *Comparison of median family income between counties that do and do not receive timber revenues in below-cost sale states, 1983*

	Counties receiving payments		Counties not receiving payments	
	Average income	Number of counties	Average income	Number of counties
Region 1				
Idaho[a]	16,445	34	15,355	10
Montana	16,736	34	16,198	23
Region 2				
Colorado	19,153	42	16,254	22
South Dakota	16,508	6	13,532	60
Region 3				
New Mexico	15,078	21	14,209	11
Region 4				
Idaho[a]	16,445	34	15,355	10
Nevada	19,367	14	17,746	3
Utah	17,818	29	—	0
Wyoming	21,465	19	20,273	4
Region 5				
California	18,033	40	21,632	18
Region 8				
Kentucky	11,158	23	15,211	97
North Carolina	14,443	23	15,593	77
Tennessee	14,270	10	14,642	85
Virginia	16,139	30	17,956	65
Region 9				
Illinois	15,486	10	19,957	92
Maine	15,233	1	15,276	15
Michigan	16,181	32	19,150	51
Minnesota	17,870	7	17,554	80
Missouri	13,322	29	15,655	85
New Hampshire	16,442	3	19,242	7
Vermont	16,972	6	15,925	8
West Virginia	14,029	13	16,659	42
Wisconsin	14,853	11	18,643	61
Region 10				
Alaska	29,422	8	22,565	15

[a]Idaho is the only state to be split into two Forest Service regions; the panhandle is in region 1, and the rest of Idaho is in region 4. The totals for Idaho are listed in *each* region.

Table 8.11. *Timber receipts and costs on national forests consistently experiencing below-cost timber sales (thousand dollars, current prices)*

	Receipts	Allocated costs	Net	Percentage loss
Region 1				
1974–78	149,990	202,255	− 52,265	26
1982	16,833	56,833	− 40,000	70
Total	166,823	259,088	− 92,265	36
Region 2				
1974–78	31,299	89,293	− 57,994	65
1982	1,937	34,026	− 32,089	94
Total	33,236	123,319	− 90,083	73
Region 3				
1974–78	25,467	46,091	− 20,624	45
1982	1,166	10,091	− 8,925	88
Total	26,633	56,182	− 29,549	53
Region 4				
1974–78	90,627	100,022	− 9,395	9
1982	4,175	31,268	− 27,093	87
Total	94,802	131,290	− 36,488	28
Region 5				
1974–78	3,422	5,730	− 2,308	40
1982	336	1,540	− 1,204	78
Total	3,758	7,270	− 3,512	48
Region 8				
1974–78	34,912	79,205	− 44,293	56
1982	8,813	25,002	− 16,189	65
Total	43,725	104,207	− 60,482	58
Region 9				
1974–78	30,683	94,800	− 64,117	68
1982	10,374	38,071	− 27,697	73
Total	41,057	132,871	− 91,814	69
Region 10				
1974–78	61,618	128,064	− 66,446	52
1982	3,390	41,583	− 38,193	92
Total	65,008	169,647	− 104,639	62
Overall totals				
1974–78	428,018	745,460	− 317,442	43
1982	47,024	238,414	− 191,390	80
Total	475,042	983,874	− 508,832	52

Sources: Barlow et al. (1980); Sample (1984).

benefits necessary to justify existing harvesting and land-use policies. Table 8.11 is based on the accounts for the national forests listed in Table 8.3, only those that consistently sell timber below narrowly defined direct costs (Table 8.11). Averaged over all these forests and the six years 1974–78 and 1982, estimated excesses of direct costs over sales revenues were 52 percent of those costs. The total loss exceeded $500 million for the six years, or about $85 million per year. By region, the percentage loss ranges from zero in the Pacific Northwest to 73 percent in the Rockies, and 69 percent in the eastern national forests.

Conclusions

This chapter demonstrates the complexities of managing forests for multiple uses in high-income countries, where diverse interests compete for the use of available resources more through political and bureaucratic means than through direct market competition. Forest planning and management must contend with interdependencies among uses, among regions, and over time – and must function in a highly charged political atmosphere. Economic efficiency is only one of the criteria that the Forest Service has been directed to apply in fulfilling its responsibilities and has until now not been a prominent criterion, despite urgings in this direction by Congress and top management. Other criteria at the forest management level include service to local communities and important users, adherence to professional forestry practices and standards, fulfillment of authorized targets, and promotion or protection of Forest Service programs.

Despite the difficulties in evaluation, the weight of evidence indicates substantial economic inefficiencies in national forest management, of which timber operations in marginal areas are one manifestation. Large areas in the Rocky Mountain, Intermountain, Alaskan, and Eastern regions that are economically unsuitable for commercial timber production continue to be used for this purpose, at a substantial economic cost unlikely to be offset by net non-timber benefits. Significant efficiency gains could be achieved if timber operations were subordinated to other objectives in such areas.

A range of remedies is available. The narrowest, which could be initiated by administrative action, would be to put minimum bid restrictions on timber sales that ensured recovery of direct, separable, production costs, with waiver provisions that required a positive showing by forest managers that non-timber benefits or other reasons justified such below-cost sales. The Department of Agriculture has moved in this direction by remanding forest plans for four western forests for further study and demonstration that broader objectives justified such sales (USDA, Office of the Secretary 1985).

A broader approach would be to interpret and implement Section 6(k) of the National Forest Management Act in such a way as to remove more of these economically unsuitable areas from the timber base prior to the development of forest management plans, thus ensuring that timber operations on those lands would be incidental to the attainment of other objectives. The effect of such action on the annual allowable cut, and thereby on the need for reexamination of other restrictions on harvesting – especially non-declining even-flow provisions – would have to be considered were this broader approach followed.

A still broader approach would attempt to deal with the underlying forces that promote uneconomic timber operations. Important among these is the predominant contribution of timber revenues, deposits, and credits in providing financial resources for a wide range of forest management activities. Also important is the weakness of the incentives forest managers have to pursue economic efficiency when it comes into conflict with other objectives (USDA, Forest Service 1986).[4] Dealing with these forces would require two general changes. The first would be to reduce the budgetary weight of timber and the distinction between marketed commodity outputs from the national forests by installing market-based user charges for a wider range of recreational and other services: fishing, hunting, camping, hiking, fuelwood extraction, and so on. While many of the benefits produced by the national forests cannot be marketed, particularly their protective functions (and this is the main reason why national forests are public institutions), many can be and are already routinely marketed by private, state, and, to some degree, national forests. Broadening the policy of charging market prices to direct beneficiaries and users of forest services would dilute the inevitable tendency of managers to give high effective priority to those activities that carry most of the budgetary freight. Recreational, wildlife, and other uses would receive more managerial emphasis as their demonstrated value and contribution to revenues increased.

Such a proposed change would undoubtedly seem heretical to users who feel themselves entitled to "free" access. However, providing recreational services from the national forests entails real cost now being borne by the general taxpayer, just as much as do below-cost timber sales. Moreover, provision of non-timber services is subject to the same efficiency test applied to the timber program: whether at the margin the benefits exceed the separable or avoidable costs. This efficiency test is even more difficult to apply to non-marketed benefits. Moreover, as a group, recreational users of the national forests are not a disadvantaged group, relative to the average user of wood or paper products.

The second change is to increase the effective incentives of forest managers to respond to relative costs and returns in supplying the multiple outputs from the national forests by making the forests more self-sufficient financially, reducing their dependence on appropriated funds, and increasing the importance of retained net revenues in their total budgets. Their present reliance on a combination of appropriated funds and retained gross revenues does not provide adequate incentives for economic efficiency. There are many ways in which this recommendation could be carried out, by budgetary and legislative restructuring, or by the creation of public corporations for national forest management. Without such a reorientation, however, improvements in operating management's concern for economic efficiency will be slower and less certain.

Endnotes

1. Assumptions embodied in Forest Service plans and planning studies assume considerably faster rises in timber prices than in costs in future periods, raising the profitability of future harvests.

2. Given its assumptions regarding price and cost trends, the Forest Service study found that efficiency would be improved if current timber operations were reduced and road investments delayed (Scheuter and Jones 1985).

3. Efficiency actually *requires* that there be trade-offs at the margin between timber and non-timber benefits in regions such as the Pacific where timber revenues cover direct costs. Otherwise, total net benefits could be increased by harvesting more timber.

4. This weakness was identified by the Forest Service's own review of below-cost sales in four national forests: "In general, the existing system is designed to reward managers for achieving goals established by Congress and the Administration, and these goals have little to do with cutting costs and raising revenues. . . . The incentive system has tended to stress target attainment and budget compliance rather than revenue enhancement" (USDA, Forest Service 1985: 1-4).

References

Advisory Commission on Intergovernmental Relations. 1978. *The Adequacy of Federal Compensation to Local Governments for Tax-Exempt Federal Land.* Washington, D.C., July.

Barlow, Thomas, et al. 1980. *Giving Away the National Forests.* Washington, D.C.: Natural Resources Defense Council, June.

Benson, Robert E., and Michael J. Niccoluci. 1985. Costs of Managing Non-timber Resources When Harvesting Timber in the Northern Rockies. Ogden, Utah: USDA, Forest Service, Intermountain Research Station.

Beuter, John H. 1985. *Federal Timber Sales*. Washington, D.C.: Congressional Research Service, Library of Congress.

Clawson, Marion. 1976. *The Economics of National Forest Management*. Baltimore, Maryland: Johns Hopkins University Press for Resources for the Future.

Deacon, Robert T., and M. Bruce Johnson, eds. 1985. *Forestlands Public and Private*. Cambridge, Massachusetts: Ballinger Publishing for the Pacific Institute for Public Policy Research.

Dowdle, Barney, and Steve H. Hanke. 1985. Public Timber Policy and the Wood Products Industry. In Robert T. Deacon and M. Bruce Johnson, eds., *Forestlands Public and Private*. Cambridge, Massachusetts: Ballinger Publishing for the Pacific Institute for Public Policy Research.

Evans, Brock, Vice-President for National Issues, The National Audubon Society. 1985. Statement before the House Interior Committee, Subcommittee on Public Lands. Washington, D.C., June 11–13.

Gorte, Ross W., and Pamela Baldwin. 1986. *The Timberlands Suitability Provision of the National Forests Management Act of 1976*. Washington, D.C.: Congressional Research Service, Library of Congress.

Helfand, Gloria E. 1983. *Timber Economics and Other Resource Values: The Bighorn-Weitas Roadless Area, Idaho*. Washington, D.C.: The Wilderness Society.

Hyde, William F. 1981. Timber Economics in the Rockies: Efficiency and Management Options. *Land Economics*, Vol. 57, No. 5: 630–637.

Hyde, William F., and John V. Krutilla. 1979. The Question of Development or Restricted Use of Alaska's Interior Forests. *The Annals of Regional Science*, Vol. 10, No. 1.

Idaho Department of Fish and Game. 1985. Statement before the House Interior Committee, Subcommittee on Public Lands. Washington, D.C., June 11.

Jackson, David H. 1986. Below-Cost Sales: Causes and Solutions. *Western Wildlands*, Vol. 12, No. 1: 11–15.

Johnson, Ronald N. 1985. U.S. Forest Service Policy and Its Budget. In Robert T. Deacon and M. Bruce Johnson, eds., *Forestlands Public and Private*. Cambridge, Massachusetts: Ballinger Publishing for the Pacific Institute for Public Policy Research.

Krutilla, John, and Michael Bowes. In press. Below Cost Timber Sales and Forest Planning. In *The Economics of Multiple Use Forestry*. Baltimore, Maryland: Johns Hopkins University Press for Resources for the Future.

LeMaster, Dennis C., Barry R. Flamm, and John C. Hendee, eds. 1987. *Below Cost Timber Sales*. Proceedings of conference sponsored by The Wilderness Society, the University of Washington, and the University of Idaho, Spokane, Washington, February 17–19, 1986. Washington, D.C.: The Wilderness Society.

Minckler, Leon. 1985. *Review of the Final Hoosier Forest Plan and EIS*. Eugene, Oregon: CHEC, Inc.

Nelson, Robert H. 1982. The Public Lands. In Paul R. Portney, ed., *Current Issues in Natural Resource Policy*. Baltimore, Maryland: Johns Hopkins University Press for Resources for the Future.

O'Toole, Randal. 1984. Cross Subsidization – The Hidden Subsidy. *Forest Planning*, May.

Sample, V. Alaric, Jr. 1984. *Below-Cost Timber Sales on the National Forests.* Washington, D.C.: The Wilderness Society, July.

Scheuter, Ervin G., and J. Greg Jones. 1985. Below-Cost Timber Sales: Analysis of a Policy Issue. Ogden, Utah: USDA, Forest Service, Intermountain Research Station.

Sog, Cindy F., and John B. Loomis. 1984. Empirical Estimates of Amenity Forest Values: A Comparative Review. Fort Collins, Colorado: USDA, Forest Service, Rocky Mountain Forest and Range Experimentation Station, March.

USDA, Forest Service. 1985. *Analysis of Costs and Revenues in Four National Forests.* Washington, D.C.

USDA, Forest Service. 1986. *Timber Sale Program Information Reporting System: Draft Report to Congress.* Washington, D.C.

USDA, Office of the Secretary. 1985. Decision on review of administrative decisions by the Chief of the Forest Service related to administrative appeals of the forest plans and EISs for the San Juan National Forest, and the Grand Mesa, Uncompahgre, and Gunnison National Forest. Washington, D.C.

U.S. Department of Labor. 1983. *Handbook of Labor Statistics, 1983.* Washington, D.C.

U.S. Forest Service Annual Report. 1985, for Fiscal Year 1984. Washington, D.C.: U.S. Government Printing Office.

U.S. Forest Service and Small Business Administration. 1983. *Small Business Timber Sale Set-Aside Program.* Washington, D.C., August.

U.S. General Accounting Office (GAO). 1984. *Congress Needs Better Information on Forest Service's Below-Cost Timber Sales.* Washington, D.C., June.

Wilkinson, Charles F., and H. Michael Anderson. 1985. Land and Resource Planning in the National Forests. *Oregon Law Review,* Vol. 114, Nos. 1 and 2.

Wolf, Robert E., Assistant Division Chief, Environment and Natural Resources Policy Division. 1980. *State-by-State Estimate of Situations Where Timber Will be Sold by the Forest Service at a Loss or a Profit.* Washington, D.C.: Congressional Research Service, Library of Congress, June.

Wyoming Game and Fish Department. 1985. Statement before the House Interior Committee, Subcommittee on Public Lands. Washington, D.C., June 11.

9   Conclusion: findings and policy implications

MALCOLM GILLIS AND
ROBERT REPETTO

Principal findings

Some low-income nations are confronting a shortage of wood; the world is not. The world is, however, facing a decline of natural tropical forests. Even as policy-makers begin to understand the value of natural tropical forests, they are rapidly shrinking and deteriorating. The studies underlying this volume were neither intended nor needed to verify these problems. Other scientists have clearly established the extent of forest decline and likely economic, social, and environmental consequences (Brown et al. 1985; Eckholm 1976; Fearnside 1982; Grainger 1980; Lanly 1982; Myers 1980, 1984, 1985; Spears 1979).

Many of these and other studies (Allen and Barnes 1985; Bunker 1980; Ehrlich and Ehrlich 1981; Plumwood and Routley 1982; Tucker and Richards 1983) have also established the principal causes of deforestation. They have identified three major outgrowths of population growth and rural poverty – shifting cultivation, agricultural conversion, and fuel-wood gathering – as threats to natural forests in the Third World, along with the impacts of large development projects. Virtually all previous studies have also focussed upon commercial exploitation, including logging and land-clearing for ranches, as major sources of the problem.

The studies contained in this volume support these findings. In addition, the authors have identified government policies that have significantly added to and exacerbated other pressures leading to wasteful use of natural forest resources including those owned by governments themselves. We emphasize the policy dimension, because changes in policy can substantially reduce resource wastage. Our emphasis on wasteful use in the economic sense does not ignore or minimize the importance of non-economic objectives underlying forest policies. Rather, our research implies that policies leading to economic waste have also undermined conservation efforts, regional development strategies, and other socioeconomic goals.

Separating the effects of government policies from those of other causes is impossible, because policy-induced exploitation interacts with other pressures on natural forests: forests opened by loggers encouraged by liberal concession terms are more accessible to shifting cultivators; government-sponsored settlements also attract spontaneous migrants; development policies that worsen rural poverty lead to more rapid encroachment on forest lands. The indirect effects of government policies are substantial. Despite this complexity, important conclusions can be distilled from the case studies of the preceding chapters.

A basic conclusion is that wastage of publicly owned natural forests has been widespread and long-standing. To an extent heretofore unappreciated, these outcomes have been the avoidable consequences of government policies.

The policies in question include not only those formulated by government agencies nominally responsible for oversight of forest utilization ("forestry policies") but a wide range of other policies designed to serve broader governmental goals ("non-forestry" policies) but which nevertheless have badly undermined the value of natural forest assets.

Forestry policies include those pertaining to the terms of timber harvest concessions, such as their duration, permissible annual harvests and harvest methods, levels and structures of royalties and fees, policies affecting utilization of non-wood forest products, and policies toward reforestation. The 12 case studies included in the foregoing chapters (including four in West African countries and three in the Malaysian states) furnish ample evidence that forestry policies have provided strong incentives for wasteful use of natural forest resources. Inadequate use of royalties and other charges to collect the economic rents potentially available to harvesters of mature timber have set off "timber booms" and scrambles for short-term profits. In most of the countries studied, including Indonesia, Malaysia, the Philippines, Ivory Coast, and Ghana, the result has been rapid, careless timber exploitation that outran both biological knowledge and administrative capacity for sustainable forest management.

Timber concessions have generally been too short in duration to allow loggers to conserve forest values even if they were inclined to do so. Harvesting methods, particularly the variants of selective cutting prescribed or allowed by most tropical countries, have undermined forest quality. The structure of timber royalties has promoted excessive mining (high-grading) of forest value and yielded too few revenues for government as resource owners. Where reforestation policies aimed at regeneration and restoration of commercial mixed forest stands have been tried, they have proved largely ineffective.

In many countries non-forestry policies have caused greater forest destruction than misdirected and misapplied forestry policies. Non-forestry policies prejudicial to forest conservation may be arranged on a continuum that ranges from self-evident to subtle. Most obvious are the effects of policies leading directly to physical intrusion in natural forest areas. These include agricultural programs that clear forest land for estate crops such as rubber, palm oil, cacao (Chapters 3, 4, and 7), for annual crops (Chapters 2, 4, and 7), and even for fish ponds (Chapter 4). Closely related are investments in mining, dams, roads, and other large infrastructure projects that incidentally result in significant, once-and-for-all destruction of forest resources (Chapters 5 and 7). Many such projects are politically driven and of questionable economic worth, even apart from the forest and other natural resource losses they impose.

Further along the continuum are tax, credit, and pricing policies that stimulate private commercial investment in forest exploitation, whether in logging or timber processing. Such policies have induced timber harvesting in excess of rates that otherwise would have been commercially profitable (Chapter 8). One step removed are policies that stimulate private investments in competing land uses, such as ranching, farming, or fish culture. The principal instruments of fiscal and monetary policies contributing to forest destruction are generous tax treatment (Chapters 3, 4, 6, and 7), heavily subsidized credit (Chapters 4, 6, and 7), and direct government subsidy (Chapters 2, 5, and 8).

Next on the continuum are land tenure policies that encourage deforestation. Of these, the most direct are tenurial rules that assign property rights over public forests to private parties on condition that such lands are "developed" or "improved." Such rules have facilitated small-farmer expansion into forested regions (Chapters 3 and 7), but in some countries have been used by wealthier parties to amass large holdings (Chapters 4 and 6). A few countries have demonstrated that this policy works in reverse, by awarding private tenures to *deforested* public wastelands on condition that they be reforested (Chapter 5).

A more indirect tenurial policy has been the centralization of proprietary rights to forest lands in national governments, superceding traditional rights of local authorities and communities. Although intended to strengthen control, such actions have more often undermined local rules governing access and use, removed local incentives for conservation, and saddled central governments with far-flung responsibilities beyond their administrative capabilities.

Finally, the furthest points on the continuum represent those policies that appear at first glance to have few implications for forest use, but which ultimately prove to be significant sources of policy-induced forest destruction. Included here are all domestic policies that further impoverish households living close to the margin of subsistence, especially in rural areas (Chapters 2, 3, 4, 6, and 7). These include pricing policies and investment priorities biased against the agricultural sector, development strategies that depress the demand for unskilled labor, and agricultural policies that favor large farms over smallholders. These policies retard the demographic transition, make rural populations more dependent on natural forests for subsistence needs, and increase the concentration of agricultural landholdings.

This emphasis on the policies of national governments responsible for their public forest lands does not imply that external agencies have been blameless in the long-term misuse of natural forests in developing countries. It has long been apparent that trade barriers protecting wood-processing industries in the United States, Japan, Europe, and Australia have usurped many of the benefits from forest-based industrialization that could have accrued to poor countries, while inducing Third World government to take strong, sometimes excessive, countermeasures (Chapter 1). And multinational timber enterprises from industrial countries were among the primary beneficiaries of log extraction in tropical nations at least into the 1970s, although these large multinationals have steadily withdrawn from the forests of developing countries. By 1980 they had completed divestment of virtually all logging operations in Southeast Asia and West Africa, their places having been taken by smaller multinationals from developing countries and by firms owned by domestic entrepreneurs (Chapters 2, 3, and 7). The image of the now-departed multinational firms, popularly identified as the principal engines of forest destruction for so long, no longer obscures the part played by domestic government policies.

There is little doubt that policies of governments have been inimical to the rational utilization of valuable forest resources. Why have these policies been adopted, and why do they persist? In many countries the policies have been deliberately intended to reward special interest groups allied with or otherwise favored by those in power. The existence of large resource rents from harvesting mature timber has attracted politicians as well as businessmen to the opportunities for immediate gain.

That is not the whole story, however. To a considerable degree, the policy weaknesses identified in this study arose despite well-intentioned development objectives. The shortcomings have been failures of under-standing and execution. Distillation of the lessons of the foregoing chap-ters provides six reasons that help to explain why government policies have erred in the direction of excessive depletion of natural forest re-sources. The six are first presented in summary form, then discussed at greater length.

1. The continuing flow of benefits from intact natural forests has been consistently undervalued by both policy-makers and the general public.
2. Similarly, the net benefits from forest exploitation and conversion have been overestimated, both because the direct and indirect economic bene-fits have been exaggerated and because many of the costs have been ignored.
3. Development planners have proceeded too boldly to exploit tropical forests for commodity production without adequate biological knowl-edge of their potential or limitations or awareness of the economic conse-quences of development policies.
4. Policy-makers have attempted – without much success – to draw on tropical forest resources to solve fiscal, economic, social, and political conflicts elsewhere in society.
5. National governments have been reluctant to invest the resources that would have been required for adequate stewardship of the public re-source over which they asserted authority.
6. National governments have undervalued the wisdom of traditional for-est uses and the value of local traditions of forest management that they have overruled.

Natural forest endowments remain undervalued in all countries stud-ied, not only by the general public but by governments as owners or as regulatory authorities and by international institutions. An asset that is undervalued is an asset that will inevitably be misused.

Forest exploitation has concentrated on a relatively few valuable com-modities, neglecting other tangible and intangible values. Natural forests everywhere serve protective as well as productive functions. Assigning money values to the protective services is much more difficult than esti-mating the market value of timber harvests. This difficulty accounts for much of the worldwide tendency to undervalue natural forest assets, but, in addition, potential production has also been undervalued. Forests in the tropics have generally been exploited as if only two resources were of any significance: the timber and the agricultural land thought to lie be-neath it. A third resource has been overlooked virtually everywhere: the capacity of the natural forest to supply a perpetual stream of valuable non-wood products that can be harvested without cutting down trees. In the tropics, these include such commodities as nuts, oils, fibers, and plant

and animal products with special uses (Chapter 2). In advanced temperate zone countries, forests' recreational value is often underestimated
(Chapter 8).

Second, the timber and agriculture products expected to flow from log
harvests and clearing of natural forests have been overvalued. The common expectation that tropical logs could be harvested every 35 years in
cut-over stands was based on over-optimistic expectations of the rate and
extent of regeneration and has been grossly unfulfilled in Indonesia, the
Philippines, and other countries studied. Assumptions about the agricultural potential from land underlying tropical forests have been even
more optimistic, and results have been even more disappointing, particularly in the Indonesian transmigration program (Chapter 2) and in the
Brazilian schemes for promoting Amazonian development through
large-scale cattle ranching (Chapter 6).

Policy-makers have usually overestimated the employment and regional development benefits associated with timber industries, infrastructure
investments, and agricultural settlements in tropical forests. Where such
initiatives have not been economically sound to begin with, they have not
induced further development or even been able to sustain themselves
without continuing dependence on government subsidies. However, one
result of such inflated expectations has been that governments, both as
holders of property rights in forests and as sovereign taxing authorities,
have often allowed even the timber that is readily valued in world markets
to be removed too cheaply. This problem is reflected in persistently low
timber royalties and license fees and unduly low – sometimes zero –
income and export taxes. Only in Sabah (and then only after 1978) have
governments been moderately successful in appropriating a sizeable
share of the rents available in logging (Chapter 3). Elsewhere, sizeable
rents available to timber concessionaires have generated destructive timber booms and pressures for widespread, rapid exploitation (Chapters 2,
3, 4, 5, and 7).

Nor have the employment benefits expected from forest utilization
been realized. The wood products industries in tropical countries have
provided some employment, to be sure, but with the single exception of
Gabon, the timber sectors in tropical wood exporting nations have typically provided jobs for less than 1 percent of the labor force, a figure only
as high as in the diversified United States economy (Chapter 8) and half as
high as in Canada. The drive to expand employment, domestic value-
added, and foreign exchange earnings has led to strong protection of
domestic processing by banning log exports and imposing high export
taxes on logs but not on timber products. Stress on forest-based indus-

trialization has also lead to tax and credit incentives for sawmills, plymills, and even pulp and paper projects that are inappropriate for short-fibered tropical hardwoods. Jobs in the forest-based sector were indeed created, but at very great cost to the nation. Large amounts of taxes and foreign exchange earnings were dissipated, and some of the domestic industries that were sheltered became inefficient claimants to more of the forests than foreign mills could ever command (Chapters 2, 4, and 7).

Along with overestimated benefits, there has been a pervasive tendency to underestimate not only the economic but also the social and environmental costs of forest exploitation. Although some of these costs have by now been well documented (Chapters 5 and 8), they have nowhere been taken as an explicit offset to putative benefits from forest use. The destruction of habitat that threatens myriad little-known species endemic to tropical forests and the displacement or disturbance of indigenous communities have been especially neglected costs (Chapters 3 and 6). In addition, the costs of "boom-town" development that often arise from intensified logging and processing activities were overlooked in the drive to promote development of lagging or backward regions (Chapters 2 and 6). In Indonesia, Brazil, and elsewhere, neither the infrastructure costs of providing for large inflows of immigrants to timber provinces in the early stages of timber booms nor the costs of maintaining excessive infrastructure in post-boom periods were viewed as offsets to the private economic benefits flowing from the opening of the natural forest. Instead, they were viewed as investments in regional development that could be financed through timber sales.

In some areas, large-scale extractive activity in natural forests has imposed heavy environmental costs. Misused, fragile tropical soils have been seriously damaged over large areas, for example. In Indonesia and the adjacent East Malaysian state of Sabah, these hitherto unforeseen costs reached calamitous heights in 1983 (Chapters 2 and 3). In that year of severe drought, fire in the moist tropical forests in both countries burned an area about one and a half times that of Taiwan. Previous droughts had brought fire, but on a far smaller scale and with minimal damage. But extensive logging in both areas had by 1983 predisposed even the wet rain forest to disastrous damage from fire, while unlogged forests suffered far milder damages. In Indonesia alone, losses in the value of the standing stock of trees exceeded $5 billion, and the costs of ecological damage are still unknown.

These miscalculations of costs and benefits demonstrate the third underlying reason referred to above: governments have proceeded without adequate biological or economic knowledge of tropical forest resource management. Little is known about the potential commercial value of all but a very few of the tropical tree species, so most trees are treated as weeds and destroyed during logging operations. Much remains to be learned about potential regeneration of currently valuable tree species and successful management of heterogeneous tropical forests for sustained yields. Without this knowledge loggers have blundered through the forests, extracting the few highly valued logs and severely damaging the rest. Even less is known of the potential value for agricultural, scientific, or medical purposes of the millions of other plant and animal species, despite clear indications from previous discoveries that unknown treasures may exist in the forests. Consequently, forest habitat is recklessly cleared for commodity production of marginal economic value.

Large-scale agricultural settlements and livestock operations have been encouraged without adequate study of land use capabilities. Painful and costly failures have driven home the lesson that the lush tropical forest does not imply the existence of rich soils beneath it. It is now recognized that most underlying soils are too nutrient-poor for sustained crop production without heavy fertilization, and that the better soils – along rivers, for example – are probably already being used by shifting cultivators. Similarly, massive conversions to monocultures and ranches have taken place without prior attention to potential problems of plant and animal diseases or to pest and weed management, with costly and sometimes disastrous consequences. Only now has serious attention been paid to the capabilities of tropical soils and the development of sustainable farming and livestock systems suited to them.

Governments have pushed ahead with forest exploitation not only in advance of ecological knowledge but even before understanding the likely consequence of the policy instruments with which they hoped to stimulate development. Several countries awarded concessions for the majority of their productive forest estates before enough time had elapsed to assess properly the adequacy of their forest management system, or the impact of forest revenue systems on licensees' behavior (Chapters 2, 3, 4, and 7). Governments have stimulated large domestic processing industries before evaluating the appropriateness of the levels of protection afforded them, their technical and economic efficiency, and the costs and benefits to the national economy of the incentives provided. Similarly, governments have gone ahead with large-scale conversions of tropical forests before adequately evaluating the economic viability and worth of the alternative uses.

Such precipitate actions are related to a fourth reason: governments have tended to grasp at tropical forests as a means of resolving problems arising elsewhere in society. Migration to forested regions has been seen in many countries as a means of relieving overcrowding and landlessness in settled agricultural regions, whether those conditions sprang from rapid population growth (Chapter 7), highly concentrated land tenures (Chapters 4 and 6), or slow growth of employment and opportunities for income generation (Chapter 7). Rather than modifying development strategies to deal more effectively with employment creation and rural poverty, or to tackle the politically difficult problem of land reform, many countries have in effect used forests as an escape valve for demographic and economic pressures.

Sale of tropical timber assets has been seen as a ready means of raising government revenues and foreign exchange. Governments have found that drawing down these resources has been easier in the short run than broadening the tax base and improving tax administration, or reversing trade policies that effectively penalize nascent export industries, although ultimately the assets and easy options are exhausted and the underlying problems remain (Chapters 4 and 7).

In addition, governments (and development assistance agencies) have failed to devote the resources and attention to the forest sector that would have been necessary to ensure proper management and stewardship. Development spending on the forest sector, for example, has been a tiny fraction of that allocated to agriculture. While most sizeable countries in the Third World have built up substantial agricultural research programs (including tree crop research in some countries), none have developed appreciable research capabilities or activities focussed on natural forest ecology and management.

Despite the enormous value of the resource, including billions or in some countries even trillions of dollars in timber alone (Chapter 6), and the large sums of money represented in the annual log harvest, governments have not built up adequate technical and economic expertise or effective management and enforcement capabilities. As countries with large petroleum and other mineral resources have found, the cost of such expertise and managerial capability is small relative to the value of the resource, and is an investment that is returned quickly in increased earnings to the national economy and the government treasury. Foregoing chapters have documented needless waste of forest resources from inappropriate policies or widespread evasion of justifiable but poorly enforced stipulations. Yet, the record of policy analysis in the forest sector is sparse, and infusions of funds and technical assistance to train, staff, equip, and monitor forest administration agencies have been inadequate.

Finally, while national governments have overestimated their own capabilities for forest management, they have underestimated the value of traditional management practices and local governance over forest resources. Local communities dependent on forests for a wide variety of commodities and services, not just timber, have been more sensitive to their protective functions and the wide variety of goods available from them in a sustainable harvest (Chapters 2, 3, 4, 6, and 7). Moreover, when provincial and national governments have overruled traditional use rights to the forests, local communities and individual households have been unable, and less willing, to prevent destructive encroachment or overexploitation. Conversely, some governments have found that restoring or awarding such rights to local groups has induced them to attend carefully to the possibilities of sustainable long-term production from forest resources (Chapter 5).

Improving the policy environment for utilization of natural forests

Those with important interests in the natural forests form a wide and diverse constituency. Included are not only public and private proprietors and enterprises based on forest resources, but also those who benefit from the protection that forests provide to soils, water, and wildlife. With respect to the tropical forests, for reasons offered in Chapter 1, this constituency is worldwide.

Conflict is frequent between competing interests over forest policy and is sometimes seen as inevitable. In particular, "conservationists" and "developers" are supposedly unalterably in opposition. When conservation interests are external to the region or nation seeking to develop, conflicts of interests can be acute. However, an important implication of this study is that such conflict is often more apparent than real. Policies that have led to wasteful exploitation of forest resources, both in developed and developing countries, have been costly not only in biological terms, impoverishing the biota and soils, but equally costly in economic terms. Uneconomic investments have been promoted, assets have been sacrificed for a fraction of their worth, and government treasuries have been deprived of revenues and foreign exchange earnings sorely needed for genuine development purposes.

Reforms of public policies toward publicly owned forests can save both natural and financial resources. Rather than a "win-lose" predicament, opportunities for policy reform present "win-win" situations for nations with natural forest endowments. For example, as Chapter 6 indicates, Brazilian tax and credit incentives for conversion of Amazonian forests to livestock pastures has resulted both in ecological and fiscal disasters: the incentives have made social waste privately profitable at great expense to the government budget. In the United States, as shown in Chapter 8, the Forest Service has subsidized timber production at a large cost to the treasury on lands economically unfit for the purpose, sacrificing potential superior recreational uses along with hundreds of millions of dollars. Since the central concern of economic policy is the efficient management of scarce resources, policies that promote wasteful use of forest resources are rarely economically justifiable.

Opportunities for policy improvement notwithstanding, there is a vein of truth in the assertion heard in the Third World that rich countries, having preempted a large share of the world's resources, wish developing countries to bear the costs of conservation. Since the clientele for tropical forest conservation is worldwide, it is in the rich nations' interests to help defray some of the costs of policy reforms affecting far-away forest endowments. For example, worldwide interests are clearly furthered when Brazil, Indonesia, Malaysia, or Gabon sets aside new areas in parks and forest reserves. Survey and exploration of natural forest resources, for wood and non-wood products as well as genetic resources, are costly but essential for better management. International interests are served by forest conservation and research; international financial and scientific support for them is therefore sensible.

We identify in the following sections a series of measures that should be undertaken by tropical country governments, by industrial country governments, and by international agencies to improve economic utilization and conservation of natural forest assets. This list is not intended to be an exhaustive agenda of needs and opportunities for slowing deforestation. Such a list would include recommendations for accelerating botanical, genetic, and ecological research, for improving agroforestry, plant breeding, and selection, and for strengthening forestry practices in responsible government organizations.

Our focus is complementary, although our concerns are the same. We have identified needs and opportunities for policy reform to alter incentives affecting decisions about natural forest resource use. For convenience of presentation, the reforms are grouped into two categories: those that are essentially the responsibility of governments as owners of forests and as regulators of activities in them, and those for which responsibility falls to agencies representing the worldwide constituency for forest conservation. Neither category is intended to be airtight; for example, the need for joint responsibility for financing reforms is particularly great. Further, the international community should help to provide better and more complete information on which policy reforms by owners can be based. This requires expanded policy research on several forest-related topics.

Policy reforms by national governments of tropical countries

In most cases, two objectives can serve as adequate guidelines for policy reform: (1) more efficient development of the *multiple* uses of natural forests; and (2) improved financial returns to governments. At first glance these objectives may seem to be too narrow, as they may appear to ignore non-economic considerations such as preservation of indigenous communities and ecosystems. However, steps to achieve these two objections will generally also contribute strongly to these non-economic goals as well, because of the complementarities discussed above. Throughgoing reform is required both in forestry and non-forestry policies.

Forestry policies. Royalties and related charges on private contractors for harvesting and sale of government-owned timber have been deficient on two principal counts. First, in all cases studied in this volume, charges have been maintained at levels well below the stumpage value of the timber. This has resulted not only in lost timber rents for the government, but in enormous pressures from business and political interests to obtain timber concessions and the large, short-run profits they offer. Combined with other flaws in timber policy, this rent-seeking syndrome has led to over-rapid, wasteful exploitation, including the harvest of timber in marginal stands, often in ecologically vulnerable sites such as slopes and critical water catchment areas. Second, in the developing countries covered in the foregoing chapters, the structure of royalties has, usually in combination with inappropriate selection systems, exacerbated loggers' proclivities for high-grading (or mining) forest stands, thereby causing needless damage to forest quality (Chapter 2, Appendix A). Sensible reform calls both for increases – sharp increases in many countries – in royalty levels and for modification of defective royalty structures. Inflexible, undifferentiated, specific charges based on the volume of timber harvested should be replaced by *ad valorem* royalty systems, based on export prices properly discounted for costs of harvesting logs and transporting them to ports. Valuation of logs for royalties should be equivalent for log exports and for those delivered to processing mills, for the measure of the opportunity cost of a log is usually its f.o.b. value. Countries should use differentiated *ad valorem* rates, with lower rates for so-called "secondary" species than for the most valuable "primary" species, to the extent that their forest services are sufficiently well-trained and administered to enforce them. But, if the forestry service is undermanned and untrained, flat-rate *ad valorem* royalties constitute the best of inferior options, and can be adequate if set at moderate levels and combined with other measures to capture resource rents.

Governments have generally proven reluctant to enact royalty reform. Of the countries studied in this volume only Sabah (1978) and Liberia (1979) have implemented sharp increases in royalty levels in recent years. China has sharply increased log prices administratively and by permitting market transactions, and proposals that were pending in China in 1987 will raise stumpage fees further. Elsewhere, including Malaysia, the Philippines, Indonesia, the Ivory Coast, Ghana, and Gabon, royalties continue well below true stumpage values and should be raised. In the United States, although royalties are bid (usually competitively) and approximate private stumpage values, the government's absorption of substantial logging costs means that timber with negative stumpage values is routinely harvested. The simplest remedy for this (Chapter 8) is the imposition of minimum acceptable bid prices high enough to recover the government's full separable costs of growing and marketing timber.

The effect of these changes would be to restrict harvesting of uneconomic forest lands, to slow the pace of timber exploitation in Third World countries to one more in line with the growth of forest management capabilities, to permit more complete utilization of the timber resources available in smaller, more compact concession areas, and thus to reduce wastage, infrastructure costs, and the forest disturbance that opens the way for secondary clearance and agricultural conversion.

Policy reforms in the administration of concession agreements would also further both conservation and development goals. Realignment of concessions policies requires changes not only in duration of concessions but in the level of license fees charged on concession areas. Prior to the Second World War, many tropical timber concessions were granted for periods of up to a century. Newly independent governments in the post-war period tended to view such arrangements as vestiges of colonialism. Consequently, the duration of concessions was steadily compressed. By 1987, concession periods were typically five to 10 years in length even for large tracts; few governments allowed concessions longer than 20 years.

Given the long growing cycles of commercial tropical hardwoods, short-term concessions are inconsistent with sensible resource management. Logging firms have no financial interests in maintaining forest productivity under such circumstances. Instead, they have grounds for repeated reentry into logged-over stands before expiration of their concessions, compounding damages arising from the initial harvest (Chapter 2).

Tropical foresters have long called for extension of concession periods to at least 70 years, so as to provide loggers with at least two rotations of 35 years each. Governments have not heeded such proposals, partly because pre-war experience with longer concessions showed little evidence of conservation practices by loggers. Therefore any movement to extend the life of concessions must necessarily be accompanied by appropriate safeguards to defend the public interest. Among the safeguards proposed has been periodic review of concessionaire performance, with renewal of logging rights conditional on adherence to prescribed practices. In practice, as the experience of the Philippines shows (Chapter 4), such arrangements are difficult to police without a substantial improvement in administrative capacity. Further, firms, especially the multinational companies, have grown wary of all long-term contracts, including timber concessions, because so many have been abrogated by governments. A complementary approach is to structure policies from the outset to provide strong incentives to firms for rational forest use.

For example, governments could use area license fees much more effectively to promote conservation, rational harvesting, and more complete rent capture. Fees based on the area awarded in concessions have been found to be extremely low in most countries (Chapters 2, 3, 4, 5, and 7). Ideally, logging concessions should be auctioned competitively, as offshore oil leases are in the United States, ensuring that governments capture virtually all the available resource rent. Auction, or competitive bidding, systems work well only where the owner has enough information about the resources in particular forest tracts to enable him to ascertain their approximate value, and where the number of bidders ensures competition. Acquiring the necessary information about resources is difficult in tropical timber stands, which are typically inaccessible and far more heterogeneous in species composition than temperate forests. Nevertheless, successful examples of auction systems have been reported for both Sarawak and Venezuela. The investment by governments in more detailed exploration and inventory of forest resources is likely to have an immediate pay-off in improved revenue capture.

Auction systems can be combined with higher reservation or minimum bid prices to guard against bid-rigging, or to discourage firms from entering regions that are ecologically sensitive or are better kept back for harvesting at some future time. Where auction systems are not feasible, concession contracts should employ much higher license fees per hectare than is now common in the tropics. Higher license fees serve two conservation goals, as well as revenue objectives: they discourage exploitation of stands of marginal commercial value, and when combined with royalty systems differentiated with respect to stumpage values, they encourage economic utilization of timber stands by providing logging firms with incentives to recover more volume and more species per hectare.

In nearly all 10 of the tropical forest cases studied in this volume, serious problems have been identified in selection systems governing harvest methods. Selective cutting systems are used almost everywhere in mixed tropical forests. Notwithstanding some evidence that careful selective logging can be done with minimal damage to residual stands, the selective cutting methods actually practiced in tropical forests yield unsatisfactory ecological results, principally because of heavy incidental damage and poor regeneration of harvested species, and inferior long-term economic results because of deterioration of the quality of the stand.

It is unfortunately true that much remains to be learned about ecologically sound selection systems in tropical forests. Virtually no one supports clear-cutting ("whole-tree utilization") as a silviculture method in tropical forests except where natural forests are to be cleared for other land uses. There is therefore little basis for recommendations on harvest systems, other than careful research in specific forest regions. But better enforcement of concession terms to avoid excess damage to soils and remaining trees is possible, and the policy reforms discussed above would provide stronger incentives to concessionaires to reduce logging damage and waste.

The most plausible alternative to current methods is one or another of the so-called uniform cutting systems, which also involve selective harvests but generally take more stems per hectare. Uniform cutting systems (Chapter 2, Appendix A) yield greater immediate volume but also entail greater costs and ecological impacts. The most sensible long-run approach to improving harvest methods is more research on the ecological, silvicultural, and economic implications of alternative selection methods, to find attractive variants of uniform cutting systems or more appropriate forms of selective logging.

Reforestation requirements in the temperate forests of North America and China are well understood, even if costs or other constraints have sometimes limited reforestation programs on public lands. The situation is different in tropical countries. If by reforestation is meant the restoration of logged-over stands to something closely resembling their natural states, with similar species frequency, age distribution, and density, then existing knowledge of the ecology of the tropical forest severely constrains the success of such efforts. If by reforestation is meant enrichment planting of primary species in cut-over stands, current policies have brought little success, despite encouraging results in Indonesia and a few other areas.

Despite the general lack of success in regenerating cut-over tropical forests, some reforestation policies have served other purposes. Indonesia's reforestation deposit ($4 per cubic meter harvested) did capture additional timber rents and helped discourage logging on marginal stands, although it was not nearly high enough to induce firms to undertake significant reforestation activity. The same can be said of similar charges in Malaysia and West Africa. Nor have regeneration programs mounted by governments as owners had much success, whether financed by earmarked forest taxes or directly from national treasuries. Most of these have been either underbudgeted (the Philippines, Ghana, Gabon), ill-designed (Indonesia and the Philippines), or rendered ineffective by institutional constraints (Sabah).

A viable set of regeneration policies in logged-over areas would have the following features: (1) prime focus on cutting methods favorable to natural regeneration, coupled with enrichment planting of native species; (2) carrot and stick incentives for firms to undertake regeneration (the carrot could be a one-year extension of the concession period for every prescribed number of hectares of the concession area in which regeneration is under way, and the stick could be a sizeable deposit of at least $5 per cubic meter of *primary* species harvested, refundable to firms as regeneration proceeds, on proof of their expenditures for this purpose); (3) more budgetary and scientific resources for regeneration research programs; (4) more support for government regeneration efforts by international lending institutions and aid donors.

Where reforestation is defined as replacement of forest cover on cut-over tracts by non-indigenous trees (such as pines) other than those grown for tree crops, some success has been recorded in tropical countries, including Indonesia, Malaysia, the Philippines, and Brazil. Plantation programs, especially for conifers, are well established in China, the United States, and other temperate countries such as Chile.

Plantations are an essential part of any program to conserve natural tropical forests, because they create an alternative source of supply to meet growing demands for wood products. Tropical areas under tree plantations (other than those established for industrial tree crops such as palm oil) have grown substantially in the post-war period. They covered an estimated 11 million hectares in tropical countries (excluding China) by 1980, and their area is expected to double by the end of the century (Spears 1983).

Policies that promote investment in tree plantations can help in reducing demands on natural forests, especially for domestic needs. A number of countries have adopted incentive programs, including tax benefits, concessionary credits, and guaranteed markets to stimulate private investment in plantations. The studies underlying this book have not evaluated such policies, because others have already done so in some detail (Berger 1980; Gregersen and McGauhey 1985; Laarman 1983; Matomoros 1982). However, impediments to plantation investments still exist in tropical countries. For example, potential foreign investors in Indonesia cannot own forest land in fee simple nor can they obtain leases for a sufficiently long period of time. Such limitations discourage investment commitments. Even with increasing investments in plantations, deforestation exceeds reforestation by large margins in most tropical countries, and depletion of timber supplies greatly exceeds additions from all sources (Chapter 1). Consequently, positive incentives for plantation investments, while important, are insufficient. Adverse policies that exacerbate losses of natural forest resources must also be changed.

Neglect of non-timber forest products is one such adverse policy, based largely on ignorance. Most tropical country governments do not even collect information on the annual value of production or export of dozens of valuable non-wood products that can be harvested without damaging the complex forest ecosystem. Of all the cases studied in this volume, only Indonesia supplies relatively complete and up-to-date data on production and export of these items, which now exceed 10 percent of gross log export value. Complicated new policies to protect non-timber products are not needed, but better information would be useful, export taxes on them should be removed, and export controls relaxed except, of course, for trade in sensitive or endangered species. If governments realize how valuable such products can be, they will view their loss with more concern.

Traditionally, local communities have been more sensitive to these benefits, while central governments have tended to regard tropical forests as immense timber warehouses. Even then, central governments have not been effective resource managers, nor have provincial or state governments done much better (Chapter 3). Transfer to private hands of permanent rights to natural forests is not a realistic option in most countries (both for constitutional and other reasons), and in Brazil, where it has taken place on a large scale, many large investors have been drawn to opportunities for short-term or speculative gains (in large part due to policy-induced distortions in investment incentives). The record of experience thus far, however, does suggest that local communal or collective ownership of forest property rights has been more consistent with conservation of *all* forest values than has central government ownership.

Accordingly, a larger share of rights and responsibilities over natural forests should be returned to local jurisdictions, where long-standing traditions of forest use exist, or to county governments (such as Indonesia's *kabupatens*) or to village cooperatives, as in China. Central governments would retain sovereign taxing power, and therefore would not give up forest revenues, although revenue-sharing with local communities is not only equitable but also effective in ensuring their interest in resource management. Reversion would mean principally that more forest management decisions, about the location, size, and length of logging concessions, and about land-clearing for agriculture, would be in the hands of local groups with a continuing stake in the multiple benefits that natural forests provide.

Non-forestry policies. The experiences examined in this volume demonstrate that government forestry policies have resulted in more resource depletion than would have occurred had governments tried to minimize their effects on private decisions over forest utilization. Not that *laissez-faire* would have been the desirable policy stance: our point is that, with market outcomes as a standard of comparison, government policies, on balance, have leaned toward more rapid exploitation rather than more conservation. Economic policies affecting the forest sector have had non-neutral effects that have exacerbated deforestation in almost every country investigated. In several, these effects have been extremely large.

Three forms of government subsidies have strongly affected forest use in many countries: (1) revenues lost by failure to collect resource rents from timber harvesting through appropriate royalties and taxes; (2) revenues foregone through income tax incentives for investment in logging, timber processing, and land-clearing activities for estate crops and cattle ranching; and (3) indirect subsidies in the form of artificially cheap credit for investments in those activities.

The timber booms in the Philippines (1950–1970), West Malaysia (1960–1975), East Malaysia, and Indonesia (1967–1980) were partially fueled by generous income tax exemptions for logging and processing firms, as well as by enormous uncollected rents. Some tax holidays were stretched illegally to a dozen years. Fortunately, tax incentives for logging firms are now relics of the past in Southeast Asia, although they are still found in a few African nations, including Gabon and the Ivory Coast. Tax holdiays largely explain the poor performance in capturing timber rents in most log-exporting countries, particularly before the late 1970s. They are unnecessary, and should be rescinded where they still survive.

Moreover, in Brazil and other countries, tax credits, provisions for loss write-offs, and generous depreciation provisions joined with tax holidays to convert socially wasteful land-clearing projects into privately profitable ones (Chapter 6). Policies offering tax subsidies for extractive activities where severe environmental costs are involved, as in logging, are particularly misguided. Taxes should reflect the social costs not borne by the private investor. Therefore, tax subsidies create perverse incentives. All tax incentives for logging and agro-conversion in tropical countries should be removed.

Credit subsidies have added to perverse incentives for forest investments in Ghana (Chapter 7), the Philippines (Chapter 4), and especially Brazil (Chapter 6). The artificially low interest rates and long grace periods available on loans for alternative land uses involving forest clearing have by themselves produced very powerful incentives for forest destruction. When coupled with generous tax subsidies, the incentives for very large-scale forest-clearing have proven irresistible, even where the alternative land uses were intrinsically uneconomic (Chapter 6). Virtually all kinds of subsidized credit programs have been proven wasteful. Credit subsidies for irreversible destruction of forest assets cannot be justified on any economic or environmental grounds, and they should be abolished.

As much as half of the mass of a log becomes sawdust, woodchips, or other residue when the log is processed. Residue can be used for fuel or to make such products as particle board but is much less valuable than lumber and plywood. Domestic processing of logs therefore offers substantial potential savings in shipping and manufacturing costs. Countries with forest endowments and low labor costs enjoy comparative advantages in forest-based industry; over time most timber processing should gravitate to the raw material source. Governments in industrial countries that have built processing industries on imported logs have tried to delay this process by trade barriers discriminating markedly against imported wood products, while governments in log-exporting countries have adopted policies to force exports of processed timber products rather than logs, and to encourage investments in sawmills, plymills, and other wood products industries.

While the net effect of these conflicting international trade policies has been to increase fiscal levies on processed and unprocessed tropical timber, and so restrict world consumption somewhat, severe distortions in investment patterns and losses in economic efficiency have also resulted. In industrial countries, capital and labor have been retained in declining industries. In exporting countries, log export bans have led to evasion, corruption, and the construction of high-cost processing facilities as a means of ensuring access to logs (Chapters 2, 3, 4, and 7). And the tax and tariff protection provided to wood-processing industries has been so high as to weaken competitive pressure and to undermine incentives to minimize costs. Consequently, very sizeable amounts of timber rents have been destroyed; in some countries (Chapters 2, 4, and 7) these losses have run into billions of dollars. Moreover, the low recovery rates in timber processing that have resulted from heavy protection have intensified demands on natural forest endowments. In time, forest-based industries established under the umbrella of heavy protection may achieve higher recovery rates, especially if the degree of protection is gradually reduced. But even then, the processing capacity induced by government incentives will retain a strong claim to raw materials supplies, because governments will not be inclined to allow closure of forest-based firms during the periodic slumps characteristic of world markets for wood products and will provide the logs to keep domestic mills going even if economic or ecological considerations argue against it (Chapters 2, 5, and 8).

Large industries have been established in many of the countries studied, including Indonesia, West Malaysia, the Philippines, Brazil, and parts of West Africa. Replacing log export quotas and log export bans in these countries with increased taxes on log exports would result in higher government revenues and send signals to processing industries of the need to modernize and raise efficiency. This trade liberalization should be negotiated in an international forum in exchange for reduced rates of protection for the timber-processing industries of log-importing countries. The result should be a gradual transfer of most tropical wood-processing industries to countries with large forest endowments, and substantial modernization and improved efficiencies of existing processing industries.

The costly lessons derived from the experiences of these countries can still be put to good use in those countries where forest-based industries have yet to emerge on a large scale, as in East Malaysia, Gabon, and others (Zaire, Cameroon, Papua New Guinea, and Burma, for example). For such nations, gradualism rather than haste is indicated in policies for promoting forest-based industrialization. Higher export taxes on logs than on lumber and plywood are superior in all respects to bans or quotas on log exports. Export taxes furnish whatever degree of protection is desired, raise government revenues, and also make redundant any income tax or credit incentives for sawmills and plymills. Similarly, log export bans and quotas are inappropriate means of preventing logging in sensitive or protected areas. Such areas should given protected status, and no logging concessions (for domestic use or exports) should be awarded.

Many ambitious forest-based industrialization programs have been founded on wishful thinking. The same can be said about large-scale resettlement programs that have moved people into forested regions. In general, it has been wishful thinking that sponsored out-migration could relieve pressures of rural poverty, land scarcity, or even environmental degradation in the areas of origin. The costs of resettlement have been too large, and the numbers resettled too small to make such a strategy viable (Chapters 2 and 6) without dealing with root problems of rapid population growth, rural inequality, and underemployment in the place of origin. Similarly, those in need of resettlement because they have been displaced by large infrastructure projects, such as dams, have often been victims of misapplied analysis that produced inflated estimates of projected returns on such investments. Most such investments have experienced serious cost and schedule overruns and have produced benefits only a fraction as large as anticipated.

The best way to ensure against such over-optimistic undertakings and the human, environmental, and economic losses they entail is to make the expected beneficiaries financially responsible for most of the costs of the projects. In all developing countries, spontaneous migrants outnumber government-sponsored migrants several-fold and typically bear all the costs of their movement as an investment toward an anticipated better future for themselves and their offspring. If resettlement programs were based on the principle of cost recovery, the need for sound planning would be reinforced, and there would be checks on the scale and pace of implementation. Similarly, if agencies responsible for large-scale development projects, such as dams, were financially responsible for recovering operating and most capital costs from project-generated revenues, the tendency toward inflated projections of investment returns would be brought down to earth.

Where resettlement programs are carried out, needless damage to natural forests can be reduced in several ways. The most obvious, of course, is *not* to locate new communities in natural forests at all. In many countries, ineffective or inappropriate land and agricultural tax policies allow large landholders, including governments themselves, to retain huge estates that are used to only a small fraction of their agricultural potential. In such countries, because smallholders typically achieve much higher per hectare yields and incomes through intensive cultivation, land redistribution with compensation can be brought about, provided that the government is able to reform rural taxes to induce large landowners to sell part of their holdings and is willing to finance land purchases by small farmers on long-term mortgage credit. In addition, there are degraded public forest lands and wastelands in many Third World countries that could be transferred to landless peasants, to be used for mixed farming systems involving agroforestry. The main problem here is that government agencies are typically as reluctant to disgorge any of their estates as any other large landholder.

If any agricultural resettlement projects are sited in forest areas, secondary and logged-over tracts should be selected in preference to intact forests. Individual holdings must be large enough to allow sustained support so that settlers are not forced to encroach on adjacent forest areas. Colonies should be based largely on such tree crops as rubber, which are more suitable to available soils, and on agroforestry systems. Further, collection, storage, and marketing facilities for non-wood forest products near resettlement sites would encourage relocated families to use forests sustainably. Finally, award of sites must irrevocably convey all property rights to the awardee, including clear rights of transfer. This not only enables resettled peoples to offer land as collateral in borrowing for improvements, but reinforces the incentive to protect and enhance value of the property.

We conclude this discussion of domestic policy measures to reduce wastage of forest resources by drawing attention to the connection between deforestation and failed development policies. Pressures on natural forests are reduced when towns and villages achieve adequate growth in rural output and employment, begin to convert from wood as an energy source to electricity and petroleum products, and experience declining population growth rates as birth and death rates decline and young workers are drawn off into expanding urban industries. Conversely, economic stagnation and impoverishment of rural and urban populations inevitably accelerate deforestion, as the experience of Ghana during the 1960s and 1970s illustrates so clearly (Chapter 7).

Not much is known about government policies that can accelerate economic development, except those that promote a stable, broadly-based political and economic system that rewards enterprise and productive investment by households and enterprises alike. Much painful experience has accumulated, however, about economic policies that retard development. Salient among these are market distortions, such as deeply overvalued exchange rates (Chapter 2, Chapter 7), negative interest rates (Chapter 6), and flagrantly distorted commodity prices (Chapter 7). These not only result in serious market disequilibria, scarcities, and rationing throughout the economy, but also create perverse investment incentives and generate speculation and corruption. Such policies also often imply heavy penalties on the rural sector, as the relative prices of agricultural outputs are lowered relative to industrial prices and investments are redirected toward the urban sector. If, at the same time, governments expand their direct role in the economy beyond their managerial capabilities, creating expensive state-operated industrial and agricultural white elephants, economic stagnation is nearly assured.

Reversal of policies clearly inimical to growth will not immediately stem forest destruction arising from the spread of shifting cultivation and from the search for fuelwood. But the recent experiences of relatively prosperous countries such as Gabon, Venezuela, and the western part of Malaysia reveal little deforestation arising from either shifting cultivation or fuelwood demands.

Policy changes by industrial countries and international agencies

Of course, economic development in tropical countries is influenced by policies of the industrial countries. The 1980s have seen stagnation and decline throughout much of the Third World as a consequence of worldwide recession in the early 1980s, extremely high real interest rates, and reductions in net capital flows to developing countries on international capital markets, as well as growing protectionist pressures in the industrial countries. While beyond the scope of this study, the connection must be pointed out between economic stagnation leading to deforestation in the Third World and policies in industrial countries that restrict outflows of capital to developing countries and inflows of commodities exported from developing countries.

More specifically, industrial country trade barriers in the forest products sector have been partially responsible for inappropriate investments and patterns of exploitation in Third World forest industries. Within the context of either the General Agreement on Tariffs and Trade (GATT), the International Tropical Timber Agreement (ITTA), or some other international forum, negotiations between exporting and importing countries should result in (a) reduced tariff escalation and non-tariff barriers to processed wood imports from the tropical countries, and (b) rationalization of incentives to forest industries in the Third World.

While international assistance agencies have become increasingly concerned with forest sector problems over the past 15 years, their involvement has primarily taken the form of support for discrete development projects. These have ranged widely, from reforestation and watershed rehabilitation projects to support for fuelwood and industrial timber plantations and to funding for wood-processing industries. Associated with them, of course, has been a significant amount of technical assistance in a wide variety of forestry-related subjects.

It must also be said that development assistance agencies have, in the aggregate, provided huge amounts of funding for projects that lead directly and indirectly to deforestation, including roads, dams, tree crop plantations, and agricultural settlements. Greater sensitivity is required in ensuring that such projects are, wherever possible, sited away from intact forests and especially away from critical ecosystems; in ensuring that such investments are, in fact, economically and ecologically sound once the non-market costs of forest losses are weighed in the balance; and in ensuring that such projects are executed in accordance with appropriate safeguards to minimize unnecessary damages.

Only to a much lesser extent have development assistance agencies been involved with the kinds of sectoral policy issues discussed in this book. The Food and Agricultural Organization of the United Nations (FAO) has indeed studied forest revenue systems and published manuals and related material for policy-makers. Multilateral and bilateral assistance agencies such as the Inter-American Development Bank and USAID have examined policies that would encourage private investment in forestry. In recent years, the World Bank has carried out forest sector reviews in several countries that address some of the broader policy issues identified in this book, and new directions in World Bank policies indicate growing attention to tropical forest values. Yet, it is clear that still more emphasis on policy reform in the forest sector is required. Especially as international capital flows become increasingly divorced from specific projects and linked to broad macroeconomic and sectoral policy agreements, the international development agencies must identify and analyze the effects of tax, tariff, credit, and pricing policies, as well as the terms and administration of concession agreements, on the use of forest resources. Working in cooperation with host country agencies, they can help to a greater extent in identifying needs for policy changes and options for policy reform.

More generally, governments of industrialized countries and international agencies should act upon the recognized community of interests among all nations in conservation of tropical forests. During 1987 the FAO, the World Bank, the United Nations Development Program and the World Resources Institute collaborated with non-government organizations, professional associations, and governments around the world on a "tropical forestry action plan" (FAO 1987). Intended not as a blueprint but as a framework for coordinated action, this plan identifies priority action needed on five fronts: forest-related land uses; forest-based industrial development; fuelwood and energy; conservation of tropical forest ecosystems; and institution-building. Its recommendations are complementary to and compatible with the policy implications of this study.

The call for action sounded by participants in that process serves equally well as a conclusion to this volume:

Above all, action is needed now. Hundreds of millions of people in developing countries already face starvation because of fuel and food shortages. The forest resource potential of these countries can and must be harnessed to meet their development needs. Properly used and managed, the tropical forests constitute a massive potential source of energy, a powerful tool in the fight to end hunger, a strong basis for generating economic wealth and social development, and a storehouse of genetic resources to meet future needs. This is the promise and the challenge (FAO 1987: 3).

References

Allen, Julia C., and Douglas F. Barnes. 1985. The Causes of Deforestation in Developing Countries. *Annals of the Association of American Geographers,* Vol. 75, No. 2: 163–184.

Berger, Richard. 1980. The Brazilian Fiscal Incentive Act's Influence on Reforestation Activity in São Paulo State. Ph.D. diss., Michigan State University, East Lansing.

Brown, Lester, et al. 1985. *State of the World.* Worldwatch Institute. New York: W. W. Norton.

Bunker, Stephen G. 1980. Forces of Destruction in Amazônia. *Environment,* Vol. 20, No. 7 (September): 14–43.

Eckholm, Eric. 1976. *Losing Ground.* Worldwatch Institute. New York: W. W. Norton.

Ehrlich, Paul R., and Anne H. Ehrlich. 1981. *Extinction: The Causes and Consequences of the Disappearance of Species.* New York: Random House.

Fearnside, Philip M. 1982. Deforestation in the Amazon Basin: How Fast Is It Occurring? *Intersciencia,* Vol. 72, No. 2: 82–88.

FAO, World Resources Institute, World Bank, and UN Development Program. 1987. *The Tropical Forestry Action Plan.* Rome, June.

Grainger, Alan. 1980. The State of the World's Tropical Forests. *The Ecologist,* Vol. 10, No. 1: 6–54.

Gregersen, Hans M., and Stephen E. McGauhey. 1985. *Improving Policies and Financing Mechanisms for Forestry Development.* Washington, D.C.: Inter-American Development Bank.

Laarman, Jan G. 1983. *Government Incentives to Encourage Reforestation in the Private Sector of Panama.* Panama: USAID, September.

Lanly, Jean-Paul. 1982. *Tropical Forest Resources.* FAO Forestry Paper 30. Rome: FAO.

Matomoros, Alonso. 1982. *Papel de Incentivos Fiscales para Reforestación en Costa Rica.* Centro Agronomico Tropical de Investigación y Enseñanza. Costa Rica, May.

Myers, Norman. 1980. *Conversion of Tropical Moist Forests.* Washington, D.C.: National Academy of Sciences.

Myers, Norman. 1984. *The Primary Source: Tropical Forests and Our Future.* New York: W. W. Norton.

Myers, Norman. 1985. Tropical Deforestation and Species Extinction: The Latest News. *Futures,* Vol. 17, No. 5 (October): 451–463.

Plumwood, Val, and Richard Routley. 1982. "World Rainforest Destruction – The Social Factors." *The Ecologist,* Vol. 12, No. 1: 4–22.

Spears, John. 1979. Can the Wet Tropical Forest Survive? *Commonwealth Forestry Review,* Vol. 57, No. 3: 1–16.

Spears, John. 1983. Sustainable Land Use and Strategy Options for Management and Conservation of the Moist Tropical Forest Eco-Systems. International Symposium on Tropical Afforestation, University of Waginengen, Netherlands, September 19.

Tucker, Richard, and J. F. Richards, eds. 1983. *Global Deforestation and the 19th Century World Economy.* Durham, N.C.: Duke University Press.

Index of topics

Printed in the United States
by Bookmasters

Printed in the United States
By Bookmasters